Flowering Plant
Origin, Evolution
& Phylogeny

D1481634

Flowering Plant Origin, Evolution & Phylogeny

EDITED BY

David Winship Taylor
Indiana University Southeast

& Leo J. Hickey
Yale University

CHAPMAN & HALL

New York • Albany • Bonn • Boston • Cincinnati • Detroit • London • Madrid • Melbourne
Mexico City · Pacific Grove • Paris • San Francisco • Singapore • Tokyo • Toronto • Washington

Copyright © 1996 by Chapman & Hall

Printed in the United States of America

For more information, contact:

Chapman & Hall
115 Fifth Avenue
New York, NY 10003

Chapman & Hall
2-6 Boundary Row
London SE1 8HN
England

Thomas Nelson Australia
102 Dodds Street
South Melbourne, 3205
Victoria, Australia

Chapman & Hall GmbH
Postfach 100 263
D-69442 Weinheim
Germany

Nelson Canada
1120 Birchmount Road
Scarborough, Ontario
Canada M1K 5G4

International Thomson Publishing Asia
221 Henderson Road #05-10
Henderson Building
Singapore 0315

International Thomson Editores
Campos Eliseos 385, Piso 7
Col. Polanco
11560 Mexico D. F.
Mexico

International Thomson Publishing - Japan
Hirakawacho-cho Kyowa Building, 3F
1-2-1 Hirakawacho-cho
Chiyoda-ku, 102 Tokyo
Japan

All rights reserved. No part of this book covered by the copyright hereon may be reproduced or used in any form or by any means — graphic, electronic, or mechanical, including photocopying, recording, taping, or information storage and retrieval systems — without the written permission of the publisher.

1 2 3 4 5 6 7 8 9 10 XXX 01 00 99 98 97 96

Library of Congress Cataloging-in-Publication Data

Flowering plant origin, evolution, and phylogeny / edited by David W.
    Taylor, Leo J. Hickey.
            p.   cm.
    Papers from a symposium held at the American Institute of
Biological Sciences meeting at Ames, Iowa, on Aug. 4, 1993.
    Includes bibliographical references and index.
    ISBN 0-412-05341-1 (alk. paper)
    1. Flowering plant origin, Fossil--Congresses.   2. Angiosperms--Origin-
-Congresses.   3. Angiosperms-Evolution--Congresses,   I. Taylor,
David W., 1958-   .   II. Hickey, Leo J.   III American Institute of
Biological Sciences. Meeting (1993 : Ames, Iowa)
QE980.A54   1995
582.13'0438—dc20   95-14824
                                                        CIP
To order this or any other Chapman & Hall book, please contact **International Thomson Publishing, 7625 Empire Drive, Florence, KY 41042.** Phone: (606) 525-6600 or 1-800-842-3636. Fax: (606) 525-7778. e-mail: order@chaphall.com.

For a complete listing of Chapman & Hall's titles, send your request to **Chapman & Hall, Dept. BC, 115 Fifth Avenue, New York, NY 10003.**

CONTENTS

Preface

Questions about the origin, early evolution, and basal phylogeny of the flowering plants have been areas of ferment in botanical research for nearly 25 years. Rapid progress toward resolving these problems continues to be made, owing, in part, to the unprecedented number of approaches now available, as well as to the accumulation of carefully documented empirical data from both living and fossil plants. During this time, a series of symposia and their published volumes have periodically summarized the progress made and focused research on many of the important issues that remain. The most influential of these publications have been *Origin and Early Evolution of Angiosperms,* edited by Charles B. Beck in 1976, *Historical Perspectives of Angiosperm Evolution* edited by David L. Dilcher and William L. Crepet as Volume 71, (Issue 2) of the *Annals of the Missouri Botanical Garden* in 1984, and *The Origins of Angiosperms and Their Biological Consequences* edited by Else Marie Friis, William G. Challoner, and Peter R. Crane in 1987. We feel that the time is ripe for a new book that summarizes the recent research from fields of paleobotany, comparative morphology, development, and structural and molecular phylogeny that bears on these problems.

The volume took shape at a symposium entitled *The Origin, Early Evolution, and Phylogeny of Angiosperms* held at the American Institute of Biological Sciences meeting at Ames, Iowa on August 4, 1993. Our contributors were invited because of their ability to provide the most current view of their fields and to place this knowledge into a broad context. In an effort to bring a greater degree of focus to the symposium and to the resulting book, we also asked the authors to consider a set of questions relating to angiosperm evolution. This book follows the same scheme as the symposium with the addition of a geological time scale to aid those who are unfamiliar with stratigraphic terminology, particularly at the level of the epoch and age divisions of the geological periods. As with all efforts to summarize a rapidly developing field, this publication is only a progress report, but one that we hope will serve to further accelerate the process of resolving what seemed only a generation ago to be intractable problems.

Finally, we wish to thank the following reviewers who gave generously of their time to ensure that the contents of this volume met the standards of the science: Pieter Bass, Robyn Burnham, Peter R. Crane, David L. Dilcher, Jeff H. Doyle, Peter K. Endress, Else Marie Friis, Dorian Q. Fuller, Richard C. Keating, Sergius H. Mamay, Richard G. Olmstead, Robert Ornduff, Edward L.

Schneider, Taylor Steeves, Susan Swenson, Alfred Traverse, Thomas K. Wilson and several anonymous reviewers. Very special thanks are due to Grady Webster who acted as an associate editor for the two chapters co-authored by the editors.

Contributors

David A. Baum
Department of Organismic and Evolutionary Biology
Harvard University
Cambridge, MA 02138

Gilbert Brenner
Department of Geological Sciences
State University of New York
College at New Paltz
New Paltz, New York 12562

Sherwin Carlquist
Santa Barbara Botanic Garden
1212 Mission Canyon Road
Santa Barbara, CA 92521

Bruce Cornet
27 Tower Hill Avenue
Red Bank, NJ 07701

Andrew Douglas
Department of Botany
Louisiana State University
Baton Rouge, LA 70803

Leo J. Hickey
Department of Geology and Geophysics
Yale University
Box 6666
New Haven, CT 06511

Gretchen Kirchner
Department of Biology
Indiana University Southeast
4201 Grant Line Road
New Albany, IN 47150

Henry Loconte
The New York Botanical Garden
Bronx, NY 10458-5126

Kathleen B. Pigg
Department of Botany
Arizona State University
Tempe, AZ 85287-1601

Kenneth J. Sytsma
Botany Department
University of Wisconsin
Madison, WI 53706

David Winship Taylor
Department of Biology
Indiana University Southeast
4201 Grant Line Road
New Albany, IN 47150

Robert F. Thorne
Rancho Santa Ana Botanic Garden
1500 North College Avenue
Claremont, California 91711

Mary L. Trivett
Department of Environmental and Plant Biology
Ohio University
Athens, OH 45701-2979

Shirley C. Tucker
Department of Botany
Louisiana State University
Baton Rouge, LA 70803

Introduction: The Challenge of Flowering Plant History

David Winship Taylor and Leo J. Hickey

Angiosperms are well known for their incredible diversity in species number, range of habitat, and morphology. Recent counts have shown that there are at least 234,000 described species of flowering plants in some 437 families (Thorne, 1992a). This number exceeds the total number of species of all other photosynthetic land plants and algae combined (Sporne, 1974; Prance, 1977). Not only do angiosperms dominate in number of species, they are also found in a far greater range of habitats than any other group of land plants. At their extremes they range from the tropics to the tundra, and from freshwater and marine environments to deserts. Not surprisingly, angiosperms also display an unparalleled range in form, from tiny plants in the Lemnaceae to huge trees in the Myrtaceae. Examples of their morphological variability can be found in stems (e.g., Carlquist, Chapter 4), flowers (e.g., Tucker and Douglas, Chapter 7), carpels (e.g., Taylor and Kirchner, Chapter 6), and pollen (e.g., Brenner, Chapter 5).

This great plasticity of the angiosperm body is the cause of contentious debate over ancestral form (compare Taylor and Hickey, Chapter 9; Loconte,

Chapter 10; Thorne, Chapter 11). No consensus on the ancestral states of many morphological traits has been arrived at, especially when no single living taxon seems to have all these states. Even determining the derived states shared by all angiosperms has been difficult because of exceptions to nearly every character, even among angiosperms regarded as primitive. In addition, a growing number of states that were thought to be uniquely angiospermous have now been shown to exist in angiosperm outgroups and are, thus, symplesiomorphies (see Taylor and Hickey, 1992, Chapter 9; Hickey and Taylor, Chapter 8). Recognition of these shared ancestral states may help in reconstructing what the ancestral angiosperm looked like but is of no assistance in determining whether angiosperms are monophyletic or what the relationships are within the clade.

Nevertheless, a number of features appear to be nearly universal in angiosperms and widespread in their putatively most basal group, the Magnoliidae. These universal synapomorphies for angiosperms emerged both from conventional and cladistical analysis of morphology (summarized in Taylor and Hickey, Chapter 9). One of these synapomorphies is the existence of sieve-tube members and companion cells. These cells are found in all angiosperms but in no other living group. Carpels and stigmatic germination are also universal and are another defining character. Other reproductive characters are the highly reduced megagametophyte with three or fewer free cellular divisions and the production of stored food for the embryo after fertilization. Several other characters have been suggested as ancestral, yet either they are not universal in the angiosperms or are possible symplesiomorphies because of their occurrence in angiosperm outgroups. Some of these nonuniversal characters include vessels with scalariform perforation plates, apocarpous carpels, wood fibers, and triploid endosperm. Examples of angiospermous characters that may be homologous with states from the outgroups include reticulately veined leaves, ethereal oil cells, and vessels.

ANGIOSPERM ORIGIN

One aspect of phylogeny that has been important for understanding angiosperm origin is their relationship to other seed plants. The consensus emerging from recent phylogenetic analyses (Crane, 1985; Doyle and Donoghue, 1986, 1992; Zimmer et al., 1989; Loconte and Stevenson, 1990; Hamby and Zimmer, 1992; Chase et al., 1993; Rothwell and Serbet, 1994; Nixon et al., 1994) clearly shows that the gnetopsids are the closest living relatives of the angiosperms, whereas the closest extinct group is the bennettitaleans. Outgroup relationships provide important clues for understanding the evolution of several distinctly angiospermous organs such as leaves (Trivett and Pigg, Chapter 2), carpels (Taylor and Kirchner, Chapter 6), and flowers (Hickey and Taylor, Chapter 8). Discovery of fossils similar to angiosperms or their sister groups (e.g., Cornet, 1986, 1989a, 1989b, Chapter 3; Cornet and Habib, 1992) can further the understanding of homologies among these groups.

As insights have continued to accumulate from work on living angiosperms, two basic hypotheses have developed regarding the form of the ancestral angiosperm: the Magnolialean hypothesis and the Herbaceous Origin hypothesis. The Magnolialean hypothesis suggests that the ancestral angiosperm was a woody, arborescent plant with large, many-parted flowers (e.g., Cronquist, 1988; Takhtajan, 1991; Thorne, Chapter 11). Based on the fossil record, Doyle and Hickey (1976, 183–191; Hickey & Doyle, 1977) expanded on Stebbins's (1974) concept and suggested that the first angiosperms were "shrubs of semixerophytic origin which entered mesic areas as colonizers of unstable habitats-the "weeds" of the Early Cretaceous" (Hickey and Doyle, 1977, pg. 73). The floral portion of the hypothesis finds its basis in the work of Arber and Parkin (1907) who viewed the transition to the angiosperm flower to be from the bisexual strobiloid reproductive organs of the bennettitaleans. Several recent phylogenetic analyses support this hypothesis in some form (Donoghue and Doyle, 1989a; Loconte and Stevenson, 1991; Loconte, Chapter 10).

The alternative Herbaceous Origin hypothesis (which includes the "Paleoherb" variants) suggests that the ancestral angiosperm was small in size and had many, small, few-parted flowers (e.g., Burger, 1981; Taylor and Hickey, 1992, Chapter 9). This hypothesis is similar to the Pseudanthial theory proposed by Wettstein (1907a). In this view the angiosperm flower was thought to be derived from a compound gymnosperm structure similar to that found in the gnetopsids (see also Hickey and Taylor, Chapter 9). A major difference between the Herbaceous Origin and the Pseudanthial hypothesis is that in the former, the ancestral plant is considered to be a rhizomatous, perennial herb instead of a tree. A growing number of phylogenetic analyses of morphology (Taylor and Hickey, 1992) and molecular sequences (Zimmer et al., 1989; Hamby and Zimmer, 1992; Chase et al., 1993; Qiu et al., 1993; Doyle et al., 1994; Sytsma and Baum, Chapter 12) support the Herbaceous Origin hypothesis. The phylogenetic results are tempered by reservations as to how well the basal portions of these cladograms can be reliably resolved (e.g., Doyle, 1994; Sytsma and Baum, Chapter 12).

TIME OF ORIGIN AND DIVERSIFICATION

The time of divergence of the angiosperms from other seed plant groups has been the subject of considerable debate. The earliest unequivocal angiosperm fossils are from rocks of Valanginian Age in Israel (Brenner & Bickoff, 1992; Brenner, Chapter 5) and show little morphological diversity. Flowers from before the Barremian also show little diversity (Friis, Pedersen and Crane, 1994). From that time on, the number (Lidgard and Crane, 1988; Crane and Lidgard, 1990) and diversity (e.g., Doyle and Hickey, 1976; Hickey and Doyle, 1977; Crane, 1987; Doyle and Donoghue, 1993) of fossils belonging to flowering plants increases. Major groups of angiosperms, including the herbaceous Magnoliidae, Magnoliales, Laurales, winteroids, Liliopsida and eudicots, all appear by the end of the Early Cretaceous. The number of species increases so that early in Late Cretaceous at least 50% of the species in the fossil flora are angiosperms (Lidgard and Crane, 1988). By the end of the Cretaceous, many extant angiosperm families have appeared (Muller, 1981; Taylor, 1990). This diversification follows a geographic pattern as well, with the earliest occurrences of angiosperms in equatorial regions and then dispersing toward the poles.

However, the alternative of a much earlier date for the origin of the angiosperms has been suggested by several molecular clock models (see summary in Sytsma and Baum, Chapter 12). Although the predicted dates vary widely and varying rates of base-pair substitution among different clades cause problems in dating the emergence of lineages, the minimum congruence of the times predicted by the various analyses suggests that the angiosperms originated around the Triassic Period. This is compatible with at least one evolutionary interpretation of the anthophyte fossil record that points to the apparent long independence of the line leading to the angiosperms (Doyle and Donoghue, 1986). Another argument for this early date is the occurrence of putatively angiospermous characters in pollen and megafossil remains from the Late Triassic and Jurassic (Cornet 1986, 1989a, 1989b, Cornet and Habib, 1992). For now, further substantiation is necessary before these claims can be accepted.

In contrast to the unprecedented variety of habitats occupied by living flowering plants, study of the paleoecology of Early Cretaceous sites provides evidence that angiosperms first lived along stream and lake margins (e.g., Doyle and Hickey, 1976; Hickey and Doyle, 1977; Upchurch and Wolfe, 1987; Hickey and Taylor, 1992). Later they appear in more stable backswamp and channel sites, and lastly on river terraces. Paleoecological studies of Late Cretaceous localities show that although angiosperms were taxonomically diverse, they were not the dominant forms but mostly understory herbs (Wing et al., 1993). These data are compatible with models suggesting that angiosperm success was in part due to vegetative characters and quick growth (e.g., Midgley and Bond, 1991).

RELATIONSHIPS OF THE BASAL ANGIOSPERMS

A growing number of phylogenetic analyses concentrate on angiosperm taxa at the base of the tree (Donoghue and Doyle, 1989a; Zimmer et al., 1989; Loconte and Stevenson, 1991; Hamby and Zimmer, 1992; Taylor and Hickey, 1992; Chase et al., 1993; Doyle et al., 1994; Loconte, Chapter 10; Sytsma and Baum, Chapter 12). Below, we briefly summarize the general consensus of these structural and molecular studies as they relate to the identity and systematic position of major groups of angiosperms.

The dicotyledons (Magnoliophyta) are the most ancestral group and are paraphyletic because the monocotyledons are derived from them. Within the dicotyledons there are several well-defined groups. One of these includes the herbaceous members of the Magnoliidae (eoangiosperms; see Hickey and Taylor, Chapter 8). The eoangiosperms are clearly paraphyletic, as some members are more closely related to monocots than are others. The Magnoliales and the winteroids also form distinct clades among the basal dicotyledons. More derivative groups include the Laurales and, finally, the eudicots. The eudicots are clearly a monophyletic clade based on both structural and sequence datasets. At its base appear the ranunculoids as well as some of the "lower" hamamelids, including Eupteliaceae, Tetracentraceae, Trochodendraceae, and Platanaceae. The final derivative angiosperm clade is the monocotyledons (Liliflorae). This group appears to be monophyletic, with *Acorus*, Araceae, and members of the Dioscoreales at the base (Dahlgren et al., 1985; Duvall, Learn et al., 1993).

IMPACT ON EARLY ECOSYSTEMS

Angiosperms have undoubtedly affected the evolution of other groups of plants and animals. The rise of angiosperms was paralleled by the decline of the ferns, as well as of all other seed plant groups, except the conifers (e.g., Lidgard and Crane, 1988). The evolutionary development of the flower became modified with the coevolution of flowers and their pollinators. This process has been documented by the mid-Cretaceous and certainly influenced the evolution of modern angiosperm families (Crepet and Friis, 1987; Friis and Crepet, 1987; Crepet et al., 1991). In addition, the evolution of the angiosperm fruit appears to have influenced patterns of herbivory, with coevolutionary feedback between plants and their vertebrate dispersers becoming especially intense during the Tertiary (e.g., Tiffney, 1984; Wing and Tiffney, 1987a).

CONCLUSIONS

The origin, evolution, and phylogeny of the angiosperms continue to be areas of keen interest and vigorous research. Recent developments have lead to the identification of universal synapomorphies for the group as well as to the realization that a number of putatively angiospermous characters are shared with their sister groups, including the gnetopsids. Recent research has not only deepened our understanding of the Magnolialean hypothesis but has also considerably strengthened a competing view that pictures the ancestral angiosperm as a rhizomatous herb with small, simple flowers. Evaluation of the merits of these hypotheses has encouraged a deeper study of character suites and raised questions about previous hypotheses.

It is clear that our knowledge of angiosperm diversification is now better documented than ever. The increasing number of well studied Early Cretaceous fossils, including flowers, shows that angiosperms diversified quickly with the major basal clades appearing by the end of the Early Cretaceous. Still not completely resolved is some evidence suggesting that the origin of angiosperms may have been earlier than their appearance in the fossil record. However, their fossils do show that the herbaceous habit has existed throughout the history of angiosperms. Finally, recent phylogenetic analyses have confirmed the existence of six major clades among angiosperms and have begun to clarify the details of relationships.

Yet considerable work still needs to be done. Careful study of living members of the Magnoliidae should continue so that additional characters will be available for phylogenetic analyses and for understanding the transformations that took place during the early evolution of the groups. New characters need to be identified, and these include characters from leaf architecture and other understudied suites. New molecular characters such as DNA rearrangements and other sequences are needed, as is the identification of unique sequence characters for all angiosperms. A crucial element in the developing synthesis is in-depth character analysis, from a morphological, molecular and developmental perspective, of states that are structurally similar in angiosperms and their outgroups. In many cases the question to be answered is simple: Are these character states evolutionary homologies or examples of adaptive convergences?

Current models of the ancestral angiosperm structure and form should be used to make predictions concerning their evolution and development, and these hypotheses should be tested. For example, suggested homologies between flowers and floral parts to gnetopsid reproductive organs can be tested by comparing the genetic control for their development between the two groups. Continued evolutionary and functional morphological investigations are needed to explain the convergence of ferns and other rhizomatous plants with angiosperms in characters such as leaf venation and vessels.

The fossil record will continue to be a rich source of novel insights. More detailed paleoecological studies are needed of early angiosperm sites. The acquisition of additional Early Cretaceous fossils is crucial to furthering our understanding of the diversity and relationships of the group in its earliest phases. These studies need to be extended into the earliest Cretaceous and the

Jurassic and should not neglect the intriguing fossils possibly related to anthophytes.

Finally, additional phylogenetic analyses of structural and molecular characters are an urgent necessity and should be extended to include putative angiosperm outgroups, particularly within the gnetopsids. This will allow an assessment of homologies between angiosperms and their plausible outgroups. Continued work on the relationships within angiosperms is also necessary. Identification of the minor groups within the Magnoliidae and the sequence of branching are issues that also need further study.

Angiosperm origin and early evolution are enigmas that have intrigued botanists for well over a century. Even so, it is an exciting time for these studies, and rapid progress is being made. The extent of the "abominable mystery" of angiosperm origin is rapidly diminishing and the pace of research is now such that we feel that, in 20 years, the issue will be no more mysterious than for any other major plant group.

A Survey of Reticulate Venation Among Fossil and Living Land Plants

Mary Louise Trivett and Kathleen B. Pigg

Leaves with reticulate venation are recognized as a hallmark of flowering plants. Dicots are usually characterized by reticulate venation and monocots by parallel venation. However, such distinctions are based primarily on the architecture of major veins, and leaves in both groups typically possess several higher orders of intercostal reticulate venation (Gifford and Foster, 1989). Because most major veins in monocots either anastomose progressively toward the apex or form arcuate anastomoses at the laminar margin, Troll (1939a) proposed the term "striate" rather than parallel to reflect the closed venation of most monocots. Reticulate venation is so ubiquitous and comprises so much intertaxonomic variation among the angiosperms that it has become an important component of the systematics of both fossil and extant taxa. Like many other vegetative features though, reticulate venation is not unique to the flowering plants. Rather, as Alvin and Chaloner (1970) point out, it appears to be a grade of morphological complexity shared by several diverse groups of living and fossil vascular plants including ophioglossalean, marattialean, filicalean, and marsilialean ferns, lycopods, cycads, *Ginkgo*, gnetophytes, pteridosperms, cycadeoids, and some fossil plants of uncertain affinities (Table 2.1).

Table 2.1. A comparison of venation characteristics of various representative taxa discussed.

Taxon	Affinity	Age	Midrib	Vein organization	Branching of secondaries	No. of orders	Orders with Anastomoses	Frequency of anastomoses	Freely ending veinlets
Ginkgo biloba	Ginkgoales	Extant	Absent	Flabellate		1	Order #1	Infrequent	Absent
Stangeria	Cycadales	Extant	Multistrand	Pinnate	Dichotomous	2	Order #2	Frequent	Absent
Welwitschia	Gnetales	Extant	Absent	Parallel	Unbranched	2	Order #2	Frequent	Absent
Gnetum	Gnetales	Extant	Multistrand	Pinnate	Dichotomous	4	Orders #2-4	Frequent	Simple-Branched
*Reticulopteris**	Cf. Medulossales	Paleozoic	Single	Pinnate	Dichotomous	2	Order #2	Frequent	Absent
Glossopteris	Glossopteridales	Permian	Multistrand	Pinnate	Dichotomous	2	Order #2	Frequent	Absent
Gangemopteris	Glossopteridales	Permian	Absent	Flabellate		1	Order #1	Frequent	Absent
Gigantonoclea	Gigantopteridales	Permian	Multistrand	Pinnate	Unbranched	4	Orders #3 and 4	Frequent	Absent-Branched
Delnortea	Gigantopteridales	Permian	Cf. Multistrand	Pinnate	Unbranched	4	Order #4	± Frequent	Absent
*Sagenopteris**	Caytoniales	Jurassic	Single	Pinnate	Dichtomous	2	Order #2	± Frequent	Absent
*Ctenis**	Cycadales	Mesozoic	Absent	Parallel		1	Order #1	Infrequent	Absent
*Dictyozamites**	Bennettitales	Mesozoic	Absent	Parallel		1	Order #1	Frequent	Absent
Drewria	Gnetales	Cretaceous	Absent	Parallel	Unbranched	2	Order #2	± Frequent	Absent
Sanmiguelia	Uncertain	Triassic	Absent	Parallel	Variable	4	Orders #1-4	Variable	Absent
*Marcouia**	Uncertain	Triassic	Cf. Multistrand	Pinnate	Dichotomous	2	Order #2	± Frequent	Absent
Furcula	Uncertain	Triassic	Cf. Single	Pinnate	Excurrent/Dichotomous	3-4	Orders #2-4	Frequent	Present
Pannaulika	Uncertain	Triassic	Single	Pinnate	Excurrent	4	Orders #2-4	Frequent	Present

* The leaf is compound and leaflet characters are reported here.

9

Modern dicots have a venation syndrome characterized by a suite of features usually including three or more well-organized ranks of progressively finer veins, abundant anastomoses within and between orders, and freely ending veinlets. In addition, dicot leaf ontogeny involves different phases and loci of meristematic activity relative to those of ferns and gymnosperms (Esau, 1965; Pray, 1955, 1960, 1962, 1963; Slade, 1957). Although primitive angiosperms have much less organized leaf architectures than modern dicots, as the Early Cretaceous record indicates, even the earliest recognizable an-giospermous leaves are essentially ordinally reticulate veined (Hickey and Wolfe, 1975; Doyle and Hickey, 1976). Such ordered hierarchical reticulate venation is much more limited in its distribution than simple reticulate patterns (Table 2.1).

Hickey and Doyle (1977) described early angiosperm leaves from Lower Cretaceous strata of the Potomac Group. They noted that leaves from the Barremian-Aptian Stage Zone I tend to be no more than 5 cm long, simple and entire, and often with little distinction between petiole and blade. The leaves have disorganized venation marked by irregularly sized and shaped intercostal areas delimited by secondary veins, indistinct differences between secondary and intersecondary veins, and poor differentiation as well as nonregular courses of higher orders of veins (Doyle and Hickey, 1976; Hickey and Doyle, 1977). Based on comparison of Zone I leaves with "first-rank" architecture (Hickey, 1973) to more advanced angiosperms and gymnosperms, Doyle and Hickey (1976) proposed a leaf reduction-expansion model for angiosperm leaf origins. They postulated a reduction trend, perhaps in response to xerophytic pressures, from a pinnately compound, dichotomously veined seed fern or cycadophyte precursor to a small, simple, pinnately veined preangiosperm leaf. Secondary expansion of the blade and elaboration of ordinal venation mediated by inter-calary growth followed the reduction phase (Doyle and Hickey, 1976). Despite disagreement regarding the precise timing of the origin of flowering plants, most workers agree with Doyle (1978) that the earliest unequivocal evidence of their occurrence comes from the Lower Cretaceous.

In this chapter, we present a survey summarizing the occurrence of reticulate venation patterns among vascular plants, with emphasis on those groups of seed plants that have been implicated as angiosperm ancestors. In addition, we use examples from extant gymnosperms as well as recent studies on the glossop-terids (Pigg, 1988, 1990; Pigg and Taylor, 1990; Pigg and McLoughlin, 1992; Pigg and Trivett, 1994) to demonstrate that different vascular anatomies may produce external venation patterns that appear identical, thereby emphasizing the need for caution when interpreting and comparing venation patterns from external morphological features alone. Because terminology for identifying and describing facets of leaf architecture is not always uniformly defined, we follow that set out by Hickey (1973, 1979).

DESCRIPTION

Extant Vascular Plants with Reticulate Venation

Pteridophytes.   We present examples of reticulate venation among nonseed plants merely to underscore its widespread occurrence among vascular plants. Although reticulate venation is relatively common among ferns, it has received much less attention than that of angiosperms. However, at a gross morphological level, variations in venation have been of significant taxonomic value (e.g., Wagner, 1979 and references cited therein). Wagner (1979) regards the many cases of reticulate venation among ferns to be the result of repeated parallelism and suggests that reticulate venation has appeared independently in this group over 50 times. Figure 2.1A–C illustrates some of the more common types of filicalean reticulate venation patterns.

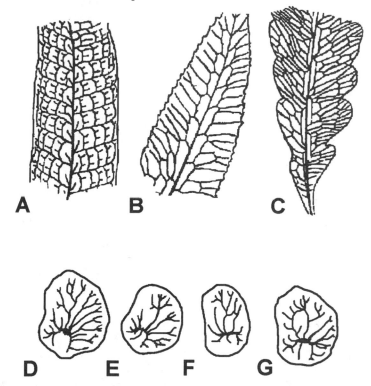

Figure 2.1.   Reticulate venation among extant pteridophytes. (A–C) Three patterns of simple reticulate venation that occur in filicalean ferns: (A) *Polypodium;* (B) *Woodwardia*; (C) *Onoclea*. (D–G) The unusual branched, occasionally reticulate venation occurring in some microphylls of the lycopsid *Selaginella schaffneri*. [A–C redrawn from Gifford and Foster (1989); D–G redrawn from Wagner et al. (1982).]

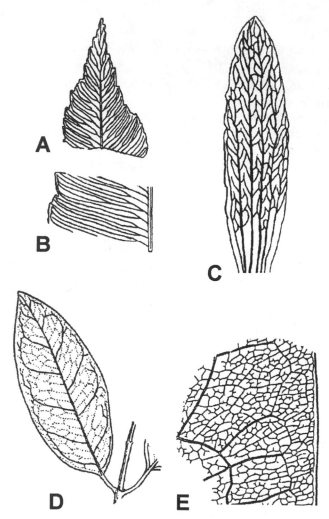

Figure 2.2. Reticulate venation among extant gymnosperms. (A,B) Cycadales. (A) Apex and (B) midregion of a leaflet of *Stangeria schizodon* with dichotomously branched lateral veins that anastomose submarginally. (C–E) Gnetales. (C) Venation pattern of *Welwitschia mirabilis* cotyledon comprising longitudinal primary veins and secondary veins that anastomose to form apically oriented chevrons. (D,E) *Gnetum*. Overall (D) and higher-order (E) venation patterns. Higher-order veins form a fine reticulum with occasional freely ending veinlets. [A, B, and E redrawn from Troll (1939a); C redrawn from Crane and Upchurch (1987); D redrawn from Gifford and Foster (1989).]

The discovery of branched venation in two species of *Selaginella* (Wagner et al., 1982) is surprising because it contradicts the concept of the lycopsid leaf as a single-veined microphyll. *Selaginella adunca*, from the western Himalayas, has a rather simple, open, ternate venation, but *S. schaffneri* (Fig. 2.2D–G), from three localities in the Central Mexican Plateau, exhibits a relatively complex, branched venation which approaches that of many ferns and gymnosperms. Approximately one-third of the leaves of *S. schaffneri* have reticulations with one to three areoles per leaf (Wagner et al., 1982). Wagner et al. (1982) do not call to question the separate origins of megaphylls and microphylls nor do they suggest that the leaves in question are megaphylls. Rather, they make the point that venation can be labile, with similar patterns achieved via markedly different evolutionary and developmental pathways, including the branched elaboration of a single vascular strand in a tiny microphyll.

Gymnosperms. The comparative morphology of venation patterns in extant gymnosperms has received even less attention than that of the ferns. However, a survey of the literature suggests that *Ginkgo* (Arnott, 1959), the cycads, particularly *Stangeria* (Troll, 1939a; Brashier, 1968), and the gnetophytes *Gnetum* (Rodin, 1966, 1967) and *Welwitschia* (Rodin, 1953, 1958a, 1958b) have reticulate venation.

Ginkgo leaves lack midribs and have a single order of open dichotomous flabellate venation that lacks freely ending veinlets (Table 2.1). However, Arnott (1959) found occasional (1–5 per leaf) anastomoses in about 10% of his more than 1000 specimens. He documented several types of anastomoses. Most common are those in which the bundle sheaths and accompanying transfusion tissue of adjacent bundles are united. In other types that lack bundle sheaths and transfusion tissue, there is complete fusion of the two veins. In still other cases, apparent anastomoses merely represent the abutting of adjacent veins, that is, "vein approximations" (Arnott, 1959). Thus, although *Ginkgo* is characterized by open dichotomous venation, Arnott's (1959) study shows that several types of reticulations occur in this form. The anastomoses in *Ginkgo* are instructive because they illustrate a variety of vascular reticulations at the anatomical level that appear identical at the gross morphological level.

Cycads typically bear pinnately compound leaves with most leaflets having a simple or open venation pattern, although some taxa exhibit more complex venation. *Stangeria schizodon*, for example, has leaflets with a well-defined multistranded midrib and pinnate secondary veins that dichotomize and anastomose to form a continuous submarginal vascular system (Brashier, 1968) (Table 2.1; Fig. 2.2A,B). Among cycads, lateral veins are essentially all one size-order and no true cross-tie veins or freely ending veinlets are found (Wylie, 1939; Brashier, 1968). The venation pattern in cycads is sometimes complicated by the presence of a well-developed, nonvascular connecting tissue that superficially resembles veins (Brashier, 1968). In *Zamia pumila*, connecting tissue extends along the edge of the leaflet, mimicking the appearance of marginal anastomoses of the lateral veins. In *Stangeria paradoxa*, tissue of this sort is associated with the marginal vascular system, making the veins appear larger (Brashier, 1968). However, cycad venation, although reticulate in some taxa, is less complex than that of even the least complex first-rank angiosperms.

Among gnetaleans, *Ephedra* has very simple venation in its reduced leaves, whereas *Welwitschia* (Table 2.1; Fig. 2.2C) produces reticulate venation in both its cotyledons and its single pair of long-lived foliar leaves. *Welwitschia* leaves have several longitudinally oriented major veins with secondary veins that interconnect to form chevrons. Chevron tips either end in the mesophyll or connect to the base of the next distal chevron (Fig. 2.2C). *Welwitschia* venation, like that of the cycads, is less complex than that of modern angiosperms but more rigidly organized than early angiosperms.

In contrast to occasional anastomoses in *Ginkgo* and the reticulum of secondary veins in some cycads and *Welwitschia*, *Gnetum* (Fig. 2.2D,E) has hierarchical reticulate venation. The following description of venation in *Gnetum* is summarized from Rodin (1966, 1967) (Table 2.1). Leaves have a prominent multistranded midrib, strongly developed pinnate secondary and tertiary veins and weakly developed quaternary veins. Regular anastomoses occur among vein orders 2–4 and simple and branched freely ending veins may be present. Additionally, *Gnetum* has "anomalous veins," that is, secondary and/or tertiary veins that do not "connect up" completely with major veins. The midrib has 4–13 (typically 5–7) parallel bundles near the base of the leaf, although, distally, the number diminishes as the bundles diverge toward the leaf margin. Pseudomonopodial divisions of the major bundles produce secondary veins in *Gnetum gnemon*. Basally, the two outermost bundles divide once to several times, with each successive subordinate branch (secondary vein) arching toward the leaf margin before undergoing a final, equal, submarginal dichotomy. Distally, the outer two longitudinal bundles arch toward the margins, ending in a submarginal bifurcation. At about the level of the submarginal bifurcation, the two adjacent longitudinal bundles begin the same progression of divisions, divergences, and bifurcations. At the apex, the remaining median bundle dichotomizes. Submarginally, derivatives of ultimate bifurcations anastomose to form a series of large "meshes" on either side of the midrib (Fig. 2.2D,E). In mature leaves, a reticulum of tertiary, occasionally quaternary, veins is present. Exterior to the coarse reticulum is a finer reticulum of tertiary veins (Fig. 2.2E). Rodin (1967) emphasized that the venation ontogeny of *Gnetum* is unlike that of angiosperms. However, Rodin was comparing *Gnetum* to modern dicots. Fossil angiosperm leaves below high second rank have dichotomous intercostal venation similar to *Gnetum* (Hickey and Doyle, 1977). *Gnetum* leaves contain features that are unusual in gymnosperms, such as laticifers and a well-differentiated mesophyll of palisade and spongy layers (Rodin, 1966, 1967), but as Hickey and Doyle (1977) noted, the dense mat of subepidermal fibers with simple pits is not typical of angiosperm leaves. Thus, *Gnetum* has a combination of gymnospermous and angiospermous foliar characters and is considered by some to be the gymnosperm with leaves most similar to the dicots (Wolfe, in Hickey and Doyle, 1977; Hickey and Taylor, Chapter 8).

The Gnetales exhibit other angiospermous characteristics including vessels in the wood, compound strobili with flower-like similarities, reduced mega-gametophyte development without archegonia formation, and in *Gnetum* and *Welwitschia*, cellular embryogeny. Recent studies by Friedman (1990a, 1990b) on *Ephedra* and by Carmichael and Friedman (1994) on *Gnetum gnemon* have demonstrated the presence of a fertilization event that, in several features, mimics the defining synapomorphy of angiospermy, double fertilization. Clad-istic analyses (e.g., Doyle and Donoghue, 1987; Rothwell and Serbet, 1994) place angiosperms as the sister group to the Bennettitales and the Gnetales.

Fossil Seed Plants with Reticulate Venation

Reticulate venation in gymnosperms was much more common during the Paleozoic and Mesozoic. Although most gymnosperms today have needle-like foliage with highly reduced venation, many fossil representatives of living groups had much more diverse foliage with more complex venation patterns. In addition, reticulate venation was widespread among seed ferns and other groups now extinct. A variety of reticulate patterns occurs in such groups as Paleozoic and Mesozoic seed ferns, fossil cycads, bennettitaleans, gigantopterids, gnetop-terids, and some *Incertae sedis* taxa. Unfortunately, despite the prevalence of reticulate venation among fossil gymnosperms, there often is a paucity of information regarding such patterns both in the fossils and the literature. Coarser-grained compression/impressions do not preserve nuances of fine struc-ture; many fossils lack preservation of anatomical structure, and there is a lack of emphasis on venation patterns compared with other characters in some studies. Whenever possible in this chapter, descriptions of venation patterns will be based on information from anatomically preserved fossil foliage.

Paleozoic Seed Plants with Fern-like Foliage

Several taxa of Paleozoic pinnules thought to have medullosan affinities exhibit similar morphologies but a suite of venation patterns. The first comple-ment of pinnules has an overall morphology typified by *Neuropteris* (Fig. 2.3A), a widely distributed form-genus with elongate tongue-shaped pinnules with cordate bases, constricted points of attachment and nearly parallel margins that taper to a rounded apex. Pinnules have a relatively prominent single-

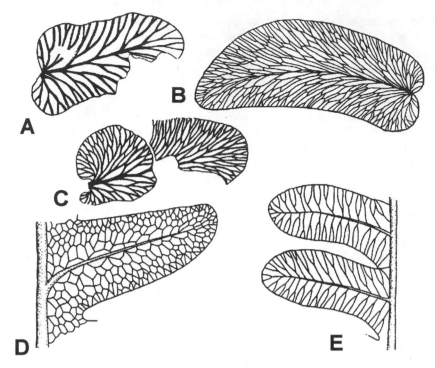

Figure 2.3. Paleozoic seed plants with fern-like foliage. (A) *Neuropteris rarinervis* pinnule with pinnate, open dichotomous secondary venation. (B) *Linopteris* pinnule with reticulate secondary venation. (C) Composite reconstruction of *Reticulopteris muensteri* pinnule with venation similar to *Linopteris*. (D) *Lonchopteris* pinnule with prominant midrib and reticulate secondary venation. (E) *Alethopteris* pinnule with open dichotomous secondary venation. [A and C redrawn from Reihman and Schabilion (1978); B, D, and E redrawn from Taylor and Taylor (1993).]

stranded midrib that does not extend to the pinnule tip and dichotomous secondary venation (Brongniart, 1822; Sternberg, 1823, 1825).

Linopteris (Fig. 2.3B) and *Reticulopteris* (Table 2.1; Fig. 2.3C) are genera of *Neuropteris*-like reticulate to pseudoreticulate veined pinnules. Like *Neuropteris*, they have a single-stranded midrib that does not extend to the tip of the pinnule and pinnately arranged dichotomous secondary venation. However, they have frequent anastomoses of the secondary veins. They lack freely ending veinlets (Figs. 2.3B, C). Gothan (1941) proposed the form genus *Reticulopteris* for pinnules similar to *Linopteris* but which were borne on fronds terminating in a single pinnule and he restricted *Linopteris* to fronds terminating in a pinnule pair. Unfortunately, many pinnules are found dispersed. *Linopteris* pinnules are a minor component at a variety of Carboniferous compression localities (Remy and Remy, 1959; Andrews, 1961; Wagner, 1964; Boersma and Broekmeyer, 1979; Laveine et al., 1989). Some *Linopteris* pinnules occur in coal balls and, based on constancy of association, are thought to belong to the medullosan frond *Sutcliffia* (Stidd et al., 1975). Interestingly, Scott and Taylor (1983) discuss the possible mimicry of net-veined early insect wings to *Linopteris* pinnules. *Reticulopteris* pinnules are far more common and have been described from many localities in Euramerica (Bell, 1938, 1962; Gothan and Remy, 1957; Crookall, 1959; Josten, 1962; Wagner, 1964; Reihman and Schabilion, 1978; Cleal and Zodrow, 1989). Reihman and Schabilion (1978) described anatomi- cally preserved *Reticulopteris* pinnules (Fig. 2.3C) with reticulate venation, whereas others have documented specimens with pseudoreticulations (Crookall, 1959). Crookall (1959) remarked on the intergradation between *Reticulopteris* and *Neuropteris* species that exhibit flexuous (e.g., *N. obliqua* cf. *R. muensteri*) to occasionally anastomosing (e.g., *N. heterophylla*) venation. Reihman and Schabilion (1978) acknowledge the possibility that *N. rarinervis* and *R. muen- steri* may be congeneric or even conspecific, based on anatomical and mor- phological similarities and consistent co-occurrence at certain localities (Bell, 1938, 1962; Gothan and Remy, 1957). With so much intergradation in the presence or extent of vein anastomoses, much remains to be done to resolve the biological affinities of this large and complex group.

Lonchopteris (Fig. 2.3D), a rarely occurring genus also thought to have medullosan affinities, produced pinnules with reticulate venation that otherwise are morphologically quite similar to more commonly occurring dichotomous- veined *Alethopteris* (Fig. 2.3E) pinnules (Remy and Remy, 1959; Boureau and Doubinger, 1975). *Lonchopteris* pinnules are broadly attached with revolute margins and decurrent bases with conspicuous midribs that extend to the pinnule apex. Secondary venation is coarsely reticulate, with veins forming a steep angle with the midrib.

The occurrence and possible intergradation of dichotomous and reticulate venation within and among taxa of neuropterid or alethopterid pinnules may simply reflect the lability of venation patterns within certain Paleozoic seed ferns. Neuropterid and alethopterid fronds are too large and complex and their last occurrence in the fossil record is too far removed from the Early Cretaceous for them to be considered direct angiosperm leaf precursors. In addition, medul- losan fertile structures, particularly the highly derived pollen organs, are not at all angiosperm-like, making the group unlikely as an angiosperm ancestor.

Late Paleozoic and Mesozoic Seed Plants

Glossopterids.   The genus *Glossopteris* (Table 2.1; Fig. 2.4) encompasses entire spatulate leaves with a distinct midrib and reticulate lateral venation. Leaves of this type dominate the Permian Gondwana impression/compression

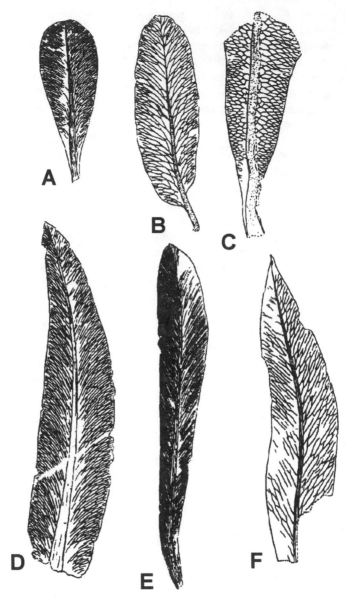

Figure 2.4.   Examples of variation in size, shape and venation patterns in *Glossopteris* leaves: (A) *Glossopteris pandurata*; (B) *G. sastrii*; (C) *G. tortuosa*; (D) *G. leptoneura*; (E) *G. angustifolia*; (F) *G. gondwanensis*. [A-F redrawn from Chandra and Singh (1992).]

floras of Australia, Antarctica, South Africa, India, and South America (Taylor and Taylor, 1993). It has been suggested that at least 200 species of *Glossopteris* have been described to date (Pigg, 1990). Although part of this taxonomic diversity is a measure of biological diversity, part reflects our present inability to delimit species of glossopterid plants and to recognize the suite of developmental and morphological forms that may comprise a single organ-taxon. Many species are based on often poorly preserved impressions, and organs typically are found dispersed rather than interconnected. It is beyond the scope of this report to examine *Glossopteris* leaves in detail; however, a generalized description of leaf morphology and venation is presented. Figure 2.4 illustrates a small sample of the variation in size, shape, and venation encompassed within the genus. More detailed information regarding individual taxa can be obtained from Chandra and Surange (1979), Anderson and Anderson (1983, 1985), Pigg (1990), Chandra and Singh (1992), Pigg and Taylor (1993) and references cited therein.

The venation of *Glossopteris* is characterized by a combination of features (Table 2.1; Fig. 2.4) which together comprise a structural organization distinct from other reticulate leaf forms (Pigg, 1988). Venation patterns were central to Melville's (1969, 1970, 1983) proposal of *Glossopteris* as progenitor to the angiosperms. *Glossopteris* leaves are consistently reticulate but have simple rather than hierarchical lateral venation (Fig. 2.4). Midribs are composed of several parallel vascular strands with the stoutness of the midrib resulting from the size of the individual vascular bundles and from increased mesophyll thickness. The ground tissue in the midrib region may be specialized as in some anatomically preserved forms (Gould and Delevoryas, 1977; Pigg, 1990). Distally, midribs either persist to the leaf apex, become indistinct distally, or dichotomize and anastomose distally.

Other genera of glossopterid leaves have different morphologies but secondary venation patterns similar to *Glossopteris*. *Gangamopteris* (Table 2.1), which is larger and lacks a well-defined midrib, occurs in Lower Permian sediments in contrast to the mostly Upper Permian *Glossopteris* (Rigby, 1984). *Belemnopteris* (Pant and Choudhury, 1977) has a sagittate base, basally trifurcating midrib, and usually coarse-meshed venation similar to some species of *Glossopteris* and *Gangamopteris*. *Mexiglossa*, a genus for Middle Jurassic leaf impressions from Oaxaca, Mexico, has venation patterns remarkably like those of *Glossopteris* (Delevoryas, 1969; Delevoryas and Person, 1975). Of the six forms and several intergradational types, four have venation patterns that are referable to species of *Glossopteris*. The relationship of *Mexiglossa* to *Glossopteris* is unclear.

Recent and ongoing investigations of anatomically preserved leaves in permineralized chert from Antarctica (Pigg, 1988, 1990; Pigg and Taylor, 1990) and Australia (McLoughlin, 1990; Pigg and McLoughlin, 1992; Pigg and

Trivett, 1994) relate external morphology to underlying anatomy and allow comparisons of leaves from various provenances of Gondwana. Two species described from Antarctica, *G. schopfii* and *G. skaarensis*, have narrow-and coarse-meshed venation, respectively, and would be recognized as species of *Glossopteris* on external morphology alone. However, their internal anatomical structure and cuticular details differ dramatically (Pigg, 1990). Leaves similar to the narrow-meshed *G. schopfii* also occur at several localities in the Bowen and Sydney Basins in Australia, attesting to the apparently widespread distribution of this general leaf form (McLoughlin, 1990; Pigg and McLoughlin, 1992). Additional investigations of anatomical material and correlations with impression/compression specimens should greatly refine our knowledge of this diverse group.

Some glossopterids bore ovulate structures that have been interpreted as angiospermous (e.g., Retallack and Dilcher, 1981b), and they are more recent than the neuropterid/alethopterid seed ferns. However, their last occurrence in the Triassic is still far removed from the Late Jurassic-Early Cretaceous earliest known angiosperms. The simple, entire glossopterid leaf-forms presumably were derived from compound seed-fern fronds, but the leaves are much larger and more complex than the hypothetical xeromorphic intermediate leaf of Doyle and Hickey (1976), whereas their single order of reticulate lateral venation is less complex but more regular than that of early angiosperms. Although the glossopterids may approximate an early grade in the reduction-expansion model for the evolution of angiosperm leaves (Doyle and Hickey, 1976; Hickey and Doyle, 1977), their basic venation is so widespread among Late Paleozoic and Mesozoic gymnosperms that, as Doyle (1978) noted, it tells us little about possible evolutionary relationships. It is unlikely that the glossopterids gave rise to the angiosperms (Doyle and Hickey, 1976).

Gigantopterids.   The gigantopterids (Fig. 2.5A,B) are represented by large (up to 80.0 cm long) leaves recovered primarily from Permian localities of China and other Asian countries (Koidzumi, 1936; Asama, 1976, 1984) and western North America, particularly Texas (Read and Mamay, 1964; Mamay, 1986, 1988). Although most gigantopterids occur as impressions, some wholly or partially permineralized specimens are known. Gigantopterids are regarded as putative pteridosperms, but the dearth of detailed information about reproductive and vegetative structures other than leaves keeps the affinities of these plants unresolved. Gigantopterids are highly variable in their frond morphology and range from simple petiolate leaves to forked and pinnately compound fronds. Simple leaves and pinnules have a pronounced midrib and three to four orders of lateral veins with at least the higher vein orders forming a reticulum.

Redescribed permineralized material of *Gigantonoclea quizhouensis* (Table 2.1; Fig. 2.5A) from the Upper Permian of southeast China (Li and Tian, 1990;

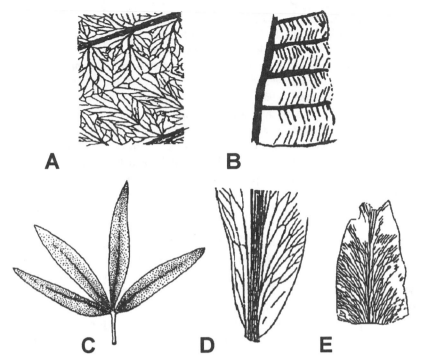

Figure 2.5. Venation patterns among the Gigantopteridales and Caytoniales. (A,B) Gigantopteridales. (A) Detail of *Gigantonoclea guizhouensis* venation showing two secondary veins with numerous tertiary veins terminating in an intercostal sutural vein. (B) Portion of *Delnortea abbottiae* venation illustrating the "herringbone" pattern of the secondary and tertiary veins suggestive of gnetalean venation. (C–E) Leaf form and venation in the Caytoniales. (C) Palmately compound leaf of *Sagenopteris phillipsi*. (D) Basal portion of *S. phillipsi* leaflet with coarse-meshed venation. (E) Apex of *S. colpodes* leaflet with fine-meshed venation. [A redrawn from Li et al. (1994); B redrawn from Mamay et al. (1988); C redrawn from Stewart and Rothwell (1993); E and F redrawn from Harris (1964).]

Li et al., 1994) reveals a probably simple, elliptical leaf about 10 cm long and 4 cm wide with a serrate margin. Secondary and tertiary veins are pinnately arranged. Quaternary veins dichotomize (equally and unequally) and anastomose to form long, simple, rectangular to triangular meshes that occasionally enclose blind veinlets. Additional long, triangular meshes are formed by the anastomosis of cladodromously ending quaternary veins. Such meshes fuse with superadjacent meshes to form more or less distinct zigzag sutural veins that lie between and parallel to secondary veins (Li et al., 1994).

Mamay et al. (1986, 1988) described an intriguing species of gigantopterid leaves from the Lower Permian of Texas. The simple, oblong-elliptical *Delnor-*

tea abbottiae (Fig. 2.5B) leaves have crenate margins and range in length from 1.2 to about 35 cm with short, stout petioles that flare to a basal abscission zone. The leaves have four orders of venation forming a precise "herringbone" pattern. Secondaries and tertiaries are unbranched and pinnately arranged; quaternaries divide sparsely and fuse with other fourth-order veins to form a dense reticulum of small meshes (Mamay et al., 1988). Unlike *Gigantonoclea guizhouensis*, sutural veins and freely ending veinlets are absent. Although *Delnortea* leaves are referable to the gigantopterids based on overall venation, they have several unusual features of interest. Thickened tissue forms a narrow band along the entire leaf margin, a feature unknown in other gigantopterids. Secondary veins terminate in the bases of the marginal sinuses, flaring slightly where they coalesce with the thickened band. The usual condition in the gigantopterids, as well as most vascular plants, is for veins to terminate in marginal teeth or lobes, not between them, although Mamay et al. (1988) call attention to the occurrence of this anomalous feature in the extant *Nothofagus gunnii*. Although they are unable to suggest a possible progenitor for *Delnortea*, they do speculate on its potential role in the ancestry of the Gnetales (Mamay et al., 1988). Drawing on palynological data, similarities in major venation orders between *Delnortea* and *Gnetum*, and strong concordance between features of *Delnortea* and the hypothetical ancestral anthophyte leaf of Doyle and Donoghue (1986), Mamay et al. (1988) present evidence for a *Delnortea-Gnetum* relationship.

Although the gigantopterids have hierarchical reticulate venation, it is a rigidly organized pattern not found in extant flowering plants (Doyle and Hickey, 1976). In addition, their venation is much more regular and complex (viz., freely ending veinlets) than the earliest known angiosperm leaves. It is unlikely that the predominantly Permian-aged gigantopterids gave rise to the angiosperms. In fact, until something is known about their reproductive structures, their phylogenetic position remains uncertain. However, it may be possible to establish a link between *Delnortea* and the Gnetales, thereby elucidating the origins of that enigmatic but apparently natural group (Mamay et al., 1988).

Caytoniales.   The Caytoniales originally were described by Thomas (1925) as a new group of angiospermous plants and are the best known of the Mesozoic pteridosperms (Taylor and Taylor, 1993). Leaf remains assignable to *Sagenopteris* (Table 2.1; Fig. 2.5C–E) are widely distributed both geographically and stratigraphically, extending from the Upper Triassic to Lower Cretaceous of Greenland, England, Sardinia, Siberia, the eastern portions of the former USSR, Japan, and western Canada (Stewart and Rothwell, 1993). The palmately compound leaves have three to six leaflets borne at the petiole apex (Fig. 2.5C). Leaflets are up to 7.0 cm long and lanceolate with a midrib that extends beyond the midpoint of the leaflet but not to the apex (Fig. 2.5D,E). Secondaries arise at a steep angle to the midrib and curve outward, ending freely at the leaflet margin. Secondary venation is reticulate with obliquely elongate areoles (Fig. 2.5D,E; Harris, 1932b, 1964).

Reproductive organs of the Caytoniales are well known and have attracted much attention due to their potential as angiosperm precursors. Some Caytoniales produced seed-bearing cupules that have been regarded as carpel (Meeuse, 1972; Krassilov, 1977; Doyle, 1978) or bitegmic, anatropous ovule homologues (Gaussen, 1946; Stebbins, 1974). They have loculate microsporangia that superficially resemble anthers. Although most caytonialean leaves have open venation, *Sagenopteris* leaves have simple reticulate venation. Doyle and Hickey (1976) suggest that palmately compound *Sagenopteris* leaves may represent a stage in their hypothesized leaf reduction trend. The evidence *in toto* supports the possibility of a Caytoniales-angiosperm lineage, a relationship suggested by the cladistic analysis of Doyle and Donoghue (1987) and bolstered by a geologic range that extends into the Lower Cretaceous.

Cycadales.   *Ctenis* (Table 2.1; Fig. 2.6D,E) is a genus of pinnately compound cycad leaves 1–2 m long that occurs at a number of localities in England, Hungary, Sweden, eastern Greenland, North America, Japan, Manchuria and Korea (Harris, 1964 and references cited therein). Leaves are once pinnate, with the typically entire leaflets broadly attached to a stout rachis. Leaflets lack a midrib but have several roughly parallel veins that anastomose, ending at or near the margin (Fig. 2.6A–C). The genus *Dunedoonia*, from the Permian of New South Wales, is a pinnately compound frond whose pinnules exhibit glossopterid-like reticulate venation. *Dunedoonia* is thought to have cycadophyte affinities (Holmes, 1977).

Bennettitales.   The Bennettitales (Cycadeoidales of some authors) are an extinct group of enigmatic Mesozoic gymnosperms that do not persist beyond the Cretaceous (Harris, 1932b, 1969). Bennettitaleans are similar to cycads in habit and overall vegetative morphology although reproductive structures are distinctly different and have figured in theories of angiosperm origins (e.g., Crane, 1986; Doyle and Donoghue, 1986b, 1987). Frond architecture in the group is so comparable to that of cycads that paleobotanists were unable to distinguish reliably between them prior to investigations by Thomas and Bancroft (1913) of extant and Mesozoic cycadophyte cuticles. Cycadeoids differ from cycads in features of epidermal cells, degree of cutinization of guard cells, orientation of stomata, and in having syndetocheilic rather than haplocheilic stomata. Although pinna venation is typically parallel to open dichotomous, *Dictyozamites* (Table 2.1; Fig. 2.6D,E) has reticulate venation. Fronds of *Dictyozamites* are once pinnate with pinnae up to 3–4 cm long with asymmetrically rounded to slightly lobed bases. Pinnae typically are crowded to overlapping on the adaxial surface of the rachis (Fig. 2.6D). Veins diverge

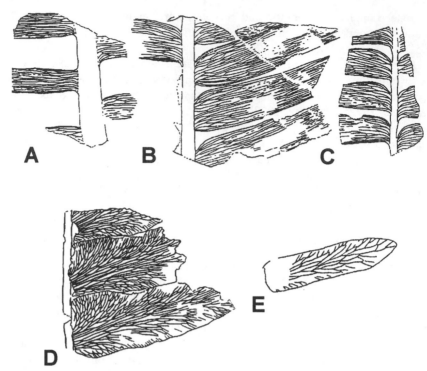

Figure 2.6. Reticulate venation in fossil Cycadales and Bennettitales. (A–C) Cycadales. (A) Lower, (B) middle, and (C) upper portions of *Ctenis kaneharai* leaves. Leaflets lack a midrib and have simple, reticulate venation. (D,E) Bennettitales. (D) Portion of *Dictyozamites hawelli* leaf and (E) leaflet with simple reticulate venation and no midrib. [A–C and E redrawn from Harris (1964); D redrawn from Taylor and Taylor (1993).]

from the constricted point of attachment, forking and anastomosing frequently (Fig. 2.6D,E).

   Ctenis (Cycadales) and *Dictyozamites* (Bennettitales) are quite similar in frond architecture and venation. However, several recent cladistic analyses have shown that the Cycadales and Bennettitales (viz., the cycadophytes) are at best only distantly related (Crane 1985, 1988a; Doyle and Donoghue, 1986b, 1987; Rothwell and Serbet, 1994) and that vegetative similarities are most likely the result of parallel evolution. Both have fairly large, compound fronds like Paleozoic seed ferns and they have only one order of reticulate venation like the glossopterids and *Sagenopteris*. Leaves of both groups are similar to the hypothetical, pinnately compound, seed-plant stage basal to the reduction-expansion model of Doyle and Hickey (1976). Cycadalean reproductive structures are distinctly gymnospermous, whereas certain bennettitalean reproductive organs have been described as "flowers" or perianth-like (Wieland, 1906, 1916) although other studies (Delevoryas, 1968; Crepet, 1974) have shown that they did not open and did not function as Wieland (1916) proposed. The structures contain seeds interpreted to be double integumented (Crane, 1985, 1986). Crane's (1988a) cladistic analysis suggests that the Bennettitales may have affinities with the Gnetales and the flowering plants, a possibility supported by the occurrence of the group in the Cretaceous.

Gnetophytes. Prior to the description of *Drewria potomacensis* from the Lower Cretaceous Potomac Group of Virginia (Crane and Upchurch, 1987), the gnetophytes were essentially unknown as megafossils, although Mamay et al. (1988) discuss a possible phylogenetic relationship between *Gnetum* and the Permian gigantopterid *Delnortea*. *Dechellyia* (Ash, 1972) and *Dinophyton* (Ash, 1970; Krassilov and Ash, 1988), enigmatic plants from the Triassic Chinle formation of the southwestern United States, and *Archaestrobilus* (Cornet, Chapter 3) from the Late Triassic of Texas, also have been suggested to have gnetophyte affinities. The following description of *Drewria* is based on Crane and Upchurch (1987). *D. potomacensis* (Table 2.1; Fig. 2.7A,B) has oblong leaves up to 20 mm long with two to three pairs of longitudinally parallel primary veins. The centralmost pair of veins extends to the acute to sharply tipped leaf apex, branching in the apical region to form a reticulum (Fig. 2.7,A). Secondary cross-veins form apically oriented chevrons (Fig. 2.7A,B. Leaves are thin with abundant subepidermal, longitudinally oriented fibers and are attached at swollen nodes by narrow-acute to decurrent bases. Phyllotaxis is opposite and decussate. The herringbone venation of *Drewria* leaves is not angiosperm-like and is most similar to that seen in the extant *Welwitschia* (Fig. 2.2C).

Triassic Plants of Uncertain Affinities

SANMIGUELIA. Sanmiguelia (Table 2.1; Fig. 2.7C) is perhaps the pre-Cretaceous fossil most often implicated in angiosperm origins. Brown (1956) originally described *Sanmiguelia* leaves from siltstone and fine sandstone impressions from the Late Triassic of Colorado and suggested alliances of the plicate leaves to the Palmae. Information regarding the plant habit (Fig. 2.7C) emerged from the discovery by Tidwell et al. (1977) of three-dimensional casts of *in situ* stems with attached leaves. Recent descriptions of *Sanmiguelia* (Cornet, 1986, 1989b) from the Upper Triassic of Texas present a presumed primitive angiosperm incorporating features of both monocots and dicots. The following description of venation patterns is based on descriptions by Cornet (1986) of a vegetative colony of *Sanmiguelia* plants in growth position at a Late Triassic locality in northwest Texas. The broadly elliptical leaves exhibit a narrow basal region that expands into a decurrent ensheathing leaf base while apically they

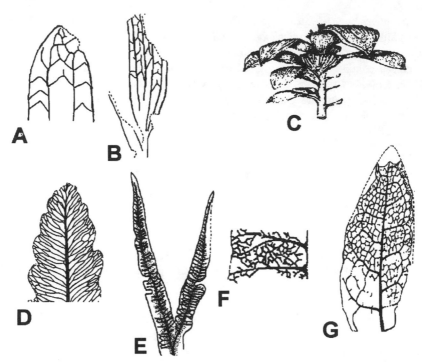

Figure 2.7. Venation patterns in gnetophytes and some Triassic plants of uncertain affinitiess. (A,B) Venation of fossil gnetophytes. (A) Apical and (B) basal portions of *Drewria potomacensis* leaves with venation similar to *Welwitschia* cotyledons. (C-G) Triassic plants of uncertain affinities. (C) Portion of *Sanmiguelia lewisii* shoot. (D) Apical portion of a *Marcouia neuropteroides* pinnule with sub-marginal reticulations. (E,F) *Furcula granulifera* (E) Bifurcating leaf and (F) higher-order venation pattern. (G) Portion of presumed trilobed leaf of *Pannaulika triassica* with four orders of angiosperm-like venation. [A and B redrawn from Crane and Upchurch (1987); C, E, and F redrawn from Stewart and Rothwell (1993); D redrawn from Ash (1972); G redrawn from Cornet (1993).]

taper to an acute to acuminate tip. Small to large leaves are borne in loose to crowded helices (Fig. 2.7C) and have up to four orders of venation. Higher vein orders show common to abundant bifurcations and anastomoses with tertiary and quaternary orders forming distinct cross-veins to the lower-ordered parallel venation. Distinct longitudinal pleats are a prominent feature of the leaves. Large leaves are 48–60 cm long and 28–31 cm wide with 24 plications. Primary veins are restricted to the basal portion of the lamina. They divide to produce smaller parallel primary veins and, following sufficient diminution, continue apically as secondary veins. Basal primary veins alternate with the plications, but more apically, primaries as well as secondaries occupy the folds. Longitudinally oriented tertiary and quaternary veins occur mostly between the smaller primary veins and comprise the majority of the veins except in the base of the leaf. In the tapering apex of the blade, numerous anastomoses occur. Although tertiary veins form only occasional cross-veins, quaternary veins form abundant cross-veins and loops. Cornet (1989b) reports vessels in the secondary xylem of the roots. Associated with the Texas plants are reproductive structures reconstructed as ovulate (*Axelrodia*) and pollen-producing (*Synangispadixis*) inflorescences (Cornet, 1989b).

Currently, the taxonomic position of *Sanmiguelia* is not clear. Some workers have postulated affinities with the monocots [viz., Palmae (Brown, 1956); Liliaceae (Tidwell, et al., 1977)] and others with the cycadophytes (Bock, 1969), whereas Cornet (1986, 1989b) has interpreted *Sanmiguelia* to be a premagnoliid angiosperm that occurred close to the dicot-monocot divergence and that combines features of both. Cornet (1986, 1989b) maintains that because of its angiospermous venation and the associated reproductive axes, *Sanmiguelia* should not be dismissed as an early angiosperm simply because of its Upper Triassic occurrence. However, until clearer documentation of the putative features and biological affinities of the suite of associated organs is presented, it is not possible to assess the morphology and reproductive biology of *Sanmiguelia*. Obviously, our understanding of this controversial taxon is far from adequate.

MARCOUIA. *Marcouia* leaves (Table 2.1; Fig. 2.7D) occur in the Upper Triassic Chinle Formation of Arizona and New Mexico (Ash, 1972), and originally were described by Daugherty (1941) as *Ctenis neuropteroides*. Most specimens consist of pinna fragments with a few pinnules attached. The most complete leaf is a palmately compound frond with four to five basally united pinnae. The pinnae are linear lanceolate and fairly large (≥ 30 cm long by ≥ 15 cm wide), with the laminae segmented into opposite to subopposite pinnules. Pinnules have a well-defined midrib that extends to 1–2 cm below the apex. Secondary veins dichotomize and anastomose several times near the pinnule margins (Fig. 2.7D). Material of *Marcouia* is so limited and fragmentary that it is difficult to asssess the taxon fully. However, its time of occurrence and known architecture have nothing to suggest intimate links to angiosperm evolution.

FURCULA. Harris (1932a) first described *Furcula* (Table 2.1; Fig. 2.7E,F) from the Upper Triassic of Greenland as a lanceolate, often bifurcating leaf with reticulate venation. *Furcula* leaves are up to 15 cm long with a strong midrib that forks in Y-shaped specimens (Fig. 2.7E). The midrib gives rise to distinct perpendicular secondary veins (Fig. 2.7E,F) spaced about 2 mm apart. In the intercostal regions higher vein-orders form a reticulum of poorly delineated areoles containing freely ending veinlets (Fig. 2.7F).

Harris (1932a) considered *Furcula* to be angiosperm-like both in its higher orders of venation and its syndetocheilic stomata, characters that led him to observe that, "If *Furcula* were a Tertiary fossil—particularly if only the undivided type of leaf were known—it would unquestionably be regarded as a Dicotyledon on the evidence of venation alone." However, because the forked laminae are more reminiscent of pteridosperms or cycadophytes, Harris (1932a) did not assign a suprageneric classification.

Although *Furcula* seems to approach the dicots in venation and cuticular structure, it has several nonangiospermous characters including a bifurcating midrib and blade, predominantly equal dichotomous secondary venation, and higher vein orders with relatively acute angles of origin (Hickey and Doyle, 1977). Moreover, the use of syndetocheilic stomata as a character of the anthophyte clade (viz., the Bennettitales, Pentoxylales, Gnetales and angiosperms) must be reevaluated in light of their occurrence in such distantly related taxa as the lycopod *Drepanophycus spinaeformis* (Stubblefield and Banks, 1978). As with other taxa discussed in this chapter, until more is known of the vegetative and reproductive biology of the *Furcula* plant, it is not possible to assess fully its contribution, if any, to angiosperm phylogeny.

PANNAULIKA.   Cornet (1993) has recently described a dicot-like leaf from a Late Triassic locality on the Virginia-North Carolina border. *Pannaulika* (Table 2.1; Fig. 2.7G), a new genus, is diagnosed as follows:

> Small leaf with entire margin and brochidodromous secondary veins arising excurrently from primary; tertiary veins form loops or arches outside secondary loops, and subaxially oriented transverse veins occur within some of the secondary loops; very regular oriented quaternary reticulate venation developed inside secondary and tertiary loops; a band of dendritic or branching quaternary venation extends from tertiary loops or reticulum to the margin of leaf. (Cornet, 1993).

The single known leaf lacks a base and most of the margin on one side. It is 3.2 cm long and 1.3 cm wide. Secondary and tertiary arches are asymmetrically developed along the wider side of the leaf and the primary vein curves slightly toward the narrow side in the basal half of the leaf (Fig. 2.7G) (Cornet, 1993). Cornet (1993) infers that the specimen may represent the left lobe of a trilobed leaf and rightly observes that should this be the case, it will be important in resolving systematic affinities. He compares the leaf to those of reticulate veined seed ferns and pteridophytes and presents major differences with both groups. Cornet (1993) finds the characters of the hypothesized three lobed palmately veined leaf to be most concordant with those of angiosperm leaves of at least rank 2. Three unusual reproductive structures were found in the same layer with the leaf. He describes one as a monocot-like axis resembling an aroid spike, one as a dicot-like receptacle with two carpel-like megasporophylls each with an orthotropous ovule, and one as a dispersed achene (similar to the carpels) containing a ribbed seed. Of course, it is not certain that any of the reproductive structures were produced by the plant that bore *Pannualika* leaves. However, should additional specimens confirm the proposed trilobed palmately veined architecture of *Pannaulika triassica* or the angiosperm-like features of the reproductive structures, it would support the presence of angiosperms as far back as the Late Triassic and give weight to Axelrod's theory (1952) of a Permo-Triassic tropical upland origin of the angiosperms.

DISCUSSION

Over the years, as botanists have sought to pinpoint the origin of flowering plants, they have advanced most major groups of fossil gymnosperms and occasionally pteridophytes as putative angiosperm ancestors. Various vegetative and reproductive characters, alone or in concert, have fueled such speculations. Leaf venation patterns have garnered particular attention because flowering plants, particularly dicots, have a venation syndrome that is relatively restricted among tracheophytes and because the early fossil record of angiosperms is heavily skewed toward pollen and leaves. But as Alvin and Chaloner (1970) point out in their reply to Melville (1969), reticulate venation alone is not an adequate indicator of phylogeny because, as seen in the present survey, it has apparently arisen independently numerous times in only distantly related groups. Reticulate venation appears to be a relatively easily acquired grade of development and may prove to be more useful as an ecological pointer (Givnish, 1976, 1979; Hickey and Doyle, 1977). If we accept the hypothesis that angiosperm evolution was driven by a xeromorphic bottleneck resulting in reduction of reproductive and vegetative structures and plant habit (Doyle and Hickey, 1976; Hickey and Doyle, 1977; Taylor and Hickey, 1992) and if the origin of angiospermy is defined *sensu stricto* as the origin of double fertilization leading to the production of nutritive endosperm, then the earliest angiosperm leaves would not have angiospermous characters unless leaf evolution preceded changes in reproduction. Therefore, depending on the sequence of evolutionary events, the protoangiosperm might have small, entire, first-rank, angiospermous leaves with disorganized venation or they might be more like simple, entire, taeniopterid-like or palmately compound *Sagenopteris*-like leaves. Although many living and fossil taxa have leaves or leaflets with reticulate venation, just a few have features that approach the more complex venation of angiosperms. Among the taxa reviewed, only *Gnetum* (extant), the gigantopterids (Permian), *Sanmiguelia* (Triassic), *Furcula* (Triassic), and *Pannaulika* (Triassic) have more than a single order of lateral veins. Some members of the gigantopterids (e.g., *Cathaysiopteris*, *Zeilleropteris*, and *Gigantonoclea*) and *Furcula* (uncertain affinities) have such angiospermous characteristics as several orders of progressively finer veins and freely ending veinlets. However, overall, the gigantopterids and *Furcula* have venation patterns that are not found in flowering plants (Hickey and Doyle, 1977) and their leaves are too large and compound or forked to be angiosperm leaf precursors or early angiosperm leaves. *Sanmiguelia* and *Pannaulika* leaves also have several angiospermous characters and they both are associated with intriguing reproductive axes that appear to be angiosperm-like. Unfortunately, the last occurrences of the gigantopterids, *Sanmiguelia*, *Furcula*, and *Pannaulika* are on the order of 100 million years before the first credible evidence for the flowering plants, making them unlikely ancestral angiosperms or preangiosperms.

Of all taxa reviewed, *Gnetum* has venation that is most similar to early angiosperm leaves (Hickey and Doyle, 1977; Hickey & Taylor, Chapter 8), although it is more organized than lowest Zone I leaves. However, we are comparing a modern leaf to early evolutionary forms. In addition, members of the Gnetales exhibit other vegetative and reproductive characteristics that are intriguingly angiospermous. Unfortunately, although it might be useful to employ a suite of gnetalean leaves in our assessment of their bearing on angiosperm evolution, the present-day Gnetales are restricted to three quite different genera and the gnetalean megafossil history is slight. To date, the only unequivocal gnetalean megafossil is *Drewria*, which is not particulary angiosperm- or preangiosperm-like in its venation. Fortunately, the palynofloral record suggests the Gnetales were a diverse group in the past. Future gnetophyte discoveries may clarify the early evolution of this group and quite possibly may provide valuable insights into angiosperm phylogeny.

CONCLUSIONS

A review of the temporal and taxonomic distribution of reticulate venation indicates that it is found in all major fossil and living groups of vascular plants except for the most ancient. Among pteridophytes, reticulate venation occurs most widely among filicalean ferns and is fairly simple and nonhierarchical. In most gymnosperms, reticulate venation also is nonhierarchical. Even in those gymnosperms with more complex reticulate venation, including progressively finer vein orders and freely ending veinlets, most (e.g., gigantopterids and *Furcula*) have patterns atypical of angiosperm leaves. Additionally, other features of their leaf morphologies are quite divergent from angiosperm, particularly early angiosperm, leaves. Other leaves (e.g., *Sanmiguelia* and *Pannaulika*) have been described as angiospermous, but they occur much earlier than the best substantiated time of origin of flowering plants. Finally, the extant *Gnetum* has relatively angiosperm-like venation, although the earliest known presumed gnetalean leaf does not. At present, the Gnetales seem to offer the most promise for resolving early angiosperm origins. However, from this review, it is obvious that venation patterns alone provide no clear-cut evidence of which, if any, known gymnospermous taxon gave rise to the flowering plants.

ACKNOWLEDGMENTS

The authors thank Sergius Mamay and an unknown reviewer for their cogent comments, Chris Trivett and Gene Mapes for their unofficial reviews, and David Winship Taylor and Leo J. Hickey, the organizers of the symposium. This work was funded in part by Faculty Grant-in-Aid B–600, Arizona State University and NSF grant BSR–9006625 to KBP.

A New Gnetophyte from the Late Carnian (Late Triassic) of Texas and its Bearing on the Origin of the Angiosperm Carpel and Stamen

Bruce Cornet

Over the last 100 years, theories on the origin of the angiosperms have shifted like the ebb and flow of the tide. But even the most recent ideas expressed at the AIBS symposium in Iowa (Angiosperm Origin, Early Evolution and Phylogeny) are not entirely new. The monocot, Amentiferae, gnetalean, and bennettitalean theories of angiosperm origin (Eames, 1961; Arber and Parkin, 1907, 1908) were proposed long before Burger (1977; 1981), Crane (1985), and Doyle and Donoghue (1986b) resurrected the anthophyte, monocot, and paleoherb theories in modified form (see also Tucker and Douglas, Chapter 7). Even concepts about the time of angiosperm origin have been cyclical, ranging from Permo-Triassic (Axelrod, 1952, 1961) to Early Cretaceous (Wolfe et al., 1975; Hughes, 1976; Hickey and Doyle, 1977) and back to Permo-Triassic (Cornet, 1986, 1989b; Cornet and Habib, 1992; Doyle and Donoghue, 1986a; Martin et al., 1989; Martin et al., 1993).

Instead of bringing us closer to resolving the morphology and origin of the first angiosperms, the study of Cretaceous angiosperms has brought us almost full circle to the realization that the Cretaceous radiation may represent only the phylogenetic history of the extant angiosperm flora, and not the history of angiosperm origin. Cladistics has extended the angiophyte clade or angiosperm phylogenetic lineage back at least to the Late Triassic and has focused interest on angiophyte sister groups for evaluating character polarity and importance. Fossil Gnetales are much less known than the extinct Bennettitales. Therefore, the discovery of a possible pregnetalean plant that occupied the same habitat as the controversial Late Triassic *Sanmiguelia lewisii* is of critical importance in the search for angiosperm ancestors.

GNETOPHYTES AND GNETALES

The Gnetales have been interpreted as transitional between gymnosperms and angiosperms at least as far back as Arber and Parkin (1908) and Wettstein (1911). This intermediate status was emphasized by the segregation of *Ephedra*, *Welwitschia*, and *Gnetum* into the subdivision Chlamydospermae (from the Greek word meaning seeds with an envelope or "cloak"). Some authors have even compared the outer envelope of the ovule of the gnetophytes with the carpel or ovary wall in angiosperms (cf. Gifford and Foster, 1989). But the notion that the Gnetales are closely related to angiosperms also cycled through a period of disfavor, only to reemerge due to new evidence favoring a very close relationship with angiosperms (Doyle and Donoghue, 1986a, 1992). The reassessment of the phylogenetic position of the Gnetales was in large part due to the contributions of Martens (1971 and references therein) and Muhammad and Sattler (1982), along with the advent and results of cladistic analysis and molecular-based phylogenetic studies (e.g. Zimmer et al., 1989).

It was not until the similarities or synapomorphies (e.g., siphonogamy, tunica-corpus, lignin chemistry, reduced megaspore wall, and granular exine) among the Gnetales, Bennettitales, Pentoxylales, and angiosperms were qualified within cladograms that their true sister-group relationships became apparent (Crane, 1985; Doyle and Donoghue, 1986a, 1986b, 1992). There was difficulty still in recognizing homologies for the carpel, anther, and second integument in the Gnetales. For example, Crane (1985) failed to recognize a cupule homolog (i.e., second integument) in the Gnetales, but both Crane (1985) and Doyle and Donoghue (1986b) recognized cupules in the Bennettitales and Pentoxylales. The angiosperm carpel was interpreted along conventional lines as derived from a simple conduplicate leaf-like megasporophyll that enclosed two marginal rows of uniovulate cupules (phyllosporous origin; see Taylor and Kirchner, Chapter 6). The carpel wall, however, was not resolved with any particular structure in nonangiospermous anthophytes. Furthermore, little comparison was made of microsporophyll morphology within the anthophytes other than to point out the apparent differences.

The reproductive morphology of the new *Welwitschia*-like plant described below provides a unique look at plesiomorphic characters and organs that have become reduced (modified) or lost in extant Gnetales, thus making homologies with angiosperm reproductive organs either unrecognizable or unconvincing, particularly when viewed through the bias of the conduplicate carpel/laminar stamen theory. In addition, these new (or lost) characters require a broadening of our concept of the Gnetales by defining gnetophytes as including the sub-clade Gnetales, and not as the complete equivalent (homolog) of the Gnetales. In this regard, the Gnetales are considered here to be the crown group, just as angiosperms are considered to be the crown group within the angiophytes (Doyle and Donoghue, 1993).

Materials and Methods

The holotype (female) for the genus was discovered on a field trip to Sunday Canyon in September, 1986. The paratype (male) and a second female reproductive axis were discovered in 1980 during the original excavation of the *Sanmiguelia* bed. All specimens come from one locality along a dirt road winding down the north wall of Sunday Canyon, just west of Palo Duro Canyon State Park, Randall County, Texas (latitude: 101° 44'; longitude: 34° 50'). The strata containing the plants occur just below a sequence of conglomerate and sandstone and appear to represent a shallowing upward interdistributary lake deposit on top of a paleosol. The *in situ* plant bed, along with transported and dropped leaves, stems, and occasional allochthonous gymnosperm cones, is restricted to the west end of a long gray mudstone lens, which is terminated westward by a down-cutting sequence of channel sandstone (see Fig. 1 in Cornet, 1986). The new taxon was found in growth position above the siltstone paleosol (Figs. 3.1, and 3.2), and as fallen spikes within the paleosol, upon which a *Sanmiguelia* colony and ferns were found rooted [see Fig. 2 in Cornet (1986) for a diagram of the plant bed; the holotype of the new taxon was found about 0.3 m behind the three vertical axes of *Sanmiguelia* connected by rhizomes shown to the left in that diagram].

Some of the specimens required degaging or breakage in order to reveal hidden parts. Most specimens, however, provided enough evidence for study and interpetation from the way they were initially exposed. Standard palynological techniques were used to secure pollen from microsporangia as well as cuticle fragments from reproductive organs. Preparations of pollen and cuticle were studied using a Zeiss binocular microscope and photographed with a Zeiss photomicroscope containing a built-in camera. Photographs of megafossils were made using a Minolta 35-mm camera with enlargement lenses and attachments.

The new reproductive organs are deposited at the Field Museum of Natural History, Chicago, and include the following:

1. PP44195—a nearly complete female spike bearing close to 90 flowers, found above the paleosol oriented with apex pointed downward and base extending into outcrop wall as if attached to a rooted plant. Excavation too dangerous to recover more of specimen because of overburden. Specimen found in 1986 in first pinkish siltstone layer above paleosol.

2. PP44196—a portion of a female spike found oriented subvertically with pith cast uncompressed and flowers radiating outwards (undistorted). Specimen found in 1980 in same pinkish siltstone layer containing PP44195.

3. PP44197—a portion of the proximal sterile part of a spike found lying horizontal at top of lacustrine mudstone lens to the east of main plant bed and about 0.7 m higher in section. Specimen found in 1986 about midway between western pinch-out of lacustrine unit and sharp bend in road as it turns north up canyon wall.

4. PP44198—a collection of three male spikes found attached to a common axis and bent around one another. Some portions of blocks containing specimens not collected, because the size and extent of specimens was not apparent until prepared in laboratory. Specimens found in 1980 within light gray siltstone paleosol, beneath *in situ Sanmiguelia* colony and associated with isolated *Pelourdea* leaves, a log, or large stem and isolated conifer/pteridophyte male cones containing bisaccate pollen.

5. PP44199—microscope slides containing acid-resistant cuticles and pollen from specimen PP44198.

Systematics

In this section the *Welwitschia*-like male and female reproductive axes are named and described as parts of a single species even though they were not found in organic attachment. The similarity of the reproductive structures is so great as to leave little doubt that they represent the same plant; the occurrence of both male and female reproductive axes in the same layers is further support that they belong to the same taxon. It is unreasonable to expect organic attachment of these organs to the same plant if that plant were dioecious. Further justification of naming both types of organs as members of the same species is dependent on morphological comparison and will follow the taxonomic description in the discussion section. The two types of axes are described under separate diagnoses of the species. The term macrocupule is introduced for clarity, because Crane (1985) and Doyle and Donoghue (1986b) used the term cupule to include extraintegumentary structures of uncertain origin that surround the ovule. Macrocupule, as defined here, is a much larger,

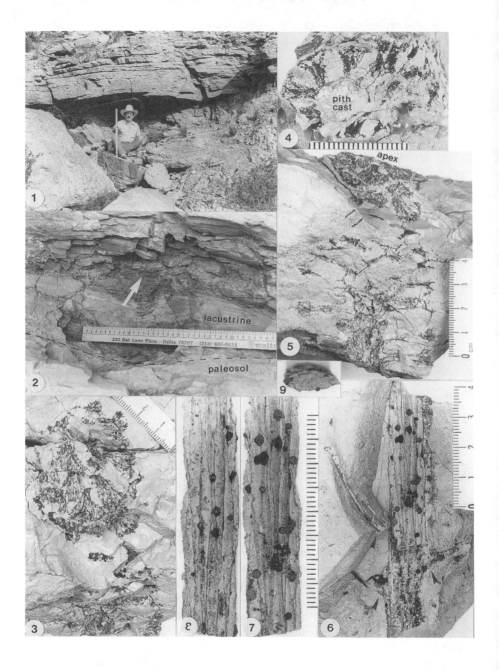

cupule-like structure that was probably derived from one or more foliar bracts. The microsporangiate "flower" of *Welwitschia mirabilis*, for example, is interpreted here as a macrocupule.

Archaestrobilus Cornet *gen. nov.*

Type species. Archaestrobilus cupulanthus Cornet *sp. nov.*

Diagnosis.   Unisexual macrocupules spirally borne on terminal axes to form strobilus-like spikes. Female spikes at least 10 cm long; male spikes at least 24 cm long. Macrocupules consist of a curled, bract-like organ with a narrow shaft and flared funnel-shaped apex. Sutures represent convergent margins of bract; sutures oriented apically (ventral). Female macrocupules contain one central elongate ovule or seed with a bulbous base (bowling-pin shaped) and gradually tapering micropylar extension; ovule surrounded by six to seven sterile scales (= interseminal scales when between ovules) with expanded flat tops that terminate below apical rim of macrocupule and isolate ovule from macrocupule wall. Ovules/seeds possess long, hair-like processes attached at micropyle apex to form a parachute dispersal apparatus. Unusual structures with short narrow pedicels and swollen heads are attached to outside of female macrocupule; these structures larger and more numerous around the macrocupule rim, decreasing to small surface ornamentation along macrocupule shaft. Three to four short

Figures 3.1–3.9. 1. *Sanmiguelia* locality at Sunday Canyon, Texas, located behind author; 3-ft. meterstick for scale; 9/86. 2. Close-up of lake bed and paleosol containing *in situ Sanmiguelia* colony, *Cladophlebis* ferns, and *Archaestrobilus*; location of holotype PP44195 shown; specimen found bent downward with apex embedded in pinkish white siltstone/fine sandstone below, with base of strobilus projecting upward (arrow, to left of specimen) into overlying gray claystone; axis then curved back down toward paleosol; burial took place over an extended period of time; meterstick used for scale; 9/86. 3. *Archaestrobilus cupulanthus* Cornet gen. et sp. nov. (holotype PP44195), female strobilus, apical view (same specimen as in Fig. 3.5); scale in millimeters for this and all subsequent figures. 4. *A. cupulanthus* (PP44196), female strobilus, showing large central pith as sediment-filled cast. 5. *A. cupulanthus* (holotype PP44195), side view showing wide main axis with macrocupules borne along its length. 6. Sterile base of reproductive axis (PP44197), bearing three 3.8 cm long lanceolate bracts in a tight spiral, followed by long internode (arrows point to bracts); pith cast in place, showing dark anastomosing and bifurcating grooves filled with lignitic material representing probable traces of primary vascular bundles that surrounded pith; isolated specimen found in same bed containing *Archaestrobilus*; bracts unlike those of *Axelrodia* and *Synangispadixis*. 7–9. Close-up of pith cast in Fig. 3.6, showing protoxylem traces, embedded in its outer surface, anastomosing and bifurcating; round dark objects are hematite nodules (after pyrite?). 8. Opposite side from that shown in Fig. 3.7. 9. Cross-sectional view of pith cast; elliptical shape due to compression.

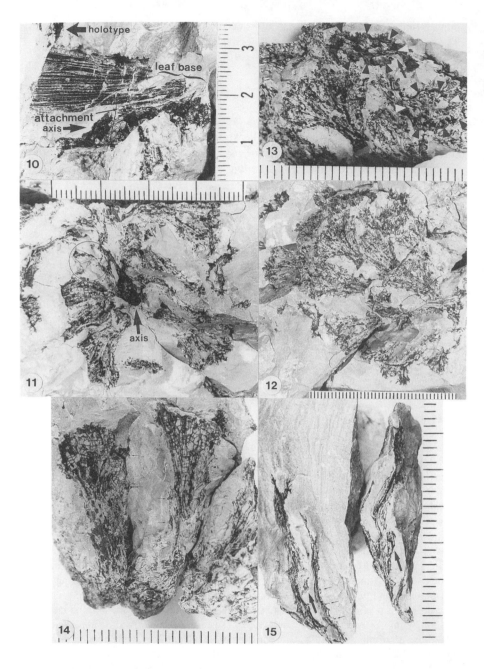

Figures 3.10–3.15. 10. *Pelourdea poleoensis* leaf, basal part showing attachment to axis (inferred from contact) and portions of inferred clasping base torn away below stem; note only two sizes of parallel veins (one large, one very small), mostly dichotomous with no apparent cross-veins; specimen on same block with *Archaestrobilus* holotype (edge of PP44195 indicated). 11. *Archaestrobilus cupulanthus* (holotype PP44195), female strobilus, showing apical macrocupules removed to expose central axis (arrow); circle around macrocupule and attached bract (see Fig. 3.18 caption). 12. *A. cupulanthus* (holotype; detail from Fig. 3.3), close-up of apical macrocupules before removal (in Fig. 3.11), which are only partially compressed and still retain some three-dimensionality (upper left); axis below center of splay (axis shown in Fig. 3.11). 3. Detail from Fig. 3.12; close-up of partially three-dimensional macrocupules, showing ring of approximately seven sterile scales (B&W arrow heads) within funnel-shaped apex, and gland-like structures borne around rim of macrocupule (visible as black dots and finger-like knobs along rim and perimeter; see Fig. 3.15 for detail); ventral suture (opening) to top of macrocupule indicated by arrow. 14. Three *Archaestrobilus* macrocupules in side view (holotype), showing flared apex, attachment to axis (below), and reduced (vestigial?) gland-like structures along shafts (small arrows). 15. Cross-sectional views of three different female macrocupules (vertical, linear black compressions with notched apices), showing open funnel-shaped apex with gland-like bodies attached, small bracts attached or positioned near base of macrocupules (large arrows), and reduced (vestigial?) gland-like structures along margin of shafts (small arrows).

narrow bracts surround base of female macrocupule. Male macrocupules without central ovule or vestigial ovule; contain thread-like appendages inside macrocupule instead of large sterile scales. Hundreds of bivalved microsporophylls are densely packed along length of shaft and around outside of funnel-shaped apex; no bracts present on outside of male macrocupule. Microsporophylls consist of two ovate bracteoles enclosing pollen sacs; three to five pollen sacs borne in cluster at end of narrow axillary stalk. Bracteole attachment opposite at base of stalk. Pollen, simple, monosulcate with thin subtectal granular-layer.

Derivation. From *archae*—Greek, meaning beginning, first cause, old; *strobilus*—Latin, meaning twisted, spiral, pine-cone-like.

Archaestrobilus cupulanthus Cornet *gen. et sp. nov.*
Holotype.   PP44195, Figs. 3.2, 3.3, 3.5, 3.11-3.19, 3.21, and 3.22.
Paratype.   PP44198, Fig. 3.4.
Type locality.   Sunday Canyon, near Canyon, Texas, U.S.A.
Stratigraphic position.   Trujillo Formation.
Age.   Late Carnian (Late Triassic).
Derivation.   From *cupula* - Latin for tub, vat, cup; *anthos* -Greek, meaning flower, in reference to flower-like reproductive units possessed by various anthophytes.

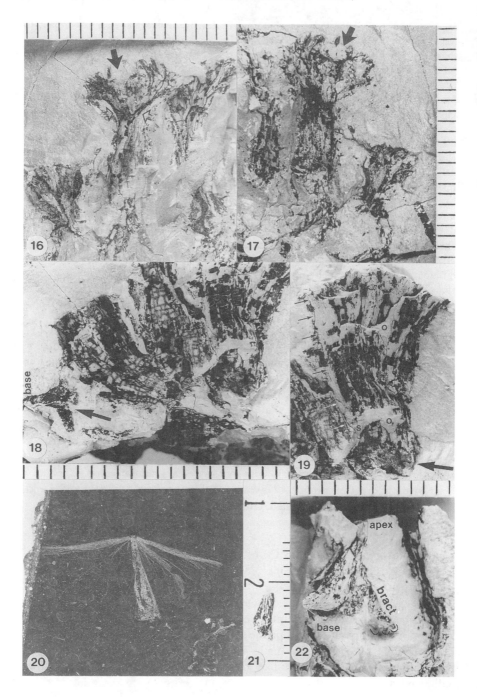

Figures 3.16–3.22.   16. *Archaestrobilus* female macrocupules (holotype), showing apical, enlarged gland-like bodies around rim (darker elliptical to finger-like knobs, complete to partially exposed), cleft (ventral suture) becoming closed below rim (closed arrow), and tops of sterile scales near base of funnel-shaped apex (scales overlap and are not distinct from macrocupule except at top), represented in central specimen by carbonaceous compression material (open arrows). 17. *Archaestrobilus* female macrocupules (holotype), counterpart to specimens in Fig. 3.16 (detail from lower portion in Fig. 3.11), showing cleft in apical rim of central specimen (arrow), representing ventral suture. 18. *Archaestrobilus* female (holotype), showing bases of three macrocupules, central axis, and one exposed digitate bract (arrow), which is bent with one digit projecting downward; base of bract (labeled) normally constricted at point of attachment, and connected to an adjacent macrocupule not present on this specimen (but visible in Fig. 3.11; circled). 19. Enlargement of female macrocupule on right side of Fig. 3.18, showing very thin to open margins of macrocupule along side (small arrows; interpreted as the ventral suture), clay-fillings within sterile scales, portions of either deformed central cavity of macrocupule and/or ovule cast (labeled O), which merges with suture traceable for entire length of that side (labeled S), and round nodule at base of macrocupule (large arrow). 20. Dispersed seed from Cow Branch Formation, Dan River/ Danville basin, North Carolina, lacustrine unit B12 (late Carnian; see Cornet, 1993). 21. Seed cast (base) from macrocupule shown in Fig. 3.22; note longitudinal carbonaceous striations imbedded in surface of seed cast that represent either remnants of a corrugated seed coat or hair-like processes similar to those on seed in Fig. 20. 22. *Archaestrobilus* female macrocupule (holotype), showing seed cast in place with a gradually tapering apex (labeled apex) and curved base (small arrow, labeled base), and one digitate bract (labelled bract along one side) attached to outside of same macrocupule above base; note constriction of digitate bract at point of attachment, and finger-like apical margin (emphasized by curved lines just beyond bract tip).

Figures.   3.3-3.9 and 3.11-3.35.

Diagnostic description (female).   Unisexual spikes borne individually at end of main axis. Distal fertile part of spike up to 10 cm long, 5.8 cm wide, containing about 100 spirally arranged macrocupules (Figs. 3.3 and 3.4). Spike axis tapers from 1.0 cm proximally to 0.4 cm distally. Sterile, basal part of spike at least 9 cm long, 14 mm wide (Fig. 3.6), bearing widely spaced, lanceolate bracts borne in subopposite pairs (1 cm apart) separated by more than 7 cm; bracts 4-5 cm long and 4 mm wide with parallel veins. Large central pith (as a cast) in sterile base of spike extends apically into lower, fertile part of spike (Fig. 3.4); compressed pith cast 11 mm wide in sterile base (Figs. 3.7-3.9), possessing loose net of 15-17 anastomosing and bifurcating protoxylem traces embedded in outer surface of pith cast. Macrocupules comprised of an axially curled (tubular), bract-like organ with a narrow shaft and flared funnel-shaped apex (Figs. 3.11-3.19). Sutures may represent convergent, unfused margins of the bract; sutures oriented apically (ventral) along spike (Figs. 3.13-3.14, and 3.19). Macrocupules 2.0 cm long on distal part of spike, 2.3 cm long proximal-

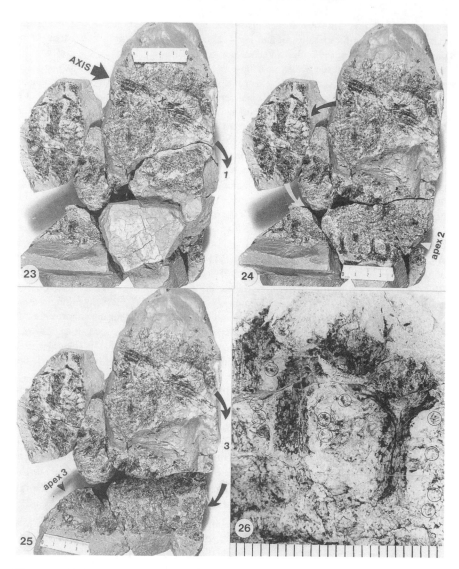

Figures 3.23–3.26. 23. *Archaestrobilus cupulanthus* (male; PP44198), top view of three male strobili attached to a common axis at X (basal attachement of axes 2 and 3 faint in photographs; also strongly implied by orientation and position; the base of male strobilus No. 1 shown by arrow pointing in direction of apex. 24. *A. cupulanthus* (PP44198), with one piece of block removed to expose male strobilus No. 2, shown by arrows as it curves from left to right, with apex indicated by white arrow head; macrocupules mostly squashed laterally; scale in centimeters. 25. *A. cupulanthus* (PP44198), with three pieces of block removed to expose strobilus No. 3, shown by

black arrows as it curves from right to left, with apex indicated by black arrow head; macrocupules mostly squashed proximal-distally; scale in centimeters. 26. Close-up of male macrocupules exposed in Fig. 3.24, belonging to male strobilus No. 2; note fragments of detached microsporophylls dispersed in matrix between macrocupules (some of which are encircled; see transfer preparation in Fig. 3.37 for apparent density of dispersed microsporophylls).

ly; macrocupule shaft 3.2-5.0 mm wide (4.5 mm average); flared funnel-shaped apex 5.0-7.0 mm long, 10.0-13.0 mm wide. Macrocupules contain a single central ovule or seed with a bulbous base and a gradually tapering elongate micropylar extension (bowling-pin shaped; Figs. 3.21 and 3.22); ovule casts about 1.2 cm long, 2 mm wide at base, about 1 mm wide at apex (Figs. 3.21 and 3.22). Ovules surrounded by six to seven sterile scales (= interseminal scales when between ovules) with expanded, flat tops which terminate 3-4 mm below apical rim of macrocupule and which isolate ovule from macrocupule wall (Fig. 3.13). Ovules/seeds possess long hair-like processes attached at micropyle apex that form a parachute dispersal apparatus (Figs. 3.20 and 3.21). Unusual structures with short narrow pedicels and swollen heads attached to outside of female macrocupule (Figs. 3.11-3.17); these structures (0.6-1.1 mm long; 0.4-0.5 mm wide heads) larger and more numerous around the macrocupule rim, decreasing to small, scattered, clavate projections (0.2-0.4 mm long; 0.1-0.2 mm heads) along macrocupule shaft (Figs. 3.14 and 3.15). Three to four 5 mm long digitate bracts surround base of female macrocupule; bracts 0.8 mm at point of attachment to macrocupule, widening rapidly to 2 mm with apical dissection into three to four short lobes (Figs. 3.18 and 3.22).

Number of specimens examined.   Two.

Diagnostic description (male).   Unisexual spikes borne in groups (of three) at the end of main axis (Figs. 3.23-3.25). Spikes fertile for almost entire length from point of origin, with macrocupules spirally borne from apex down to near base of spike. Spikes variable in length, ranging from 13 cm to 24 cm long (estimated original length up to 33 cm long); spike width ranges from 5.2 cm to 6.3 cm for the longest specimen. Central axis width decreases from 1.1 cm proximally to 0.6 cm distally. Male macrocupules without proximal bracts (naked; Figs. 3.26 and 3.27); without central ovule or vestigial ovule. Filament-like appendages present inside macrocupule instead of large sterile scales (Fig. 3.33). Ventral suture open on funnel-shaped apex, but apparently closed (fused?) on shaft (Figs. 3.29-3.31). Macrocupule length relatively uniform: about 2.2 cm long. Proximal shaft 1.5-3.0 mm wide; flared, funnel-shaped apex 6.0-9.0 mm long, 10.0-10.2 mm wide, increasing to 17-20 mm on macrocupules squashed proximal-distally. Hundreds of bivalved microsporophylls densely packed along length of shaft and around outside of funnel-shaped apex (Figs.

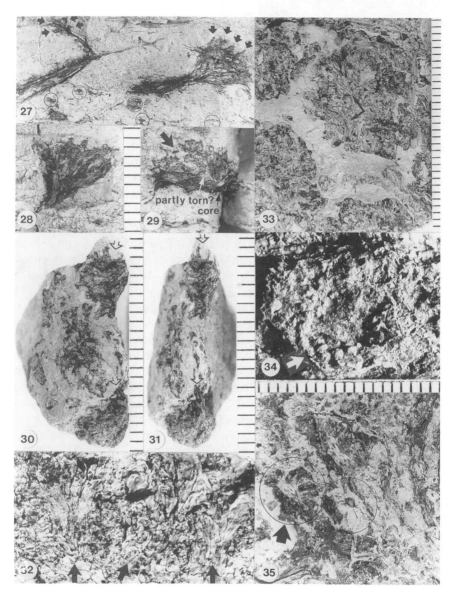

Figures 3.27–3.35. 27. *A. cupulanthus* (male; PP44198); two male macrocupules of strobilus No. 2, one in side view (right) and one in cross section (left); note bivalved microsporophylls still attached on parts of rims (arrows) but missing along stalk; note also bracteoles and microsporophylls in matrix surrounding macrocupules (circled). 28. Counterpart to macrocupule on right in Fig. 3.27. 29. Funnel-shaped apex to a third male macrocupule from strobilus No. 2 (from between two blocks), viewed from its base

(central sediment-filled cavity labeled core); note cleft (ventral suture; large arrow) open near rim; irregularity of cleft due to superimposed bracteoles (one of which is situated within cleft); cleft appears to be become a tear further down funnel-shaped apex where shaft begins, implying that cleft (suture) did not continue down shaft. 30. Funnel-shaped apices of two male macrocupules of strobilus No. 2 (PP44198); open arrows point to sutures aligned on same sides of macrocupules; alignment of sutures in ventral (apical) orientation. 31. Same two male macrocupules as in Fig. 3.30 viewed from below, showing orientation of ventral sutures; top suture spread apart by matrix; bottom suture closed with lips raised (open arrows); alignment of sutures in ventral (apical) orientation. 32. Three male macrocupules from strobilus No. 2 (PP44198), compressed laterally (arrows at bottom); attachment points of bivalved microsporophylls on macrocupule impressions appear as black (compression-filled) depressions where macrocupule wall pulled away, leaving microsporophyll bases exposed; scale/same as Figs. 3.30 and 3.31. 33. Male macrocupules from strobilus No. 3 (PP44198), squashed proximal-distally, showing thin thread-like sterile scales within central macrocupule (arrows), and concentration of numerous bracteoles to microsporophylls on outsides of macrocupules (three-dimensionally extend into matrix; demonstrated to be outside on other specimens). 34. Oblique light view of raised (three-dimensionally preserved) bivalved microsporophylls around a proximo-distally compressed macrocupule in strobilus No. 3 (PP44198); note cluster of clam-shaped bracteoles and their crowded distribution where they can be seen clearly (white arrow) due to shadow contrast. 35. Axial region of male strobilus No. 2 (PP44198), near apex, showing bases of two male macrocupules (circled); wall of macrocupule can be seen as made up of carbonaceous, plate-like thickenings curved outward and distally from base of macrocupule (large arrow; details of base in right circle not as well preserved); within matrix-filled center of macrocupule base on left, a narrow, incomplete oval of black structures can be seen (small arrow within left circle), that may represent inner ring of thin thread-like sterile scales visible in Fig. 3.33.

3.33 and 3.34). Microsporophylls consist of two ovate bracteoles enclosing pollen sacs (Figs. 3.34; 3.37); bracteoles about 2.0 mm in diameter (Fig. 3.37). Three to five pollen sacs borne in cluster at end of narrow axillary stalk (Figs 3.36-3.39); stalk 1.2 mm long, cylindrical to clavate with constricted base; pollen sacs round to ovate, 0.3 mm diam. to 0.3 mm x 0.5 mm; dehiscence by longitudinal slit (Fig. 3.36). Bracteole attachment opposite at base of stalk. Pollen simple, monosulcate with thin, subtectal, granular-layer (Figs. 3.40-41); pollen size range: 19.2 mμ × 9.6 mμ to 30.4 mμ × 16.0 mμ (av. length 24.5 mμ; av. width 13.6 mμ: 20 grains measured in four pollen masses).

Number of specimens examined. Three.

DISCUSSION

The morphology of the reproductive organs, for the most part, is apparent and photogenic; however, some organs are quite different from comparable organs

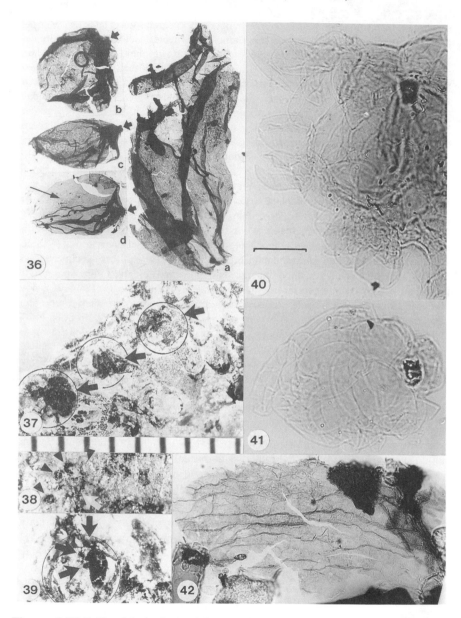

Figures 3.36–3.42. 36. *Archaestrobilus cupulanthus* (male; PP44198) pollen-apparatus, recovered from basal portion of male strobilus No. 2 in Fig. 3.24; pollen apparatus consists of a central, inflated(?) stalk with a constricted base (a: 1.2 mm long) and round to oval pollen, sacs (*b–d:* 1.5–1.7 mm long) that were attached apically and possibly down one side of stalk (see Fig. 3.39); a few faint pollen grains (identified by morphol-

ogy) appear within lowest pollen sac (*d:* long arrow); pollen-sac attachments are on right side and visible as nipples (short arrows); pollen-sac sutures longitudinal (aligned with long axis of sac) and difficult to see in photographs due to folds (suture on b along top margin; suture on *c* along bottom margin). 37. Transfer preparation of acid-resistant pollen-sac apparati (arrows) each encompassed by (overlapping) outlines of a pair of bracteoles (encircled; overlapping pairs most apparent in upper right circle); individual pollen-sac outlines can be discerned, but stalk obscured. 38. Pollen-sac apparatus exposed on surface of rock (white arrow: PP44198, strobilus No. 2), with at least four roundish pollen sacs visible in an arc around stalk (black arrow heads); stalk incomplete but black fragments of it are visible; pollen sacs to another apparatus exposed to right. 39. Transfer preparation of one pollen-sac apparatus, showing stalk with at least three small pollen sacs attached on left side (arrows); outline of bracteole on one side encircled (raised margin of bracteole cuticle just inside circle); slightly enlarged compared to Fig. 3.37. 40. Pollen mass (PP4419) recovered from bulk maceration of matrix from specimen PP44198; note monosulcate pollen around perimeter of mass; pollen ranges from 19 to 30 μm in length; scale bar 20 μm. 41. Pollen mass (PP44199), possibly from a small pollen sac, recovered from bulk maceration of matrix from specimen PP44198; same scale as Fig. 3.40. 42. Acid-resistant cuticle (PP44199) of bracteole belonging to bivalved microsporophyll, recovered from bulk maceration of matrix from specimen PP44198; matrix came from area containing abundant microsporophylls (male strobilus No. 2), and only bracteoles and pollen-sac apparati have visible cuticle preserved under microscope examination.

described in the literature (e.g., the microsporophylls), and model-dependent interpretation had to be avoided as much as possible. Some characters, although observed on a number of macrocupules, could be distinguished from artifacts of preservation only in one or two specimens having that character (e.g., suture) and showing no evidence of damage during transportation and burial. Reconstruction of some organs (e.g., pollen-sac apparatus) was dependent on several lines of evidence, any one of which might by itself be equivocal because these organs are so different. Black-and-white photographs of organs were selected for their clarity and resolution of detail, but those goals were not always achieved, requiring additional labeling and notation on the photographs to highlight or encircle what is evident on the specimens.

The male and female spikes are constructed similarly, each possessing hundreds of spirally arranged, unisexual macrocupules of similar size (Fig. 3.43C). The reproductive axes are described as strobili or spikes and not as cones, because the reproductive units or macrocupules are radially symmetrical, not bilaterally flattened as in a cone. The male spikes tend to be longer than the female spikes and were apparently borne in clusters (three per cluster), whereas the female spikes were found as individual specimens (compare Figs. 3.3-3.5 and 3.23-3.26). Although the complete length of the spikes could not be determined, based on the reconstruction of the compound spikes in their blocks of matrix, the male axes were about three times the length of female axes. Unlike the unisexual reproductive organs of *Sanmiguelia* (from the same beds), which were also not found attached to the same specimen and which are quite different in morphology, the strobili of *Archaestrobilus* are very similar in overall construction, even down to corresponding details of the macro-cupules. It is because of these similarities that they are described as a single natural taxon and given the same name.

Female Macrocupules

The macrocupules are radially symmetrical and trumpet shaped (Fig. 3.44I). The presence of ventral sutures was determined from a few informative speci-mens where the sutures were spread apart or filled with sediment (Figs. 3.13 and 3.19, arrows). Their apical or ventral orientation (Fig. 3.43C) was deter-mined by position relative to the central axis, and the occurrence of macro-cupules were in succession with similarly oriented sutures. Such an orientation would be expected if the macrocupule is derived from a single bract.

Seeds.   The morphology of the seeds was determined by seed casts, one of which was removed from its macrocupule (Figs. 3.21 and 3.22). No nucellus or megaspore membrane is present. Thin, hair-like processes can be seen along the side of the seed. Similar and better preserved seeds were found associated with *Pelourdea* leaves in coeval strata of the Dan River basin, North Carolina (Fig. 3.20). The hairs appear to originate from the seed apex and flank the seed down to near its base (derived from an outer integument?).

Sterile scales.   Based on the amount of compression material surrounding the ovule cast and the width of the macrocupule shafts, it is clear that tissue filled the space between the macrocupule wall and ovule (Figs. 3.13 and 3.16-3.19). No cuticles are preserved, either on the outside of the macrocupules or within them. The presence of sterile scales surrounding the ovule is based on a few informative specimens at the apex of the spike that were only partially compressed (Fig. 3.13). In these obliquely distorted three-dimensional macro-cupules, the tops of the sterile scales (pads) can be seen to form a floor near the base of the funnel-shaped apex (Figs. 3.13 and 3.16). The exact number of pads appears to be about six or seven, based on rings of scales that are clearly demarcated by borders with surrounding scales and structures (Fig. 3.13). The outer walls of the scales are preserved as carbonaceous compression material, whereas their centers decayed and became filled with clay (Figs. 3.13 and 2.19); distinction between clay filling spaces between scales and clay replacing cellu-lar structure within scales was determined in longitudinal sections where car-bonaceous compression material was observed capping or completing the apical

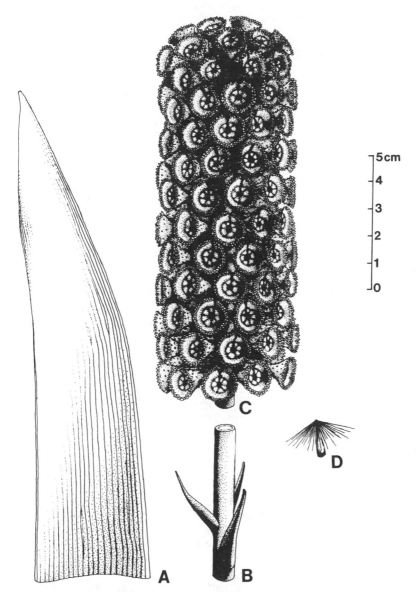

Figure 3.43.   Reconstruction of *Archaestrobilus cupulanthus* and isolated organs related by morphology and association. (*A*) Associated leaf of *Pelourdea poleoensis*; most dichotomies occur near leaf base; cross-veins unknown. (*B*) Associated sterile lower part of strobilus. (*C*) Female strobilus (holotype). (*D*) Dispersed seed (Carnian, Dan River basin, NC) showing floater apparatus; same type of seed recovered from holotype.

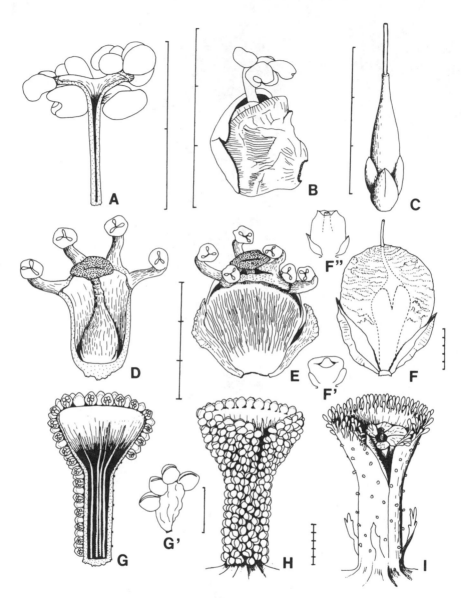

Figure 3.44. Comparison of gnetalean megasporangiate and microsporangiate organs with those of *Archaestrobilus*. (*A*) Cross section of *Ephedra distycha* male macrocupule. (*B*) Camera lucida drawing of *E. distycha* male macrocupule protruding from paired bracts. (*C*) *E. trifurca*, female macrocupule with basal pairs of opposite and decussate bracts. (*D*) Cross section of *Welwitschia mirabilis* male macrocupule (modified from Martens, 1971). (*E*) Drawing of *W. mirabilis* male macrocupule enclosed by basal pairs

of opposite and decussate bracts (modified from Martens, 1971). (*F*) Drawing of *W. mirabilis* female macrocupule ("wing integument") and basal pair of opposite bracts (modified from Martens, 1971). (*F'*). Immature female macrocupule showing origin from a closed ring structure and not from a pair of fused bracts (after Martens, 1971). (*F"*). Immature female macrocupule further along in development, showing cupular origin of "wing integument." (*G-H*) Male macrocupule of *Archaestrobilus cupulanthus*. (*G*) Cross section showing reduced sterile (interseminal) scales inside, and bivalved microsporophylls outside macrocupule. (*G'*) Pollen sac unit showing four pollen sacs attached to inflated stalk. (*H*) Outside view of male macrocupule showing suture open only on flared apex and crowded distribution of bivalved microsporophylls. (*I*) Female macrocupule of *A. cupulanthus* showing gland/food body-like organs crowded around rim (homologous organs reduced in size along below rim), suture peeled back to expose ring of sterile scales surrounding a central ovule with extended micropyle, and several digitate bracts attached around the base of the macrocupule.

outline of individual scales (e.g., Fig. 3.16). In longitudinal sections of macro-cupules, the sterile scales appear to be contiguous with their neighbors and with the macrocupule wall (Figs. 3.18 and 3.19). Whether or not they were fused with the macrocupule wall or were freestanding could not be determined. The absence of cuticles that would distinguish individual structures could imply that they were joined or fused. The term "sterile" is used rather than "interseminal," because there is only one functional ovule. Clay typically replaces or fills the space occupied by the seed.

Glands/Food Bodies.   The unusual structures attached to the outside of the macrocupule may have functioned as glands and/or food bodies; they increase in size around the funnel-shaped apex, reaching their largest size around the mac-rocupule rim (compare Figs. 3.14 and 3.15 with Figs. 3.16 and 3.17). Their morphology on the rim is similar to the clavate stalks that support pollen sacs (Figs. 3.36 and 3.39), but they are not enclosed by a pair of bracteoles (Fig. 3.37).

Bracts.   The bracts attached near the base of the macrocupules are difficult to demonstrate because of their small and delicate nature. The only two that were uncovered in their entirety are spatulate with a digitate apex (Figs. 3.18 and 3.22). Other bracts were viewed in cross section (Fig. 3.15). The number per macro-cupule (three to four) was based on the size and apparent spacing of the bracts, as no more than two could be clearly seen on any one macrocupule, and they were not attached opposite one another as they should have been if paired.

Male Macrocupules

The male macrocupules, like the female ones, are radially symmetrical and trumpet shaped (Fig. 3.44H). The shafts to the macrocupules are narrower than the shafts of female macrocupules due to the absence of a central ovule. The presence of ventral sutures was determined from a few informative specimens where the sutures were spread apart and the space filled with sediment (Figs. 3.29-3.31). Their apical or ventral orientation was determined by their position relative to the central axis. Below the funnel-shaped apex, crowded micro-sporophylls obscure the suture (Fig. 3.44H). Macrocupules compressed prox-imo-distally indicate that the suture was probably closed along the shaft, be-cause it separated (tore?) only on the funnel-shaped apex (Fig. 3.29).

Sterile Scales.   Equivalents or homologs to the sterile scales of the female macrocupules appear to be present within the male macrocupules. Their narrow filament-like morphology is apparent in only a few macrocupules that were squashed proximo-distally (Fig. 3.33). A partial ring of these scales is visible at the base of one macrocupule (Fig. 3.35). The number of such processes or sterile scales could not be determined. They are distinct from tracheids or tracheid bundles by their much larger size (i.e., width), color (tracheids appear black, whereas the scales appear as light-colored impressions sometimes with a dark vascular trace in the middle), and position within a hollow macrocupule.

Bivalved Microsporophylls.   Crowded clam-shell-shaped microsporophylls attached to the outside of the male macrocupule are similar in size from macrocupule base to apex (Figs. 3.34 and 3.44H). The ones around the macro-cupule apex and rim appear to be more stout or better attached, because they persist even when microsporophylls along the shaft have been removed by abrasion during burial or by insect herbivory. In specimens in which the microsporophylls are not attached, bracteoles can be seen distributed in the matrix between the macrocupules (Figs. 3.26, 3.27, 3.30 and 3.31). Bracteole cuticle was recovered that shows the faint outline of epidermal cells (Fig. 3.42).

Pollen-sac apparatus.   Pollen sacs are attached to the ends of cylindrical to club-shaped axes or stalks (Figs. 3.36-3.39; 3.44G, and 3.44G'). A few speci-mens were observed on broken surfaces or in transfer preparations where three to four pollen sacs were still attached or arranged distally on the stalk (Figs. 3.37 and 3.38). The size of the pollen-sac apparatus is very small and, unfor-tunately, was too fragile to be recovered intact. In transfer preparations, pollen-sac apparati were found within the concavities formed by a pair of bracteoles (which are outlined by their acid-resistant cuticles) and situated opposite their adaxial sides (Figs. 3.37 and 3.39), but only one specimen clearly shows (immature?) pollen sacs situated around the apex of the central stalk (Fig. 3.39). Maceration produced numerous specimens of stalks and pollen sacs, but none attached to one another (Fig. 3.36). Attachment scars on the pollen sacs were detected, along with torn margins on the stalks, confirming observations with the binocular microscope. The stalk cuticle is similar to that of the pollen sacs, implying a common derivation (Fig. 3.36), perhaps from a modified pollen sac. Carbonaceous material frequently coats the outside of the stalk cuticle, whereas no carbonaceous residue was found inside the compressed cuticle envelope (compare Figs. 3.36 and 3.39). The bulging and irregular, folded sides of the stalk cuticle and absence of internal residue suggest that the stalk was hollow and possibly inflated at pollen maturity. Inflation or elongation may have been necessary to elevate the pollen sacs above the bracteoles. One side of the stalk appears to be incomplete on all recovered specimens, implying either that the stalk is inverted or that pollen sacs were attached down one side (Figs. 3.36 and 3.44G').

Pollen.   Pollen masses were recovered in bulk maceration of microsporophylls. The size and shape of these masses (Figs. 3.40 and 3.41) is identical to the size and shape of the pollen sacs. Only one type of pollen is present in these masses: simple, monosulcate with a thin translucent wall (Figs. 3.40 and 3.41) and infratectal granules, which are barely visible as tiny beads within some specimens. Some pollen sacs still contain this type of pollen (Fig. 3.36, arrow), but poor contrast makes the pollen difficult to see in photographs.

Strobilus Base.   The sterile basal part of the female axis (Figs. 3.6-3.9) was not found attached, but based on its morphology, size, and pith cast, it is thought to belong to *Archaestrobilus* (Fig. 3.43B). The three long narrow lanceolate bracts on this specimen are unlike the short bracts borne by *Axelrodia*, or the leaves of *Sanmiguelia*, the only other seed plant found preserved *in situ* at this locality. One bract bears the impressions of a larger central vein flanked by several small, parallel veins (barely visible in Fig. 3.6). The close spacing of the bracts in a tight spiral, followed by a long internode (Fig. 3.43B), is similar to the manner in which the male spikes are borne in what appears to be a trichotomy of determinate branches. It cannot be determined whether the sterile base represents the lower part of the compound male strobili or the base of the solitary female spike. In either case, its morphology is distinctive and its bracts similar to the leaves of *Pelourdea*, which also occurs as *in situ* specimens (Fig. 3.10).

INTERPRETATION

Biotic Implications

The close similarity in morphology between male and female macrocupules implies that they are homologous. The presence of microsporophylls on the outside (male) and an ovule on the inside (female) implies a derivation from a bisexual ancestor (cf. *Welwitschia*). The similarity in morphology and position of the pedicellate structures on the outside of the female macrocupule to the stalks of microsporophylls also implies homology. The presence of such structures on the female macrocupules indicates that they may have had a function in attracting insects looking for food or pollen. The carbonaceous material coating the outside of the pollen-sac stalks of the male macrocupules has a resinous look about it, as do the swollen heads of similar but reduced structures around the rim of the female macrocupule. No chemical analysis of these organs has been made, however. The concentration of modified append-ages around the rim of the female macrocupule is probably a modification to enhance insect attraction to the tops of macrocupules where the micropyle was exposed. These observations are the vasis for interpreting the sterile organs on the outside of female macrocupules as glandular food bodies. The swollen heads of the sterile scales around the ovule may have provided a convenient landing pad for insects and a place for them to stand while feeding on the surrounding garden of possible food bodies.

The dense aggregation of bivalved microsporophylls around the entire outer surface of the male macrocupules would be difficult to explain in an anem-ophilous plant. On the one hand, the close spacing of the macrocupules on the axis would prevent wind from having access to pollen released below the apical rim. The paired bracteoles are unusual for an anemophilous plant, in that they would baffle effective wind velocities between the macrocupules. On the other hand, the bracteoles may have served to protect the pollen from insect herbivory until maturity. The spaces between the macrocupules were wide enough to allow small beetles, for example, to live and forage there. The bracteoles and pollen sacs on one male spike were largely removed from below the rims of most of the macrocupules, whereas on another, they remain intact. The removed bracteoles occur disseminated through the matrix between the macrocupules, and torn attachment points can be seen on macrocupule shafts. Perhaps insect foraging is responsible. Wind was probably effective in dispersing pollen from sacs around the rim, but copious amounts of pollen that had to have been produced by such large spikes could only be justified as an efficient use of energy if insects were the prime pollinators. Thus, *Archaestrobilus cupulanthus* is interpreted as ambophilous.

Association with Pelourdea

At the Sunday Canyon locality isolated leaves of *Pelourdea poleoensis* were found in the paleosol underlying the *Sanmiguelia* root zone and in thin shales interbedded within the sandstone sequence overlying the *Sanmiguelia* colony (Cornet, 1989). A long strap-shaped leaf with poorly preserved parallel ve-nation matching the overall morphology of *Pelourdea* was found next to the blocks of paleosol containing the male spikes. Leaves of both taxa are frequently reported from the same localities (Ash, 1987). On the back side of the holotype (female spike: PP44195) occurs a portion of a stem with an attached fragment of a (clasping) leaf base and connected lower part of a well preserved *P. poleoensis* leaf (Fig. 3.10). This leaf fragment shows the typical crowded parallel venation of *P. poleoensis* described by Ash (1987), as well as minor second-order venation between the primary veins. Bifurcations are more numerous than anastomoses, but subperpendicular cross veins are lacking. The apparent absence of cross veins on siltstone impressions could be the result of preservation (Daugherty, 1941; Stagner, 1941; Ash, 1987), but this specimen is preserved well enough to show that cross-veins are not present in the lower part of the leaf.

Even though this stem is not oriented in the same direction as the reclining spike, the association of *Pelourdea* leaves with *Archaestrobilus*, in the absence of any other in situ candidates, implies a relationship. That is why a *Pelourdea* leaf is shown next to *Archaestrobilus* in Fig. 3.43.

Affinity

The presence of a single ovule at the center of unisexual macrocupules, which appear to be arise from the axil of a single bract, and the presence of male macrocupules bearing stalked clusters of three to four (possibly six) pollen sacs (Figs. 3.44G-I and Fig. 3.45F) are strongly reminiscent of the Gnetales, particularly *Welwitschia* (Fig. 3.44A-F; cf. Martens, 1971; 1977; Crane, 1985; Crane, 1988b; Gifford and Foster, 1989). Other possible synapomorphies with the Gnetales include the following: elongate parallel-veined leaves/bracts (cf. *Welwitschia*, if *Archaestrobilus* belongs to *Pelourdea*); a distinctive pith surrounded by a loose net of anastomosing and bifurcating protoxylem bundles in the inflorescence axis (this may also be a plesiomorphic character for the anthophytes); ancestral bisexual macrocupules (cf. *Welwitschia*: Fig. 3.44D); sterile scales surrounding the ovule (cf. rudimentary appendages surrounding ovule in the male macrocupule of *Welwitschia*: Fig. 3.44D; Martens, 1971); male strobili borne in clusters of three (i.e., compound strobili; cf. *Welwitschia*); and monosulcate pollen with optical indications of infratectal granules (cf. *Welwitschia*). The fact that the bracts on the sterile base of the spike(s) are borne in a tight spiral of three may be significant, inasmuch as this phyllotaxis is intermediate between spiral and whorled/paired (a synapomorphy for extant Gnetales). Clearly, the greatest overall similarity is with *Welwitschia*.

The spiral arrangement of macrocupules on the reproductive axes (Fig. 3.43C), radial symmetry of both male and female macrocupules (Fig. 3.44H-I), sterile scales or bracteoles that surround a central ovule (Fig. 3.44I), unisexual macrocupules, sterile, bractless microsporophylls (glands/food bodies: interpretation) on the outside of the female macrocupule (Fig. 3.44I), spiral arrangement of determinate leaves on the stem (if *Archaestrobilus* belongs to *Pelour-dea*), parallel leaf and bract venation with few cross viens (Fig. 3.10), simple monosulcate pollen with no ribs or plications, and a large pith with anastomosing and bifurcating protoxylem bundles may all be plesiomorphic characters, indicating that *Archaestrobilus* is much more primitive than the extant Gnetales. Its pollen most closely resembles that of *Sahnia laxiphora* (Pentoxylales) under transmitted light (Osborn et al., 1991), consistent with an anthophyte affinity. However, the cupular morphology of its reproductive organs, the presence of only one functional ovule per macrocupule, pollen sacs clustered together (not synangia – free from enclosing bracteoles), and compound strobili may be the most important characters for determining its affinity.

I therefore suggest that *Archaestrobilus cupulanthus* is a stem gnetophyte, because of the number of characters which unite it with the Gnetales and which distinguish gnetophytes from all other anthophytes. The absence of decussate and opposite leaves currently excludes *Archaestrobilus* from the Gnetales but not from the gnetophytes, which are more broadly defined here by the recognition and circumscription of *Archaestrobilus*. Consequently, inclusion of *Archaestrobilus* and any other non-gnetalean seed plant in the gnetophytes is based mostly on reproductive rather than vegetative similarity, just as the inclusion of *Sanmiguelia* in the angiophytes is based on its overall suite of characters, not on its having all the characters (e.g., leaf morphology) of Cretaceous or extant angiosperms.

Homologies among Archaestrobilus, Ephedra, and Welwitschia

According to Crane (1985) and Doyle and Donoghue (1986a, 1986b, 1992; Hickey and Taylor, Chapter 8), *Ephedra* is the sister taxon to *Welwitschia* plus *Gnetum*, because it has the most plesiomorphic characters. The advances shared between *Welwitschia* and *Gnetum* over *Ephedra* are vein anastomoses in the leaves, reduction of the male gametophyte, a tetrasporic megagametophyte with free nuclei serving as eggs, and a feeder in the embryo (Doyle and Donoghue, 1986a). The presence of double fertilization in *Ephedra* (Friedman, 1990a, 1990b) could also represent a plesiomorphic character for the anthophytes. *Archaestrobilus* possesses characters that may be plesiomorphic even for *Ephedra*, such as radially symmetric floral units that are spirally arranged rather than decussate or whorled, and bracts that are grouped but still spirally arranged rather than opposite or whorled. The grouping of male spikes may be similar to that of the bracts in being arranged in a tight spiral, indicating that the characteristic synapomorphy for the Gnetales of opposite or whorled appendages is probably an advancement over the condition in stem gnetophytes (Doyle and Donoghue, 1986b).

Homologies among floral structures have been confused by misinterpretation and the previous absence of fossils that could clarify homologies among extant Gnetales. The structure enclosing the ovule in *Welwitschia* is a prime example: The wing or "aile" has been variously interpreted (Martens, 1971, p. 137), with Martens concluding that the wing consists of a pair of fused bracts. That interpretation was accepted by Crane (1985, 1988b). It probably stems from the observation that the male flower has two pairs of alternating and decussate bracts, the innermost pair of which most closely resembles the wing around the ovule on the female flower. Yet the data and illustrations in Martens (1971) contradict his own interpretation. Nowhere does a cross section or illustration of an immature female flower show two separate bracts (e.g., Martens, 1971, Figs. 66, 67, and 69). Only the diagramatic cross section (his interpretation) in his Fig. 63-6 shows two bracts. In all other cases, the actual specimens show a macrocupule or ring structure that grows upward to enclose the central ovule. Similarly, the outer envelope or "perianth" of the ovule in *Ephedra* was interpreted as arising from the fusion of a dorsal-ventral pair of bracts by Crane (1985, p. 759). That interpretation may stem from misinterpretation of Martens' (1971) floral diagrams for *Ephedra distachya*. Even when ovules are borne in pairs, each contains a single ring primordium that surrounds the ovule (Martens, 1971, Fig. 21).

If there is any indication of two separate bracts in the morphology of the ovuliferous wing of *Welwitschia*, the division between bracts would have to be placed in a dorsi-ventral position where the macrocupule wall is thinnest and its apical margin slightly cleft on the ventral side [contrary to floral diagrams in Martens (1971) and Crane (1985)]. That separation, which was suggested by Chamberlain (1966, Fig. 374B), would create two pairs of nonalternating conduplicate bracts – a symmetrical paradox. Here is a problem where a fossil can be used to distinguish between homology and analogous symmetry caused by bilateral (dorsi-ventral) flattening of the flower.

The existence of a tubular, or ring macrocupule in the Gnetales is well demonstrated by the male flower of *Welwitschia* (Fig. 3.44D-E). Immature female flowers of *Welwitschia* also show a tubular macrocupule before much dorsi-ventral flattening has occurred (Fig. 3.44F'-F"). The outer envelope of the *Ephedra* ovule is also macrocupular in construction (Fig. 3.44C; Chamberlain, 1966; Gifford and Foster, 1989). The male flower of *Ephedra* consists of a miniature macrocupule, enclosed by a pair of opposite bracts (Fig. 3.44A-B). Based on a reexamination of the male flowers of *E. distachya*, I discovered that the male macrocupule bears up to six pairs of pollen sacs (rarely three sacs per cluster) on and below the rim of a flared but dorsi-ventrally flattened funnel-shaped apex, with a central opening that becomes mostly occluded below the base of the funnel at the center of the shaft. Such morphology indicates a possible derivation from a hollow, radially symmetrical macrocupule, like that of *Welwitschia* and *Archaestrobilus*. The major difference between the male macrocupules of *Welwitschia* and *Ephedra* and those of *Archaestrobilus* is one of size (see Fig. 3.44).

Size reduction may account for the decrease in number of microsporophylls from hundreds in *Archaestrobilus* to as few as six in *Welwitschia*, and a similar decrease in the number of pollen sacs per microsporophyll from four to six in *Archaestrobilus* to two in *Ephedra* and *Gnetum*. Even with extreme size reduction, the pollen sac clusters remained stalked [this is not necessarily the case for all fossils of possible gnetalean affinity, e.g., *Piroconites* (van Konijnenburg-van Cittert, 1992)]. Size reduction and enclosure by strobiloid bracts may also account for the loss of bracteoles that enclose the pollen sacs of *Archaestrobilus* (they become redundant), as well as the loss of microsporangia from much of the shaft in *Ephedra* and all of the shaft in *Welwitschia*. The presence of bracts subtending the male macrocupules in *Ephedra* and *Welwitschia* may be an apomorphy (transposition from the female macrocupule) if the naked condition in *Archaestrobilus* is basic. Alternatively it may indicate separate derivation of the *Ephedra* and *Welwitschia* lineage(s) from a common ancestry with *Archaestrobilus*, which still had functional bisporangiate macrocupules. That possibility would make *Archaestrobilus* the sister taxon of *Ephedra* plus *Welwitschia*. The macrocupules of *Archaestrobilus* obviously became dimorphic and unisexual very early in the evolution of the gnetophyte clade.

The microsporangiate structures (anthers sensu Martens, 1971) of *Gnetum* are much simpler in construction than a ring of pollen sacs borne on reduced macrocupules and enclosed by paired bracts that occurs in *Ephedra* and *Welwitschia* (Chamberlain, 1966; Martens, 1971). They consist of a narrow stalk bearing two terminal pollen sacs; this structure is enclosed by a tubular perianth with an apical slit or suture. On the one hand, if this tubular perianth were derived from the fusion of paired bracteoles like those of *Archaestrobilus*, *Gnetum* would be the only extant member of the family to retain vestiges of such bracteoles. On the other hand, if these structures are highly reduced versions of the macrocupules plus paired bracts of *Ephedra* and *Welwitschia*, the homology would be at a higher structural level. In *Gnetum gnemon*, aborted ovules are borne in a whorl around a central axis above the microsporophylls. If the outer envelope or tubular perianth that encloses the ovules is homologous with the macrocupule of *Archaestrobilus* and the microsporangiate structures are only slightly modified versions of bivalved microsporophylls, the only major differences from *Archaestrobilus* would be the restriction of microsporophylls to the bases of macrocupules in *Gnetum* and the whorled arrangement of those macrocupules.

Homologies Among Anthophyte Androecia

Cladistic analyses based either on phenotypic or on genotypic characters have been inconsistent and, therefore, inconclusive about interrelationships between anthophyte sister groups (cf. Crane, 1985; Doyle and Donoghue, 1986a, 1986b; Zimmer et al., 1989). But one fact remains consistently clear: The Bennettitales, Gnetales, Pentoxylales, and angiosperms had a common ancestry sometime before the Late Triassic. The Gnetales are relatives of angiosperms and Bennettitales that underwent drastic floral reduction and aggregation in response to wind- pollination (Doyle and Donoghue, 1986a). The number of similarities or synapomorphies (plesiomorphies for the group) involve nonpreservable characters: siphonogamy, tunica-corpus, lignin chemistry, reduced megaspore wall, and granular exine. Additional characters that unite the Bennettitales and Gnetales are stalked or sessile ovules, whorled microsporophylls, and a micropylar tube. The three living genera of Gnetales share opposite leaves (originally probably spirally arranged and linear), multiple axillary buds, vessels, loss of scalariform pitting, simple or reduced microsporophylls, a single terminal ovule, a second integument apparently derived from the macrocupule of the ovulate flower, and striate pollen (lost? in *Gnetum*: Doyle and Donoghue, 1986a).

To this list of characters which unite the anthophytes, in particular the Gnetales, Bennettitales, and angiosperms, can be added pollen sacs borne in aggregates of two or more (Figs. 3.45C,F,I and 3.44A,B,D,E,G,G',H), paired bracteoles that enclose the pollen sacs or are fused with them to form synangia (lost in Pentoxylales by reduction?; Figs. 3.45C,F,I and 3.44G,H), microsporophylls borne on perianth parts or on the macrocupule derived from perianth parts (lost in Pentoxylales by reduction?; Figs. 3.45A,D,G), and simple tectate-granular, monosulcate pollen (this last character was assumed for the Gnetales before the discovery of *Archaestrobilus*).

Figure 3.45 shows how the basic microsporophyll units of angiophytes (Fig. 3.45C), gnetophytes (Fig. 3.45F), and bennettitaleans (Fig. 3.45I) could be derived from an ancestral sporangiophore subtended by a pair of bracts or bracteoles. In gnetophytes, the pollen sacs remained integral on a filamentous axis. Even when the number of pollen sacs and axial length is reduced, the bracteoles remain free and unfused to that axis or to the sacs. In *Sahnia* (microsporangiate flower of *Pentoxylon*), the axis either remained elongate and branched (vestigial plesiomorphy?) or became so. If bracteoles were present in early developmental stages, they are not present in mature specimens (Crane, 1985). In bennettitaleans the axis was lost and the pollen sacs became fused to the adaxial sides of the bracteoles. This condition was apparently modified in some early Bennettitales, in which the pollen sacs are not clearly borne in bivalved synangia; for example, *Bennettistemon, Haitingeria, Leguminanthus*, and *Leuthardtia* (Crane, 1986). Thus, in the Pentoxylales and perhaps also the Bennettitales, bracteoles were lost in some taxa early in anthophyte history, whereas the supporting axis was retained in most taxa where pollen sac fusion with the bracteoles did not occur.

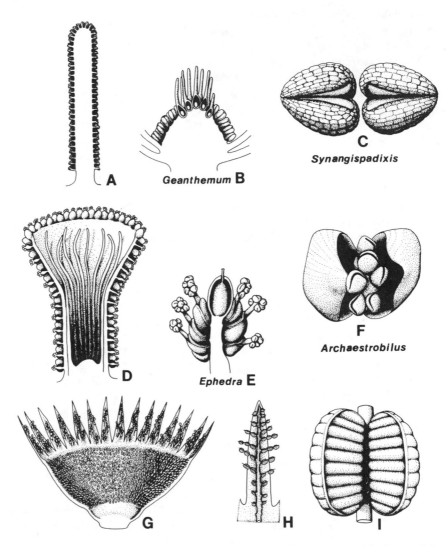

Figure 3.45.   Comparison of anthophyte microsporangiate organs with those of *Archaestrobilus*. (*A*) *Synangispadixis tidwellii* male flower, longitudinal section. (*B*) *Geanthemum* bisexual flower, longitudinal section, showing sessile stamens attached below collective bases (receptacle) of carpels/macrocupules. (*C*) *S. tidwellii*, tetraloculate anthers without connective, drawn to show origin from a pair of conduplicate bracteoles. (*D*) *Archaestrobilus cupulanthus*, longitudinal section showing distribution of bivalved microsporophylls on outside of macrocupule and reduced sterile (interseminal) scales on inside. (*E*) *Ephedra campylopoda*, showing reduced bisexual inflorescence and male macrocupules extended beyond enclosing paired bracts. (*F*) *A. cupulanthus* bivalved

microsporophyll, showing axile position of stalked pollen-sac apparatus encompassed by pair of bracteoles; an analogous morphology (higher order of complexity) is duplicated by *Ephedra* male flowers (E). (*G*) *Weltrichia sol*, longitudinal-half section through male flower, redrawn from Crane (1985, Fig. 10D). (*H*) *Weltrichia spectabilis*, detail of branched bivalved microsporophylls attached to a perianth lobe; redrawn after Stewart and Rothwell (1993, Fig. 3.25.7C). (*I*) *Weltrichia sol*, detail of bivalved synangium; redrawn after Crane (1985, Fig. 10E).

Information on the seeds and the pollen organ of the *Eucommiidites*-producing plant indicates a possible anthophyte affinity (Pedersen et al., 1989a). Those characters that support this relationship are ovules with two integuments (*Erdtmanispermum*) and spherical heads (*Erdtmanitheca*) containing hundreds of radiating pollen sacs borne in probable synangia. Because of the morphology of its seeds and its triaperturate (modified polyplicate?) pollen, this taxon has been compared to the Gnetales. It is, therefore, important to recognize diagnostic differences between *Erdtmanitheca* and *Archaestrobilus*. The apparent fusion of numerous elongate pollen sacs with a bract-like organ and their position inside that organ (Pedersen et al., 1989a) compares more with the condition found in the Bennettitales than it does with that of the Gnetales and *Archaestrobilus*, where the pollen sacs are free and unfused to associated bracteoles or bracts. There is greater similarity between *Erdmanitheca* and *Synangispadixis* male flowers in which the macrocupule has been reduced to a supporting axis (pseudoaxis).

In angiophytes (represented by the stem angiophyte *Sanmiguelia lewisii/ Synangispadixis tidwellii*) the pollen sacs became fused to the adaxial sides of the enclosing bracteoles, and the pollen sac cuticle was reduced, like that of most Bennettitales (Fig. 3.45C). In addition to the formation of synangia, the bractioles became plicate or conduplicate. In folding, they almost completely enclosed the pollen sacs. With conduplication and space limitation, the number of pollen sacs was reduced from many to two functional sacs per bracteole. The reduced median pollen sacs were not completely lost but were apparently modified to form a septum that separates the two outermost sacs. That interpretation would explain why the septum disappears with pollen maturity as do the pollen sac walls [no pollen sac cuticle can be demonstrated for *Synangispadixis* (Cornet, 1986, 1989b)]. Each conduplicate bracteole with paired pollen sacs is identical in construction to an angiosperm anther. I propose that the presence of an ephemeral septum separating two pollen sacs per bracteole plus the paired nature of the bracteoles (paired anthers give tetraloculate condition for angiosperms) represents the hallmark synapomorphy (symplesiomorphy) for all angiosperms (Cornet, 1989b).

If there is any single reason why *Sanmiguelia* cannot be placed in the gnetophytes or in any other recognized nonangiophyte sister group, this is the most important one. Not everyone may agree with this assessment: Doyle and Donoghue (1993), for example, did not recognize or accept the paired nature of the biloculate anthers and compared *Synangispadixis* with ginkgophyte strobili bearing sporophylls with two pollen sacs. Crane (personal communication, 1993), however, concurs with my interpretation.

Homologies Among Anthophyte Gynoecia

Recognition of rudimentary scales or reduced bracteoles at the base of the aborted ovule in the male flower of *Welwitschia* (Fig. 3.44D: Martins, 1971) has important implications: I interpret the scales surrounding the ovule in *Archaestrobilus* to be homologous with the rudimentary scales in *Welwitschia* (Fig. 3.44D). Their similarity in shape and position to interseminal (sterile) scales around ovules in bennettitalean flowers cannot be ignored (compare Fig. 3.46D,G,I). Some ovuliferous macrocupules of *Archaestrobilus* were found with nodular structures at the base of some of the sterile scales (Fig. 3.19, arrow; Fig. 3.46D). They may well represent aborted or rudimentary ovules and not just sterile scales with mineral deposits, and thus be derived from a multiovulate bennettitalean-like flower (cf. Takhtajan, 1969; Ehrendorfer, 1976). However, the presence of such accessory ovules is not necessary for the elucidation of homologies, only helpful.

The nucellus in *Gnetum* is surrounded by three tubular envelopes (Martens, 1971; Crane, 1985). The middle envelope may be derived from the coalescence or fusion of sterile scales (analogous to the fusion of paired bracteoles around the anther stalk?). If this interpretation is correct, only *Ephedra* would lack a structure homologous with the middle envelope of the *Gnetum* ovule and/or the rudimentary scales of *Welwitschia*. The possible relationship of the middle envelope to plesiomorphic organs is the kind of understanding that can be derived only from the fossil record. Taken together with the possible plesiomorphic morphology of *Gnetum* microsporophylls (see above), Doyle and Donoghue's (1986b, Fig. 14d) entertainment of a cladistic tree where *Gnetum* is basal on the gnetophyte branch is reasonable. For those who would interpret the dicot-like leaf (*Pannaulika triassica*) and flowers from the Late Triassic (Cornet, 1993) as gnetalean, the above possibility becomes viable possibility. In opposition, one must consider mosaic evolution (*Gnetum* does not have to be advanced in all characters) and the much longer tree length (101 steps vs. 31-47 steps from the work Doyle and Donoghue, 1986b) required to place *Gnetum* before the other Gnetales. In addition, Doyle and Donoghue (1992) and Loconte and Stevenson (1990) regard leaf vein reticulation as not homologous in angiosperms and *Gnetum* (see also Trivett and Pigg, Chapter 2).

Crane (1985) and Doyle and Donoghue (1986b) interpret female bennettitalean and pentoxylalean flowers as consisting of aggregates of uniovulate cupules separated by sterile scales (Fig. 3.46H,I). If the organization of cupular ovules and sterile scales is basic to the anthophytes (see also Taylor and Kirchner, Chapter 6), then the macrocupule in *Archaestrobilus* (and by homology the outer envelope or wing surrounding the ovules in extant Gnetales) may be homologous with the perianth of the bennettitalean flower (Doyle and Donoghue, 1986b: character 35). Such an interpretation contradicts that of Crane (1985), who homologizes the macrocupule of *Welwitschia* (his inner bracteoles) with the "perianth" of *Ephedra* and *Gnetum.*

My interpretation has the flowers of extant Gnetales and those of *Archaestrobilus* constructed on an identical bauplan with all bracts and envelopes accounted for by plesiomorphic characters (organs sometimes missing in extant taxa). The bennettitalean flower (exemplified by *Williamsonia harrisiana*: Fig. 3.46G), in this interpretation, represents the plesiomorphic condition for the anthophytes (cf. Doyle and Donoghue, 1986a). The only significant difference between the flowers of anthophytes is the reduction in the number of sterile and fertile parts.

Origin and Homologies of the Angiosperm Carpel and Stamen

With the acceptance of gnetopsids as the closest living sister group to the angiosperms, a conflict arose between the traditional phyllosporous interpretation of the carpel and the stachyosporous explanation of the bracteole-bearing axes terminated by an ovule in gnetophytes (Stevenson, 1993; Doyle and Donoghue, 1992). Complicating this is the interpretation by Taylor (1991; see also Taylor and Kirchner, Chapter 6) that the most basic of angiosperm carpels is ascidiate with a single basal, orthotropous ovule. Cornet (1986) attempted to explain the morphology of the *Axelrodia* carpel (Late Triassic) by considering a similar origin via the enclosure of a terminal ovule-bearing branch by a subtending clasping bract or leaf. If the macrocupule of *Archaestrobilus* is a reduced uniovulate version of the bennettitalean flower, then the macrocupule is probably homologous with the inner whorl of bracts in, for example, *Williamsonia* (compare Figs. 3.46D and 3.46G). The presence of a ventral suture on the macrocupule of *Archaestrobilus* implies that it is derived from either a large encircling bract or from two or more bracts, all of whose margins became fused except along one side. Because the suture does not seem to have importance in dehiscence (the ovule was enclosed by sterile scales) and the suture is apical in orientation, the former interpretation is preferred because it is a by-product of an origin from a single bract. Continued interpretation of the outer envelope of the gnetalean ovule as being derived from a pair of bracts is not supported by development or by *Archaestrobilus.* The angiosperm carpel can be either conduplicate with an extended suture or ascidiate with no suture in

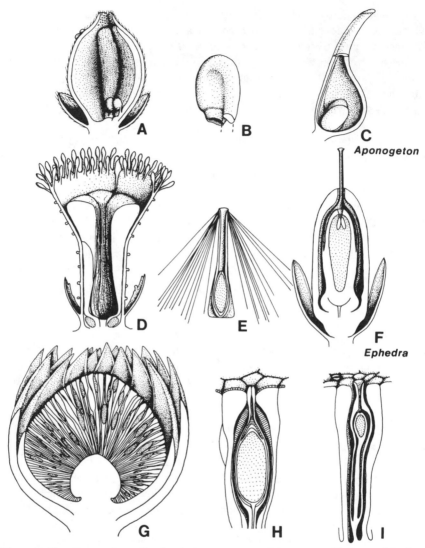

Figure 3.46. Comparison of anthophyte megasporangiate organs with those of *Archaestrobilus*. (*A*) *Axelrodia burgeri* carpel/macrocupule, longitudinal section showing two anatropous ovules attached to base of a pair of sterile scales, one on either side of ventral suture. Papillae of "transmission tissue" on ovary side of sterile scales; papillae and hairs around stigmatic apex (based on Cornet 1986, 1989b). Gland-like organs as round bumps near apex on left side, and portion of bract-like perianth also shown. (*B*) *A. burgeri* anatropous ovule showing two integuments (based on Cornet, 1989b). (*C*) *Aponogeton* (monocot) carpel with pair of basal ovules shown for comparison. (*D*) *Archaestrobilus cupulanthus*, longitudinal section showing macrocupule with gland-like bodies around

rim, ventral suture, ring of sterile scales (some with possible aborted ovules) surrounding central cavity for a single seed, and some of the digitate bracts around the base of the macrocupule. (*E*) Reconstruction of *A. cupulanthus* seed, based on specimens from Dan River/Danville basin, North Carolina (Fig. 3.20). (*F*) *Ephedra viridis* female flower, longitudinal section with emphasis on homologies with *Archaestrobilus* flower in (*D*); redrawn from Chamberlain (1966, Fig. 356). (*G*) *Williamsonia harrisiana* female flower, showing gynoecium with numerous cupulate ovules and sterile scales surrounded by bracteate perianth; redrawn from Crane (1985, Fig. 10A). (*H*) *Cycadeoidea morierei* ovule, longitudinal section showing interseminal scales enclosing a cupulate ovule with apical micropyle; redrawn from Stewart and Rothwell (1993, Fig. 25.10B). (*I*) Cycadeoid ovule with interseminal scales, redrawn and modified from Stewart (1993, Fig. 25.3A).

primitive living angiosperms; similarly, the loss of the suture in extant Gnetales is interpreted as an apomorphy, with no more or less significance than its closure in angiosperms. The evolution of the angiosperm carpel from such a macrocupule is considered here to be the most parsimonious interpretation (see also Cornet, 1989b).

If the angiosperm carpel and gnetalean macrocupule evolved from a single bract, its larger size and subtending position to one or more ovules helped it to exclude other lower or outside bracts in a phyllotactic spiral or whorl. Development of such an enlarged bract over others in the perianth may be the result of size reduction, where the gynoecium was reduced from many ovules to one or two. In the Bennettitales (Fig. 3.46G) bract size and shape in the perianth remained similar. The constriction of the macrocupule apex to protect two or more ovules inside created the carpel and led to the evolution of the stigma in angiophytes (Fig. 3.46A). In the Gnetales, a single ovule with an extended micropyle obviated the necessity for a stigma when the macrocupule apex became constricted as in *Ephedra* (Fig. 3.44C).

In both clades, the ancestral condition (also for all anthophytes) was a bisexual macrocupule: Microsporophylls were attached to the outside of the macrocupule, whereas megasporophylls were contained within it. Differentiation into unisexual flowers probably evolved as a means to increase outcrossing before carpel closure and the development of sporophytic incompatibility. Unlike the Gnetales, retention of two or more ovules (or the potential for more than one ovule) and a shift from prefertilization to postfertilization development of the ovule resulted in additional specializations in angiosperms: (1) interseminal (sterile) scales modified to form a pollen tube transmission tissue that connects stigma with ovules (Fig. 3.46A; Cornet, 1989b, Fig. 9); (2) ovules that became anatropous so that the micropyle would face the transmission tissue (Fig. 3.46B); and (3) development of bilateral rather than radial symmetry for the arrangement of paired ovules and transmission tissue tracts within the ovary (see also Taylor and Kirchner, Chapter 6). In the case of *Axelrodia*, the transmission tissue (rows of micropapillae aligned with axis of the carpel) developed on the ovary side of paired sterile scales, which became attached to the carpel wall, one on either side of the ventral suture (Cornet, 1989b, Fig. 9). If these interpretations are correct, that is, sterile scales are derived from aborted ovules, and transmission tissue from sterile scales, then pollen tubes in angiosperms follow ovule-derived structures from stigma to micropyle (Cornet, 1989b).

Evidence in support of some of these interpretations comes from the Chloranthaceae, considered to be among the most primitive living angiosperms (LeRoy, 1983b; Endress, 1987a; Todzia, 1988). Cornet (1989b) interpreted the angiosperm stamen to be homologous with the entire male flower of *Synangispadixis*, but that seems to involve too much reduction and modification. A more parsimonious interpretation is that the only thing lacking in the tetraloculate anthers of *Synangispadixis* is the connective [the anthers of *Synangispadixis* are joined at their base by a cellular connection that may be the precursor to the connective and filament (Cornet, 1986, 1989b)]. Transposition of stamens from male to female flowers resulted in the secondary evolution of a bisexual flower (perhaps a contributing factor in mid-Cretaceous diversification?). Thus, the position of anthers/stamens borne on the outside of the carpel in *Sarcandra* is viewed as not exceptional: That is where they belong on a bisexual macrocupule. Subsequent refinement moved the stamens to the base of the carpel/macrocupule (cf. *Gnetum* and Fig. 3.45b). As further evidence that the male flower of *Synangispadixis* is relevant, consider the male flower of *Hedyosmum*: It consists of an axis bearing hundreds of sessile stamens. It does not have to be interpreted as the product of extreme floral reduction (see LeRoy, 1983a, 1983b). It could represent the male counterpart of the bracteate carpel (as does the *Synangispadixis* male flower), for which the axis (i.e., pseudoaxis) is the carpel homolog. In this interpretation, the only significant distinctions between the male flowers of *Hedyosmum* and *Sanmiguelia* (*Synangispadixis*) are the evolution of a connective between the anthers, and the evolution of tectate-columellate pollen with multiple apertures. Thus, stamens are not homologs of carpels, but homologues of ovules plus transmission tissue (if derived from interseminal scales). Consistent with this interpretation, the connective or laminar part of the stamen evolved *de novo*, as did staminodial petals (Walker and Walker, 1984; cf. Friis et al., 1991).

CONCLUSIONS

Archaestrobilus cupulanthus gen. et sp. nov. from the late Carnian Trujillo Formation of Texas possesses combinations of characters shared with extant

Gnetales, such as large strobili borne in clusters, groups of three bracts attached on the reproductive axis in a tight spiral and separated by long internodes (almost whorled), and complex, unisexual, cupulate flowers. The presence on the outside of the female macrocupule of gland-like structures that resemble the stalks that bear pollen sacs on the male macrocupule, and the presence of sterile filamentous appendages inside the male macrocupule that resemble the sterile scales around the ovule, imply an origin from a bisexual macrocupule. The presence of simple, ovoid monosulcate pollen, radially symmetric rather than bilaterally flattened macrocupules, fully developed sterile scales around the ovule, and paired bracteoles that enclose the pollen organs indicate a plant more primitive than extant Gnetales, but one that belongs in the gnetophytes and is probably a stem gnetophyte.

The association of *Archaestrobilus* with *Pelourdea poleoensis* leaves and stems suggests that they belong to the same plant. This latter taxon was herbaceous (also indicated by the occurrence of reproductive axes in a semiaquatic habitat), and was unlike angiophytes (see Ash, 1987) in being non-rhizomatous with long, spirally-arranged, strap-shaped leaves that pos-sessed only two sizes of parallel veins compared with four sizes in *Sanmiguelia* (Cornet, 1986). Large strobili probably borne terminally on primary rather than secondary axes, and orthotropous seeds with long tapering micropyles add to the characters that nest *Archaestrobilus* in the gnetophytes. Probable insect involvement in pollination and the presence of an apical tuft of hair-like appendages on the seed that served in wind dispersal are also consistent with a stem gnetophyte/anthophyte affinity.

Similarities in the construction of unisexual flowers between *Archaestrobilus* and *Sanmiguelia* (see also Cornet, 1989b) indicate a common ancestry for both taxa and stress the stachyosporous rather than phyllosporous origin of the angiosperm carpel and stamen. Comparison with *Sarcandra* and *Hedyosmum* (Chloranthaceae) reveals a similar floral organization and confirms the primi-tive status of these taxa among living angiosperms.

ACKNOWLEDGMENTS

The author thanks David Winship Taylor and Leo J. Hickey, the organizers of the symposium, for inviting me to participate. He also thanks Elizabeth B. Cornet (my mother) for help in collecting the specimens in 1986, BonnieLee Cornet (deceased 1991) for her unforgettable support during the early stages of this study, and Patricia Huff-Cornet for her unconditional support in making the completion of this chapter possible. I acknowledge the support of Paul E. Olsen, who encouraged me to make this contribution and supported me through his grants at Lamont-Doherty Earth Observatory during various stages of this study.

Wood Anatomy of Primitive Angiosperms: New Perspectives and Syntheses

Sherwin Carlquist

The concepts of I. W. Bailey and his students Frost (1930a, 1930b, 1931) and Kribs (1935, 1937) concerning wood anatomy of dicotyledons as well as the work of Cheadle (1943) on xylem of monocotyledons were developed independently of evolutionary concepts in other fields such as embryology. Bailey was not unaware of developments in other fields, but he and the Harvard xylem students did not incorporate those developments in their work. There were advantages as well as disadvantages in the development of xylem data independently of work in other fields. In fact, if xylem evolution is influenced by long-term and short-term strategies for optimizing the water economy of a plant, xylem patterns ought to evolve entirely independently of patterns in, for example, pollination biology or defenses against herbivores. Ultimately, however, syntheses between xylem-evolution patterns and patterns from other fields are desirable. In particular, cladistic analyses based on a large number of morphological characters and cladistic analyses of DNA data invite comparison. There are some apparent conflicts between the traditional data on wood evolution and the newer cladistic results, and the purpose of this chapter is to examine what these apparent conflicts are and to point out possible methods of resolution for them. All of the discrepancies cannot be resolved, but there may be some value in highlighting what the discrepancies are, the concepts that may permit resolution, and the kinds of data that are needed. Trends of evolution in dicotyledon woods are analyzed in some detail here because unless we understand the nature of evolutionary specialization, we cannot develop a clear concept of primitive angiosperm woods. Those who wish a sourcebook describing wood anatomy of most of the primitive dicotyledon woods mentioned here will find Metcalfe (1987) useful.

MAJOR TRENDS OF XYLEM EVOLUTION AND THEIR PROBABLE SELECTIVE VALUE

The trends can be defined by correlation with the phyletic decrease in fusiform cambial-initial length. This decrease in length was discovered by the comparative data of Bailey and Tupper (1918), then applied to the phenomenon of storying in woods (Bailey, 1923), morphology of vessel-elements (Frost 1930a, 1930b, 1931), types of rays (Kribs, 1935), and axial parenchyma types (Kribs, 1937). Some additional features potentially correlated with fusiform cambial-initial length were considered by Metcalfe and Chalk (1950, p. xlv) and Carlquist (1975, 1992a). What has regrettably not been realized concerning these trends are the following:

1. The statistical correlations demonstrated are true, not just because there is a mathematical association among the features but because there is a functional association.
2. The apparent irreversibility of the trends is based on a functional correlation that has led to progressively greater division of labor in wood cells and progressively greater conductive efficiency and safety. Reversion of one character state without simultaneous reversion of the others would likely result in functionally suboptimal patterns. Irreversibility is not absolute, there are fluctuations that can be recognized without discarding the concept.
3. The statistical correlations of Frost, Kribs, and so forth, would not occur if evolution of each of these characters took place once or even a few times. Rather, a quick glance at the data of these authors shows that the evolution of each of these features has been **highly polyphyletic**; homoplasy is, therefore, quite common in these character state changes. Identification of homoplasy in wood studies requires analysis of a large family, or dicotyledons as a whole. One recent study of a large family, Rosaceae (Zhang, 1992) supports the concept of extensive homoplasy of characters. In a cladistic analysis of a large segment of dicotyledons (Magnoliidae), Loconte and Stevenson (1991) give consistency indices for characters studied: the low consistency index for xylem characters is indicative of a high degree of homoplasy.

Identification and Probable Adaptive Significance

The major trends of xylem evolution will be reviewed below. Probable ecological and physiological causes for the trends will be hypothesized.

1.   Shortening of fusiform cambial initials. The reasons for shortening of cambial initials are controversial. One should remember first that vessel elements elongate little or not at all compared to the length of the fusiform cambial initial from which they were derived; vessel-element length is, therefore, an accurate indicator of this trend. Zimmermann (1983) claimed there was no functional reason for the trend at all. However, Slatyer (1967) reported that even in vessels with simple perforation plates, air embolisms tend to be confined to a single vessel. This was confirmed by the results of Sperry (1985) and Carlquist (1988a, p. 317). If vessel elements are short, confinement of embolisms in them would represent minimalized interruption of the water columns. Shorter vessel elements would, therefore, be expected in species from drier areas. This was demonstrated in the case of Asteraceae (Carlquist, 1966) and has been repeatedly observed since.

2.   Increase in F/V ratio. The length of primitive vessel elements should be only slightly less than that of the accompanying imperforate tracheary elements (usually tracheids in the most primitive woods). The ratio of length of imperforate tracheary elements (F of various authors, referring to fibers) to accompanying vessel elements (V) is between 1.00 and 1.20 for primitive dicotyledons and rises to about 4.00 in the most specialized woods (Carlquist, 1975). The reason for increase in intrusiveness of imperforate tracheary elements may be the greater mechanical strength achieved by longer elements. The correlation between greater length and greater strength of imperforate tracheary elements has been demonstrated by Mark (1965), Wardrop (1951), and Wellwood (1962). Both increased fibrillar angle in longer elements and greater surface area (leading to greater bonding between adjacent elements) appear basic to the explanations.

3.   Perforation-plate morphology specialization. Frost (1930a, 1930b) hypothesized that the primitive vessel element is like a scalariformly pitted tracheid (e.g., Fig. 4.1-4.3), with the end wall highly inclined, long, with many scalariform perforations (see Fig. 4.4-4.6). Transition from a vesselless to a vessel-bearing condition involves loss of pit membranes in perforations; many primitive dicotyledons show incomplete lysis of these pit membranes (Fig. 4.5 and 4.6), as has been documented by various workers (see Carlquist, 1992a). Specialization can be conveniently measured by decrease in number of perforations (or bars between the perforations), although decrease in end-wall angle is also an index of specialization, this end-wall angle is less readily measured.

One can trace to Haberlandt (1914) the idea that loss of bars on the perforation plate lowers impedance to water flow. This idea, widely accepted, has been challenged, for example, by Schulte et al. (1989). These authors claim no gain in conductivity for vessels with simple (as opposed to scalariform) perforations plates at **nonpeak** rates of flow, but they do concede impedence of the bars at **peak** flow. The selective value of a structure often depends on performance under extreme

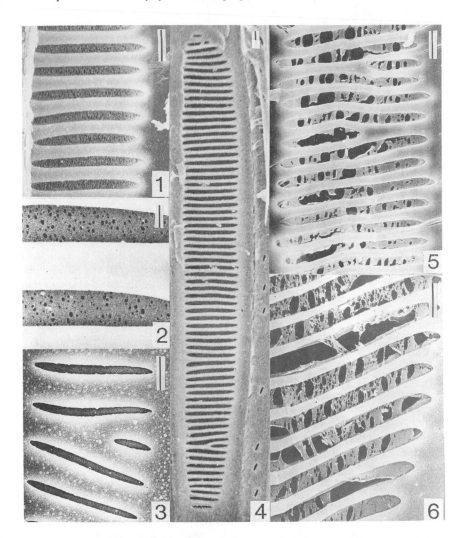

Figures 4.1–4.6. Scanning electron micrographs of radial sections of dicotyledon wood, showing vesselless dicotyledon wood (1–3) and features of primitive vessels (4–6). (1) *Tetracentron sinense*, end wall of earlywood tracheid, showing pits. (2) *T. sinense*, two pits from end wall of earlywood tracheid, showing porose nature of pit membrane. (3) *Trochodendron aralioides*, lateral wall of earlywood tracheid. (4) *Aextoxicon punctatum*, complete perforation plate of vessel element; pits in tracheid visible to right. (5) *Illicium tashiroi*, showing holes in the very finely porose pit-membrane remnants. (6) *I. cubense*, holes in the coarsely porose pit-membrane remnants. Bars at upper right in each figure indicate magnification: Figs. 4.1, 4.3, 4.5, 4.6, bar = 5 μm; Fig. 4.2, bar = 1 μm; Fig. 4.4, bar = 10 μm.

conditions, so peak flow-rate impedence is, in my opinion, very much a factor. Schulte et al. (1989) also claim that in a model made with plastic piping, introduction of impedances equivalent to bars in a perforation plate failed to reduce water flow rates. This model seems faulty, because the piping used is so much larger than actual vessel elements: to correspond with actual vessels, one should use a fluid more viscous than water in the artificial model. There is an enormous amount of comparative data showing phyletic reduction of bars in circumstances where lowering of impedance is likely. For example, there has been accelerated reduction in bar number in lianoid phylads that begin with scalariform perforation plates (Carlquist, 1975).

4.   Lateral wall specialization. Frost (1931) demonstrated a trend from scalariform to transitional to opposite to alternate circular pits. The selective reason for this trend appears to be the gain in mechanical strength in the vessel wall (Carlquist, 1975). This gain may not improve the safety of the conductive system, but rather represents improvement in mechanical strength of woods. The likelihood of this explanation is shown by some succulent Crassulaceae in which libriform fibers have been phyletically replaced by axial parenchyma; in these species, pits in vessel walls are likewise reduced in mechanical strength by great increase in pit size. Scalariform pitting has been retained by lateral vessel walls in some families (e.g., Magnoliaceae, Rhizophoraceae, and Vitaceae), but there may be mitigating factors for retention of a primitive character state in these families.

5.   Vessel outline as seen in transection. Vessels in dicotyledons are thought to be primitively angular in outline, with rounded outline a specialized condition (Frost, 1930a). A possible explanation for this trend is considered here. When a vessel is angular in transection, it typically has scalariform or transitional pitting that tends in its horizontal dimensions to correspond with vessel faces: pits on vessels rarely extend from one vessel face to another. Opposite and alternate, circular, lateral wall pits on vessels are much shorter in diameter than a vessel face, so no correlation between a vessel face and horizontal pit diameter exists. Also, vessels are often moderately narrow in primitive woods; if so, they contact fewer imperforate tracheary elements than do the wide vessels of more specialized woods. Fewer cell contacts with vessels would tend to be correlated with a more angular form.

6.   Shift in imperforate tracheary type. The imperforate tracheary elements shift from tracheids to fiber-tracheids to libriform fibers (these terms are used in accordance with the IAWA Committee on Nomenclature, 1964 because they accord best with functional significance; British and continental European literature, respectively, tend to use different sets of terms). Metcalfe and Chalk (1950) would describe the shift as beginning with "fibres densely provided with fully bordered pits" and ending with "fibres with simple pits." This trend was identified by Metcalfe and Chalk (1950, p. xlv) in a table in which they correlated it with stages in simplification of perforation plates.

The functional reason for this trend is found in the progressive division of labor that begins with vessel elements similar to tracheids (both serving both for conduction and mechanical strength) and ends with conductively more efficient vessel elements and mechanically strong libriform fibers.

7.   Axial parenchyma distribution. Axial parenchyma has progressed from diffuse into various degrees of aggregation (Kribs, 1937). Lesser degrees of aggregation (diffuse-in-aggregates) are more primitive than greater degrees (paratracheal abundant).

The functional explanation for this trend has to do with transportation and storage of photosynthates in the living cells of the wood. Aggregated axial parenchyma cells provide fewer and more massive channels for photosynthate translocation (which interconnect with similarly specialized rays, see below). There is also a net tendency for a shift from apotracheal to paratracheal parenchyma in the Kribs (1937) scheme. This corresponds with the function of paratracheal parenchyma to serve as a source for input of sugars into vessels, thereby renewing the conductive stream (Van Fink, 1982; Mooney and Gartner, 1991); this sugar input corresponds to seasonal extremes, such as vernal renewal of flow in xylem in *Acer* (Sauter et al., 1973).

8.   Evolution in ray types. Kribs (1935) subdivided rays into types based on (1) presence of both multiseriate and uniseriate rays, as opposed to loss of one or the other; (2) heterogeneity versus homogeneity of a ray, that is, both procumbent and upright cells as opposed to procumbent cells only; and (3) presence of long, uniseriate wings on multiseriate rays as opposed to loss of the wings.

The effect of these evolutionary trends is to produce horizontally oriented sheets of parenchyma tissue that consist wholly of procumbent cells. These are maximally efficient at radial translocation of photosynthates in the wood. Presence of borders on tangential walls of procumbent cells, much more frequent than reports indicate (because workers look at pits in face view instead of in sectional view), underline this conductive capability. As compared to diffuse axial parenchyma and heterogeneous rays, homogeneous rays and larger aggregations of axial parenchyma provide fewer but more massive points of contact between the vertical and axial parenchyma systems, both optimally constructed for rapidity of photosynthate translocation.

9.   Storying. Increase in storying is correlated with shortening of fusiform cambial initials (Metcalfe and Chalk, 1950, xlv). Bailey (1923) showed that nonstoried woods have pseudotransverse divisions of fusiform cambial initials to increase cambial circumference, whereas the plane of division changes to vertical in woods with storied woods.

There has been little comment in the literature on the likely explanation for this trend. Long fusiform cambial initials are extremely narrow, the lumen not much wider than the nucleus. Spatially, accommodating a tangentially oriented nuclear division and a vertically oriented cell plate is difficult under these circumstances. In a short fusiform cambial initial (and shorter fusiform cambial-initials also tend to be wider, in my experience), these kinds of cell division may be accomplished more readily. Interestingly, storying has been reported in gymnosperms—in the genus with the shortest fusiform cambial initials, *Ephedra* (Carlquist, 1989).

10.   Other possible trends associated with shortening of fusiform cambial initials. Some wood features appear more commonly in specialized woods but are not associated with the major trends of xylem evolution. Such features, which are not apparently statistically correlated with shortening of fusiform cambial initials, include large versus minute pits on lateral walls of vessels, absence versus presence of helical thickenings or other helical wall sculpture, absence versus presence of vestured pits in vessels, absence versus presence of interxylary phloem or other types of variant cambial activity, dense versus sparse vessels, and wide versus narrow vessels (Carlquist, 1991a), and tracheid dimorphism that results in vasicentric tracheid presence (Carlquist, 1988b). All of these represent a change in character state that likely represents a selective advantage in a phylad, at least at some moment in time. The possible selective values of these features and their systematic distribution are discussed elsewhere (Carlquist, 1988a). The failure of some of these features is likely due to the infrequency with which the feature has been evolved in dicotyledons. For example, vestured pits have likely evolved in ancestors of one order, Myrtales, but otherwise only in a scattering of genera of particular families. Although an apomorphy, vestured pits do not show statistical correlation with shortening of fusiform cambial initials. On the other hand, a feature such as wide versus narrow vessels can originate an almost infinite number of times, sometimes merely as a phenotypic modification. The frequency in dicotyledons is not, by itself, a criterion for association with fusiform cambial-initial length shortening.

One feature, solitary vessels (as opposed to vessels in contact in groupings as seen in transection), has been suspected of being a feature related to the major trends of xylem evolution (Pieter Baas, personal communication, 1988). Metcalfe and Chalk (1950, p. xlv) included it in their table comparing wood features to four degrees of perforation-plate specialization. However, the percentage of solitary vessels does not decrease in a progressive fashion in these four groupings (35%, 10%, 15%, and 5%). Solitary vessels (as defined by number of vessels per group below 1.20) is governed by the occurrence of true tracheids as a background tissue in the wood, or of abundant vasicentric tracheids (e.g., *Quercus*), regardless of ecology (Carlquist, 1984). To be sure, the percentage of species with tracheids ("fibres with distinctly bordered pits") decreases progressively in the table of Metcalfe and Chalk (1950, xlv): 73%, 45%, 31%, and 7%. However, a low degree of vessel grouping may also be found in specialized woods (simple perforation-plates, libriform fibers) in notably mesic localities (Baas and Carlquist, 1985). There is an appreciable number of species with libriform fibers in habitats sufficiently mesic that vessel grouping falls below 1.20. Tracheids deter vessel grouping because they form a conductive tissue that can serve when vessels embolize. Libriform fibers are not a conductive tissue, so that in woods with libriform fibers, vessel grouping offers redundancy in conductive pathways, a redundancy valuable when embolism formation occurs (Carlquist, 1984). If embolisms are unlikely to occur in a wood with libriform fibers, there is minimal selective value for vessel grouping.

Why Trends Occurred and Primitive Conditions Still Exist

Evolution of more specialized character states can mostly be attributed to evolution of a wood system adapted to management of water stress and seasonality of water availability—but in markedly diverse ways, as noted above. There are many ecological opportunities for plants with capabilities for growing successfully in these conditions and surviving the extremes they provide. One can even say that in all likelihood, ecological niches with more severe conditions and with more extremes are always unsaturated, because there is a higher extinction rate under these conditions than under relatively uniform conditions of temperature and moisture availability. These considerations tend to account for the advance into more specialized wood expressions. One should note that there is clear evidence that there has been a steady increase in advanced wood character states over time; the evidence from fossil woods is now relatively abundant (Wheeler and Baas, 1991). There are certainly some phylads that have developed specialized character states sooner than others: the woods in a particular geological stratum are at various levels of specialization, not one. This is not incompatible with an increase over time in the percentage of specialized character states. However, in wood evolution, as in so many evolutionary scenarios, one wonders why primitive character states have persisted at all.

One reason for persistence of primitive character states in wood may be lack of mutations for a character state advance in a particular phylad. This explanation can be demonstrated most readily in characters that are relatively infrequent in angiosperm woods—vestured pits, for example. If vestured pits are of selective value—as they do seem to be—they could be expected in more numerous groups, but genetic information for their formation is evidently not invented readily. A polyphyletic nature of shift from primitive to advanced in a particular character state is to be expected under these conditions. One must keep this in mind because, as we will see, many wood characters must be, by their systematic distribution, highly polyphyletic. This has implications for cladistics, where programs seek parsimony.

The features that are associated with shortening of fusiform cambial-initials are interrelated. If fusiform cambial-initials are not shortened in a particular phylad, storying does not originate. If simplification of perforation-plates permits rapid flow rates that are likely in a more highly seasonal environment, a similar increase in conductive efficiency in photosynthates through ray and axial parenchyma tissues is to be expected also. Rapid and sudden flushing of leaves and flowering tend to occur in environments that are seasonal (e.g., temperate forests, deserts, dry tropics). These events require a rapid increase in water flow rates and a rapid increase in photosynthate mobilization. In this example, the evolution of specialized axial parenchyma and rays might typically be expected to be synchronous with evolution of improved conductive efficiency in vessels, but not to precede it by very much. If invention of genetic information for a character-state change in several wood features is only moderately frequent, simultaneous advance in states of these various characters is even less frequent. Thus, lack of mutations that must be coordinated or in particular sequences to be adaptive may account for persistence of primitive character expressions in wood.

A further reason for persistence of primitive character states in wood may be the tendency for other aspects of the vegetative apparatus to evolve more readily, offering compensation. For example, leaf size and morphology apparently can change quite readily, so that a more xeromorphic leaf structure could easily compensate for a primitive xylem configuration. Microphylly in Bruniaceae and Grubbiaceae, both of which have notably primitive wood, might be examples of this. Cutinaceous stomatal plugs in mature leaves of conifers and Winteraceae are additional examples.

Yet another reason for persistence of primitive states in wood is the continuous availability of habitats that are relatively free from extremes that would require more specialized wood structures. Favorable mesic habitats are not as uniform as one might expect. Betulaceae and Cornaceae have notably primitive woods, but many species are found in areas of considerable winter cold. No Betulaceae or Cornaceae live in dry situations, so the adaptations of these families to counter winter cold permit them to occupy areas that are clearly mesic during the growing season. Primitive woods would not, however, be expected in areas that are cold in winter but with dry periods during the growing season.

Retention of tracheids in dicotyledon woods is notable, because one finds them in a series of woods in which perforation plates are simple. Examples of co-occurrence of these two character states may be found in woody Cistaceae, Epacridaceae, Proteaceae, Rosaceae, and other families and genera prominent in Mediterranean-type climates (see listing in Carlquist, 1985). The simple perforation plates are likely adaptive to peak flow conditions when soil moisture is still high and temperatures rise in late spring. The tracheids are adaptive in the hottest part of the year, when vessels become embolized; at these times, tracheids are likely to retain their water columns, supplying water to leaves in the same pathways that the vessels did. Evergreen leaves characterize the four families named in these habitats, so having a subsidiary conductive system of tracheids that can provide water to leaves when vessels have ceased to function is a requirement.

PHYLOGENETIC ISSUES RELATED TO CLADISTIC EVIDENCE

Vessellessness: Primary or Secondary?

The Baileyan concept (Bailey and Tupper, 1918; Bailey, 1944b) that vessellessness is primitive in dicotyledons has been widely accepted but recently challenged by cladistic constructions based on morphology (as opposed to molecular evidence), notably Young (1981) and Doyle and Donoghue (1986b).

The cladograms produced from molecular evidence (e.g., Qiu et al., 1993) pose similar problems, although the challenges from these latter cladograms are not explicit.

In none of these cladograms do the vesselless genera emerge in a single clade; they are separated from each other by vessel-bearing genera. This is hardly a surprise, because for several decades, leading phylogenetic systems have, for example, placed *Tetracentron* and *Trochodendron* in Hamamelidales, whereas the other vesselless families (Amborellaceae and Winteraceae) are treated as magnolialean (e.g., Thorne, 1986), although Endress (1986b) finds Trochodendrales to be intermediate between Magnoliales and Hamamelidales. The idea that *Sarcandra* is vesselless (Swamy and Bailey, 1950) has been disproved (Carlquist, 1987; Takahashi, 1988; Zhang et al., 1990). The distribution of vessels in *Sarcandra* is not entirely clear: they definitely occur in root secondary xylem of *S. glabra* (Thunb.) Nakai (Carlquist, 1987), occasionally in stem metaxylem of *S. glabra* (Takahashi, 1988), and are claimed to occur throughout the plant body of *S. hainanensis* (Pei) Swamy and Bailey (Zhang et al., 1990). These data in Zhang et al. (1990) are entirely from light microscopy; although they show that there is differentiation between end walls and lateral walls of the tracheary elements, one wants electron microscopy to demonstrate whether or not pit membranes do occur in what appear to be perforations. *Bubbia* (Winteraceae) tracheids can have scalariform end walls different from the alternate pitting on lateral walls, but pit membranes are present in the end walls (Carlquist, 1983). With the subtraction of *Sarcandra* from the list of primitively vesselless dicotyledons, instances of vessellessness do not appear so highly polyphyletic. *Tetracentron* and *Trochodendron* constitute monogeneric families that are often placed together as Trochodendrales. In the system of Qiu et al. (1993), Amborellaceae and Winteraceae fall close to each other.

The Vessel-Tracheid Continuum

The end walls of *Tetracentron* tracheids have porose pit membranes (Carlquist 1988a, p. 113; Carlquist 1992a). This is also true of *Bubbia* tracheids (Carlquist, 1983). There is only a slight difference between these conditions and the perforations of vessels of *Illicium* and other primitive dicotyledon woods in which fragments of the pit membranes characteristically remain in the perforations (Carlquist, 1992a). One could, on the basis of this evidence, reasonably argue that even today, we have some instances that there is a continuum between tracheids and vessel elements. A problem that must be dealt with is the fact that there are other differences between vessel elements and tracheids in these primitive woods: vessel elements are shorter, wider, and have lateral wall pitting that may or may not be like that of tracheids. If vessels were secondarily lost, one would have to create a monomorphic tracheary element by loss of several vessel features, not just one. Is this possible? We have at present possibly only one pertinent example, *Sarcandra glabra*, which is unusual in having some stem metaxylem vessel-elements (identical to tracheids in all respects other than loss of membranes in some pits) and vessels in root secondary-xylem (wider than tracheids and like vessels in other Chloranthaceae). Certainly the stem secondary-xylem tracheids of *S. glabra* are clearly monomorphic. As a potential example of secondary loss of vessels or origin of vessels, the genus *Sarcandra* deserves further study, but it may or may not be representative of what has happened elsewhere in dicotyledons.

Scenarios for Loss of Vessels and Their Probability

The difficulty in imagining loss of vessels in a phylad of dicotyledons may not be so much whether it is possible to create a secondary xylem of monomorphic tracheary elements from ancestors with wood-containing vessels that are intermixed with tracheids. Rather, the difficulty may be in explaining the ecological and other circumstances related to a hypothetical loss. For example, one can most easily imagine a vessel-loss scenario in a very primitive wood in which the vessel elements are so primitive there is very little difference between them and tracheids (e.g., vessel elements in Figs. 4.4–4.6). Such a hypothetical instance is so close to the origin of vessels earlier in the phylad that in effect, nothing is explained: it is just as easy, with a very primitive vessel condition, to imagine an additional instance of origin of vessels from vesselless ancestors as it is to imagine that the wood with primitive vessels is on its way to losing vessels. This is barely a character state reversion at all, in other words.

Because tracheids are of selective value under conditions of extreme cold and/or dryness (embolisms can be localized in individual cells, instead of disabling the entire length of a vessel), entry into very extreme conditions might conceivably promote vessellessness. There are a few instances in *Ephedra* in which vessels have very nearly been lost (Carlquist, 1988c), but the species of *Ephedra* in which this has occurred grow under the most extreme conditions of alpine cold and drought that the genus is capable of tolerating. Vessels have been reduced in abundance in some xeric cacti and Crassulaceae (Bailey, 1966; Gibson, 1973) but not extinguished (contrary to the implication of Young, 1981). These instances tell us that in order to lose vessels in a phylad by shifting it into extremely dry and/or cold conditions, it would have to be in almost unimaginably extreme habitats. But vesselless dicotyledons are all in quite uniformly mesic habitats, so the phylad would then have to change to the opposite extreme to explain the cases of vessellessness we have in dicotyledons. The likelihood of this happening is nil, and if vessels have been lost, we must search for some alternative scenario.

Another possible scenario is offered by Nymphaeaceae. Nymphaeaceae have no secondary xylem, but the primary xylem is vesselless. One can imagine the ancestors of Nymphaeaceae either with or without secondary xylem, and one can imagine that if they did have secondary xylem, vessels were present or they were absent in the secondary xylem: we have no evidence for these possibilities. Let us suppose one wishes to think of vessels in secondary xylem of pre-Nymphaeaceae. One can take into account Bailey's (1944) refugium hypothesis that primary xylem has characters more primitive than secondary xylem in a given plant and that vessels originated in secondary xylem and spread to primary xylem. Supporting data are presented by Bierhorst and Zamora (1965). By loss of secondary xylem in such an instance, one achieves vessellessness from vessel-bearing anestors. Bierhorst and Zamora (1965) found that in at least the first half of primary xylem, only tracheids are present in Aquifoliaceae, Buxaceae, Caprifoliaceae, Chloranthaceae, Cunoniaceae, Ericaceae, Pittosporaceae, and Styracaceae—all families with long scalariform perforation-plates in secondary xylem of some species. Unfortunately, Bierhorst and Zamora (1965) did not study the families on which one would most like primary xylem data: Aextoxicaaceae, Eupomatiaceae, Illiciaceae, Schisandraceae, Paracryphiaceae, and Sphenostemonaceae, for example (these families are mentioned because they have very primitive vessels: see Figs. 4.4-4.6). Let us suppose that the primary xylem of one or more of these phylads consists wholly of tracheids. Then, if that phylad changed in habit and ecology and thereby lost secondary xylem, it would be vesselless. Should that phylad then proceed to develop secondary xylem again, the secondary xylem might be vesselless. The example of *Sarcandra* is pertinent, because it has a sympodial habit: cane-like stems, each of which has a minimum of secondary xylem. Assuming that Takahashi's (1988) finding of some vessels in primary xylem is atypical of *Sarcanda glabra* stems and that primary xylem in this species is more typically vesselless, one can imagine a *S. glabra* ancestor has lost secondary xylem, then goes on to produce it at a later time, and uses the vesselless template in the primary xylem, extending it to the secondary xylem by a kind of paedomorphosis. The difficulty in such a scenario is that Nymphaeaceae probably lack secondary xylem because of adaptation to an aquatic habit, and related to this is a sympodial growth pattern in which stems elongate but never increase in diameter. If one is using a pattern based on Nymphaeaceae, one would have to imagine a thoroughgoing shift into an aquatic habitat, followed by a change into a woody terrestrial habit again. Possibly an herbaceous habit other than one associated with an aquatic habitat could be basic to loss of secondary xylem, but probable instances have not yet been presented.

The occurrence of vessels in secondary xylem of roots of *Sarcandra* but not in the secondary xylem of stems is suggestive of a monocot pattern, which may, in turn, be related to the sympodial habit of *Sarcandra* and *Chloranthus*

(Carlquist, 1992b). Nymphaeaceae have no vessels in roots; *Nelumbo* (Nelumbonaceae) has vessels in roots but not elsewhere (Kosakai et al., 1970). Vessel origin in monocotyledons began in roots (Cheadle, 1943). *Sarcandra* and Nymphaeaceae may be useful templates for showing us how monocotyledons and their vessel patterns appeared (Carlquist, 1992b). These two groups may be less useful as templates for constructing hypotheses about secondary vessellessness in dicotyledons, because of the multiple ecological and habital shifts that would be required.

We cannot at present cite any examples in extant angiosperms that suggest a scenario for secondary vessellessness that would clearly show us how such a change might originate. Chloranthaceae are the family that offer potential examples, but what occurs there may or may not be a pattern that has occurred elsewhere in dicotyledons. If the scenarios cited above do not offer satisfying modes for origin of secondary vessellessness, we would do well to consider the hypothesis below as an alternative.

The Polyphyletic Hypothesis of Vessel Origins in Dicotyledons

There has been an implicit belief by those who have constructed or interpreted cladograms that vessels have originated only a few times and that loss of vessels is the best hypothesis to explain vessellessness in Trochodendrales (incorporating both Trochodendraceae and Tetracentraceae), Amborellaceae, and Winteraceae. This is viewed as a parsimonious explanation for the appearance of these three groups in clades separated from each other by vessel-bearing families. But is the most parsimonious explanation the correct one? Bailey and others were well aware that the vesselless angiosperms do not form a single, close-knit group from which all other angiosperms can be derived, yet he believed vessellessness to be primitive in angiosperms, and retained from this primitive condition in the three groups named. He did not attempt to resolve this apparent contradiction. If one looks at the Frost–Kribs data and similar datasets relating phyletic advance of wood features in dicotyledons, one sees smooth progressions from primitive to specialized. This is important, because if advance in particular xylem features occurred only a few times, one might expect to see clustering of data (e.g., origin of homogeneous rays in a particular order and, therefore, the occurrence of this ray type in the families derived from that order but not in other orders). If one studies wood of numerous dicotyledons in relation to systematics, one sees no clustering of data: in group after group, for example, we can see **within the limits of a family or a genus** the evolution of a simple perforation plate from a scalariform perforation-plate. Simplification of the perforation plate of vessels has probably occurred in at least 100 families or genera independently, perhaps even several hundred times.

This has never been questioned, because the data are so clear and obvious on this point. The number of times simple perforation-plates originated from scalariform perforation-plates has not been estimated because plant anatomists tend not to have a particular phylogeny of angiosperm families committed to memory (and certainly not a phylogeny of genera within a family), and there has been a multiplicity of available phylogenies, in any case. The literature is at hand for such an estimate, should one want to do it. Let us, however assume that simple perforation-plates have originated from scalariform perforation-plates in at least 100 phylads. Is there any reason to believe that a preceding trend, origin of vessels with scalarifom perforation-plates, has occurred much less frequently? **Therefore, numerous independent origins of vessels, as well as numerous independent events of specialization in any particular vessel or other wood feature, are to be expected in dicotyledons.** This may not be parsimonious, but it appears much more probable than a limited number of origins of vessels or a limited number of events of specialization from scalariform to simple perforation-plates. One does not see a claim in cladistic literature that evolution is 100% parsimonious, but workers in cladistics do not give any estimate of how nonparsimonious evolution in a particular group might be. If primitive character states grow progressively less common in the fossil record (Wheeler and Baas, 1991), one can imagine much larger numbers of vesselless species at the time of origin of various orders that have primitive vessels. Certainly, vesselless angiosperm woods were once much more widespread than they are today, judging from the data of Suzuki et al. (1991); available data is very scanty.

Are Vessels Ever of Neutral or Negative Selective Value?

Hypotheses that involve secondary loss of vessels are conceivable only in relatively primitive woods that also have tracheids as a conductive cells type. Woods in which fiber-tracheids or libriform fibers are the background cells in which vessels are encased have only vessel-elements as conductive cells, because fiber-tracheids and libriform fibers are nonconductive. Consequently, one can have loss of vessels only in relatively primitive woods in which tracheids are the imperforate tracheary element type, unless one hypothesizes not only loss of vessels but also simultaneous invention of tracheids *de novo*. Let us imagine, for any possible remaining hypotheses, therefore, a wood in which vessels are accompanied by tracheids.

The only way in which a vessel-element is less good than a tracheid where water conduction is concerned is in the inability of vessel-elements to confine an air embolism to a single cell. This disadvantage would disappear if a phylad entered a constantly mesic area, in which freezing never occurred. Vesselless dicotyledons are all in mesic areas (freezing occurs in some habitats of Winteraceae, as well as in many habitats of *Tetracentron* and *Trochodendron*, but in none of these habitats does temperature likely fall more than about 5–8 C below freezing). Consequently, can one hypothesize loss of vessels because they are of neutral selective value?

I do not know of any instance in which evolution of a phylad into a mesic habitat has been accompanied by a development of fewer or no vessels, short of an extreme reduction in an aquatic plant (e.g., *Lemna*), and those who talk about secondary loss of vessels in angiosperms are not really concerned with instances like *Lemna*. One might expect such a phylesis in, say, Myrtaceae of bogs where water is constantly available (Myrtaceae have tracheids in addition to vessel-elements). However, in none of these Myrtaceae can one find any perceptible diminution of vessel density. In fact, it is difficult to imagine that in a mesic locality vessels would not have any positive selective value: days of peak transpiration even where water is constantly available to roots would favor vessels at least as much as tracheids. Localities that are wet but experience severe frost might be expected to favor tracheids, but even in these, vessels persist in all taxa and become more frequent (greater number per square millimeter of transection), in fact (Miller, 1975).

WOOD OF ANCESTRAL ANGIOSPERMS: COMPATIBILITY WITH WOOD OF SUPPOSED RELATIVES

The evidence seems massive from comparative studies in xylem of angiosperms that vessels primitively had scalariform perforation-plates and scalariform lateral wall pitting and that tracheids of supposedly primitively vesselless dicotyledons also have scalariform pitting (e.g., *Tetracentron, Trochodendron*, end walls of tracheids in *Bubbia* and *Zygogynum* of Winteraceae; pitting is near scalariform in wide tracheids of *Amborella* also). On the other hand, some recent cladistic analyses, including some based on morphology (Crane, 1985; Doyle and Donoghue, 1986b; Hickey & Taylor, Chapter 8) and some on DNA analysis (Qiu et al., 1993) are placing Gnetales close to angiosperms, often as the sister group. Before discussing the probability of this concept, the differences between gnetalean and angiosperm wood should be examined. Recent studies show that gnetalean wood is much more diverse than hitherto appreciated (Carlquist, 1988c, 1989, 1991b, 1995). We can still generalize that tracheids of Gnetales have large circular-bordered pits as well as vessels in which the perforation-plates have circular (foraminate) perforations like the pits of tracheids (Fig. 4.7–4.10). The claim of Muhammad and Sattler (1982) that scalariform perforation-plates can be found in *Gnetum* does not hold when large

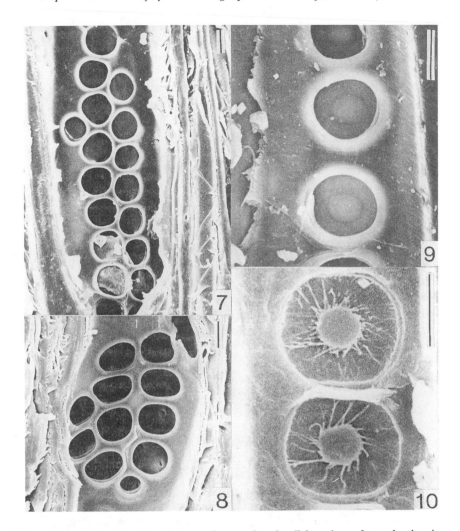

Figures 4.7–4.10.   Scanning electron micrographs of radial sections of vessels, showing the foraminate nature of the perforation plates in Gnetales. (7) *Ephedra major*, entire perforation plate. (8) *E. kokanica*, entire perforation plate. (9) *E. equisetina*, perforations, at end of a perforation plate, which have failed to lose pit membranes (these may be regarded as transitional to lateral wall pitting); these perforations show tori typical of *Ephedra* pit membranes, but the margo portion of the membranes is occluded. (10) *E. gerardiana*, perforations (or pits transitional to lateral wall pits) in which pit membranes have been retained. The tori are shown well; the strands of the margo connecting the torus to the edge of the pit cavity are evident in both pits, and the spaces between the margo strands are obviously large, but some strands have been lost by sectioning. Magnifications indicated by bars at upper right; Figs. 4.7 and 4.8, bars = 10 μm; Figs. 4.9 and 4.10, bars = 5 μm.

samples are examined: perforation plates that are anything like scalariform are extraordinarily rare, and the picture that one previously has been given of foraminate perforation plates (Thompson, 1918) proves to be very pervasive indeed (see Carlquist, 1995).

Unfortunately, the terms "scalariform" and "large circular" when referring to lateral wall pitting of vessels or pitting of tracheids are used as though these were concepts simply of shape, but they are much more than that: they represent two divergent methods of pit construction. The large circular-bordered pits of gymnosperm tracheids are not uniform, but the common type consists of a dome-shaped pit cavity and a pit membrane with a central circular torus; the torus is attached to the edges of the pit cavity by margo threads (Frey-Wyssling, 1976, p. 61), rather than a dense meshwork of microfibrfils. The margo threads have relatively large spaces between them: the spaces can pass colloidal gold particles but not India ink particles (Frey-Wyssling, 1976). As the pits age, the margo strands tend to develop even wider spaces between them (Frey-Wyssling, 1976). The spaces between margo strands may be large enough to pass small air bubbles, although larger air bubbles will be sieved out; this presumably gives tracheids a degree of conductive safety. Should the pressure between two adjacent tracheids become markedly different (e.g., one tracheid fills with air), the torus will be deflected so as to close the pit aperture—this is the obvious function of the torus, and it works because the torus fits into a circular pit aperture of circular-bordered pit. This prevents air leakage into the water column system and minimizes damage to the water column system if air embolisms do occur.

The gymnospermous pit described above definitely occurs in *Ephedra* (see Carlquist, 1991b, p. 61). *Gnetum* tracheids and lateral wall vessel pits have been claimed to lack tori, and while this appears to be true of *G. gnemon* (Carlquist, 1995), a torus and margo can definitely be demonstrated in pits of tracheary elements of some of the lianioid species of *Gnetum* (Carlquist, unpublished data).

On the contrary, the pits of angiosperm tracheids or lateral wall pits of angiosperm vessels mostly lack tori, and all lack a margo like that found in gymnosperm tracheid pits. The pit membranes do not have slender microfibrillar strands that radiate from a torus, but rather form a dense meshwork in which very small pores may be found (*Tetracentron*: Carlquist, 1992a; *Bubbia*: Carlquist, 1983). These are illustrated here (Figs. 4.1, 4.2; compare to Fig. 4.10). Air bubbles likely do not pass through these pores (see Carlquist, 1988a, p. 108), and pits do not aspirate by deflection of the pit membrane. The membrane is likely less flexible than is the gymnospermous pit with its suspension of the torus by strands. Also, a scalariform pit membrane would be more difficult to deflect and would require a scalariform torus in any case. There are a few reported instances of tori on pits of tracheary elements in angiosperms (e.g.,

Wheeler, 1983), and aspiration of pits in angiosperms can rarely be observed (Elisabeth Wheeler, personal communication 1993), but these scattered instances do not seem to represent ancestral conditions at all because they occur in a very few highly specialized woods. Therefore, I believe the contrast between gymnospermous pits and angiosperm pits in tracheary elements remains phyletically valid and involves much more than size and shape—a difference in a conductive safety system.

In addition, the circular bordered pits of Gnetales can be found superimposed on the gyres of helically banded primary xylem elements (Bierhorst, 1960, 1971). This peculiar condition is also characteristic of conifers and Ophioglossales, but is absent in angiosperms (Bierhorst, 1960, 1971; Bierhorst and Zamora, 1965). The closest approaches to this condition in angiosperms, notably *Juglans* (Muhammad and Sattler, 1982), do not represent the same phenomenon at all, according to the terminology and interpretations of Bierhorst and Zamora (1965). If primary xylem in vascular plants tends to be a refugium of primitive conditions, as claimed by Bailey (1944b), superimposition of circular bordered pits onto the helical gyres of primary xylem tracheary elements would represent a specialization, a pervasive introduction of the circular bordered pit-pattern into the entire xylem.

Scalariform pitting of tracheids characterizes, among living taxa, Cycadales. Doyle and Donoghue (1986) are incorrect in saying *Stangeria* and *Zamia* are the only cycads that have scalariform pitting on secondary xylem tracheids. *Stangeria* and *Zamia* have scalariform pitting throughout their secondary xylem, whereas the remaining genera have scalariform pitting in the metaxylem and earlier secondary xylem, yielding gradually to alternate circular pits later in ontogeny (Greguss, 1956). Bennettitales have scalariform pitting in secondary xylem (Bose, 1953; Nishida, 1969). Scalariform pitting is figured for the metaxylem of the aneurophytes (progymnosperms) *Callixylon* and *Protopitys* (Arnold, 1929; Henes, 1959; Scheckler and Banks, 1971) and for the secondary xylem of *Protopitys* (Henes, 1959). Pentoxylales have scalariform pitting in secondary xylem, with some occasional deviations (Vishnu-Mittre, 1957). Additional instances in fossil and living ferns can be cited, but are probably less relevant than instances of scalariform pitting of tracheids in seed plants. Doyle and Donoghue (1986) dismiss occurrence of scalariform pitting in primary xylem as of no significance, but if primary xylem tends to be a refugium for primitive expressions in vascular plants at large as Bailey (1944b) claims, instances of scalariform pitting in metaxylem become of great importance. Instances of scalariform pitting in metaxylem may be found in such seed plants as the seed ferns *Heterangium, Lyginopteris,* and *Medullosa* (Scott, 1923), *Glossopteris* (Pant and Singh, 1974), *Rhexoxylon* of the Corystospermatales (Archangelsky and Brett, 1961), and *Cordaites* (Scott, 1899). In secondary xylem, Doyle and Donoghue (1986) score scalariform pits as specialized compared to circular pitting (their character 20), but they take the reverse view for metaxylem (character 19). Their view is based on the absence of scalariformly pitted tracheids in secondary xylem of progymnosperms (*Protopteridium* excepted). I believe the Baileyan refugium theory applies not just to angiosperms but to all vascular plants. Circular bordered pits may have been evolved polyphyletically in tracheary tissues in this case—and that seems entirely likely. Circular bordered pits likely confer greater mechanical strength, a correlation shown by the fact that scalariform pitting in early secondary xylem is continued throughout the wood of the two genera, *Stangeria* and *Zamia*, without upright trunks, but scalariform pitting eventually yields to circular bordered pits in secondary xylem of the other genera, which have trunk-like stems.

One logical possibility is that angiosperms may have originated from less woody ancestors, in which little secondary xylem was present, but with scalariform pitting throughout. Upon attaining woodier forms, the scalariform pattern was carried forward to the additional secondary xylem, by the process I have called paedomorphosis (Carlquist, 1962). This is a concept different from Bailey's refugium hypothesis, which says that more primitive character states tend to occur in primary xylem as compared to secondary xylem. When paedomorphosis occurs, the primary xylem patterns are carried into secondary xylem so that both primary and secondary xylem tend to be alike in character state expressions. Takhtajan (1976) has advocated a paedomorphic origin of woods (as well as other features) in dicotyledons ancestrally.

A paedomorphic origin of woods with scalariform pitting of tracheary elements would accord well with the paleoherb hypothesis, discussed in great detail by Taylor and Hickey (1992). An herbaceous or minimally woody angiosperm, like *Chloranthus* or *Sarcandra*, would have scalariform pitting on primary-xylem tracheary elements and might have never had any genetic information (or it might have lost such information) for production of elements bearing alternate bordered pits. If such a paleoherb phylad became woodier, continuation of the scalariform pitting patterns by paedomorphosis could result in scalariform pitting throughout the wood.

One can imagine various ways in which scalariform pitting could be present ancestrally in primitive angiosperms without lessening their mechanical strength. One possibility is shown by *Trochodendron* and *Tetracentron*, in which earlywood tracheids are clearly scalariformly pitted on radial walls (Figs. 4.1–4.3), but the latewood tracheids, by virtue of much narrower radial wall faces, bear circular pits—there is not enough space for scalariform pitting. Thus, the latewood tracheids possess the maximal strength configuration, probably sufficient for a tree. With the origin of vessels, division of labor immediately produces narrow tracheids with circular bordered pits. If angiosperms were primitively vesselless, invention of vessels in various phylads provided not merely a conductively efficient cell type but also the tracheid, which has good wall-strength characteristics and, by their abundance in a wood, render the lateral wall pitting of vessels in a wood of relatively small importance as a means of conferring strength to woods. Yet another possibility is that extraxylary fibers, by their great strength, compensate for lower strength of tracheids with scalariform pittings. Certainly, this is shown abundantly in tree ferns. Early dicotyledons of limited height might have had extraxylary fibers sufficent to compensate for weakness of scalariformly pitted tracheids—evidence from the fossil record would be important here. The latewood tracheid solution of *Tetracentron* and *Trochodendron* was not available to vesselless dicotyledons in relatively nonseasonal environments, and it is noteworthy that Winteraceae, although they have scalariformly pitted tracheids in metaxylem, have circular bordered pits on tracheids in secondary xylem (scalariform pitting on end walls of some tracheids in a few species). The compensatory presence of extraxylary fibers certainly seems to be the strategy followed by monocotyledons. So obvious that one never sees it stated is that the xylem of monocotyledons has no division of labor, like that of dicotyledons, into conductive and mechanical systems, so the presence of extraxylary fibers as a counterbalance to the exclusively conductive function of xylem cells in monocotyledons is almost inevitable.

A sympodial plant like *Chloranthus* or *Sarcandra* could have been basic to monocotyledons (Carlquist, 1992b), and such a growth form is conceivably basic to dicotyledons as well. That would be in line with the "paleoherb" concept of angiosperm origin, although even if one endorses "paleoherbs" as basic to large assemblages such as monocotyledons and, in dicotyledons, Piperales, one can imagine that a sympodial plant of limited size and woodiness, like *Sarcandra* or *Chloranthus*, was basic to dicotyledons. The sympodial habit can shift to monopodial without great difficulty. If sympodial habits are, as I believe, basic in Piperales and Chloranthales, one can nevertheless see no vestiges of sympodial construction in trees such as *Hedyosmum* of the Chloranthaceae.

The divergence between gnetalean woods and angiosperm woods is quite marked, in any case, and detailed study only reinforces the differences. These differences led Thompson (1918) and Bailey (1944b) to claim independent origin of vessels in Gnetales and angiosperms. Crane (1985), in his cladistic scoring, accepts this concept. Doyle and Donoghue (1986) do not score the phyletic status of vessels in angiosperms, although they do score presence of vessels in Gnetales as a specialized character state.

Given that wood of Gnetales is much like wood of a conifer (or some other gymnosperms, such as *Ginkgo*), except for its vessels and multiseriate rays, are Gnetales really close to angiosperms, or, if they are, how have they attained such different wood? This is a question neither easily solved nor dismissed. The cladogram of Crane (1985) shows Gnetales only a single character closer to angiosperms than to the line leading to Pentoxylales and Bennettitales. The various cladograms of Doyle and Donoghue (1986) have the merit of entertaining various hypotheses, their various cladograms incorporating particular conceptual constraints. On the basis of vegetative characters alone (and surely many vegetative characters are more highly homoplasic than more deeply seated life-cycle features), Gnetales fall close to angiosperms, but on the basis of macroreproductive characters, Gnetales fall far from angiosperms. Doyle and Donoghue (1986b) ably show the diversity of results obtainable by cladistics, depending on change of a few parameters.

Many cladists would probably agree that phylogenetic placements are uncertain when we are dealing with groups that include fossil representatives for which many important facts remain unknown (e.g., cellular histology of reproduction). The evidence from analysis of DNA sites, when projected cladistically, can involve some ancient groups that are extant, such as Gnetales and Cycadales, but it cannot involve fossil groups. By default, this might make Gnetales look much closer to angiosperms than they really are. Surely anyone familiar with cladistics would agree that the topology of a tree based on DNA data would change markedly if we were able to study DNA of fossil groups such as Bennettitales, Caytoniales, Pentoxylales, and so forth.

The summary of consensus trees offered by Qiu et al. (1993) on the basis of *rbcL* shows Gnetales close to angiosperms. However, when Qiu et al. (1993) generate a tree based on the specifications that Nymphaeales are the sister group of all other angiosperms, Gnetales group closely with conifers, and angiosperms fall as close to Cycadales as to the gnetalean-coniferophyte line. The results of Hasebe et al. (1992), who also used *rbcL*, differ from those of Qiu et al. (1993). In the cladogram of Hasebe et al. (1992), one branch leads to ferns, another leads to seed plants, and this latter branch divides much as in the classical systems, into gymnosperms and angiosperms. In the clade of Hasebe et al. (1992), Gnetales are closer to Cycadales than to angiosperms. Monophyly of both Gnetales and angiosperms is shown in this cladogram. Zimmer et al. (1989), using ribosomal RNA, have constructed a cladogram in which conifers, cycads, and *Ginkgo* are closer to angiosperms than are Gnetales. These results are noteworthy, because they used no fewer than four species of Gnetales, from all three of the families.

Further molecular results are certainly to be welcomed, but, clearly, the topology of cladograms that cannot include important fossil groups must necessarily leave us quite unsatisfied. We must keep in mind the possibility that placements for gymnosperm groups in such cladograms may result from such default more than from positive evidence of relationship. Where wood of angiosperms is concerned, wood of angiosperms is much more similar to that of Bennettitales, Pentoxylales, and even Cycadales than to that of Gnetales. Eventually, these facts must be explained. The comparative studies of wood in all groups of vascular plants are likely to yield more results of importance.

CONCLUSIONS

Concepts of evolution of dicotyledon wood are reviewed in order to see what the basis for these concepts are and, therefore, how these concepts might relate to recent results from cladistically projected analyses of DNA data and other datasets. The concepts from workers such as Bailey, Cheadle, Frost, and Kribs show that the following features are statistically correlated with each other: shortening of fusiform cambial initials, origin of vessels, simplification of scalariform perforation plates, change from scalariform to opposite to alternate pitting on lateral walls of vessels, change from angular to rounded shape of vessels in transection, shift from tracheids to fiber-tracheids to libriform fibers, change from diffuse axial parenchyma to various types of axial parenchyma aggregation, change in ray histology, most notably from heterogeneous to homogeneous ray structure, and change from nonstoried to storied cambia. These trends were held by Bailey to be irreversible in broader outlines. The present chapter analyzes the functional nature of the character states and their selective basis. The idea that these changes have not shown extensive reversion is believed to be that there are functional correlations among the characters, so that reversion of any one feature would lead to a less optimal functional mode.

The question of whether vessellessness is primary or secondary in woody dicotyledons has arisen in recent cladistic analyses. Two possible scenarios for secondary vessellessness are reviewed. In one of these, vessels are lost during evolution into extreme drought and cold, as they have been (incompletely) in cacti and in *Ephedra*. This scenario would require survival of extreme conditions yet complete reversion to mesic conditions in order to account for the occupancy by contemporary vesselless dicotyledons of exclusively highly mesic sites, and this is regarded as highly unlikely. In a second scenario, a primitive phylad of dicotyledons would lack vessels in primary xylem only; loss of secondary xylem in such a phylad would result in secondary vessellessness. This scenario might have occurred in ancestors of Nymphaeaceae, accompanied by adaptation to aquatic habitats, but such a scenario is unlikely if a phylad remains woody. An alternative way of explaining the occurrence of vesselless taxa in more than one clade of dicotyledons is the assumption that there have been numerous origins of vessels in angiosperms. This is possible if we assume a high degree of homoplasy for character stage changes in the wood character associated with shortening of fusiform cambial-initials, such as those associated with vessel-element specialization. If such vessel specializations were highly homoplasic, vessel origins (in a way, the precursor events to vessel specializations) are also likely to be polyphyletic. Vessels may have originated numerous times in dicotyledons.

Scalariform pitting of tracheary elements is a primitive character state in angiosperms, according to results of wood anatomists such as Bailey and Frost. Scalariform pitting is certainly widespread in the metaxylem of vascular plants, although it occurs in the secondary xylem of fewer major phylads. If primitive angiosperms were herbs, in accord with the paleoherb hypothesis, metaxylem would be expected to have scalariform pitting of tracheary elements. Phylesis toward greater woodiness, if it featured paedomorphosis, would extend scalariform patterns into secondary xylem. The paleoherb hypothesis is conceivable in terms of known patterns of wood character state distributions.

Scalariform pitting in tracheary elements of angiosperms is in marked contrast to the circular pitting of gymnosperm tracheids (and vessels, in the case of Gnetales). This contrast is frequently presented merely in terms of pit size and shape, but that is misleading. The gymnosperm pit not only has a torus but also a pit margo with pores much larger than those in angiosperm pit membranes. The gymnosperm pit is a mechanism for conductive efficiency and safety, quite different from those in the pits of angiosperm tracheary elements. Although a scattering of dicotyledons have tori on pit membranes, the gymnospermous pit margo does not accompany these tori. These and other observations underline the marked difference between wood of Gnetales and that of angiosperms. Gnetales and angiosperms have been placed close to each other in some recent cladistic analyses, both on the basis of DNA evidence and other datasets. However, other analyses, involving both kinds of evidence, place the two groups well apart. If evidence from DNA of fossil groups were available, topology of cladograms would likely change appreciably. In any case, gnetalean wood and angiosperm wood are quite different.

Evidence for the Earliest Stage of Angiosperm Pollen Evolution: A Paleoequatorial Section from Israel

Gilbert J. Brenner

A new model is proposed for the origin of the monosulcate condition commonly found in angiosperm pollen from the Lower Cretaceous. This model disagrees with the widely accepted view that the sulcus in angiosperm pollen is a haptotypic feature that is inherited from their gymnospermous ancestry. The pollen types on which this new model is based were found in core samples from two wells of late Valanginian to early Aptian age (see Appendix) during a study of the palynology of the Lower Cretaceous of Israel (Brenner and Bickoff, 1992). The progression of early Cretaceous angiosperm pollen-types described in this chapter begins with inaperturates in the late Valanginian and ends with tricolpates and monoporates in the Aptian.

An understanding of the structure and general morphology of the earliest angiosperm pollen grains in the geologic record may contribute not only a better understanding of the group's botanical ancestry, but may give us a better grasp of the systematic relationship between the extant taxa. How one identifies the oldest angiosperm pollen is a difficult problem, not easily resolvable. A prototype, different from what is being proposed in this chapter, was put forth by

Walker and Doyle in 1975, and Walker and Walker in 1984. Figure 5.3 illustrates the essential idea of their model, which was based on the comparative study of living plants. Briefly stated, this model suggests that the primitive angiosperm pollen grain is large- to medium-sized, boat-shaped, smooth-walled (or weakly sculptured), with a homogenous or granular interstitium (infratectal layer), atectate, and with an endexinal layer (considered a new layer in angiosperms) that is either missing or poorly developed under the apertural area. In living plants, this type of angiosperm pollen can be found today only in the primitive members of the Annonaceae, Degeneriaceae, and Magnoliaceae.

The following list (with comments) defines some of the terms used in this chapter to describe pollen morphology. For a complete glossary of palynological terms refer to Traverse (1988).

Ektexine (Faegri and Iversen, 1975): the outer layer of the two layers of the exine of spores and pollen. The ektexine is underlain by the endexine. The ektexine may have surface sculpture and internal, complex structure. This term when used for angiosperm pollen is synonymous with "sexine." The endexine in angiosperm pollen is considered a new layer that evolved in the Cretaceous, initially beneath the foot layer or nexine, and in the position of potential aperture development.

Foot layer: a basal layer in the sexine that is often indistinguishable from the nexine.

Infratectal layer: zone within the sexine that develops structure. Structural elements are overlain by a tectal layer.

Intectate: lacking a tectum. Structural elements, such as columellae or granules are free-standing.

Nexine (Erdtman, 1952): the inner layer of the exine. Under light microscopy and scanning electron microscopy (SEM), the nexine is usually unsculptured and homogeneous in structure. When no distinct foot layer of the sexine can be distinguished (especially in fossil pollen), the term "nexine" is used for the inner layer of the exine beneath the infratectal structures. The term is not used for spores, only pollen grains.

Sexine (Erdtman, 1952): the outer layer (above the nexine) of the exine in both gymnosperms and angiosperms that contains complex structures such as granules or columellae. The columellae may be rooted in a foot layer. "Sexine" is synonymous with "ektexine."

Tectum: the outer layer of the sexine that forms a roof over the structural elements of the sexine.

Figure 5.3 is a diagram outlining the essential aspects of Walker's (Walker and Doyle, 1975; Walker and Walker, 1984) concept of angiosperm pollen evolution. The upper part of the diagram illustrates the type of gymnosperm pollen that may have preceded the first angiosperm grains. The grain is mono-sulcate, large, boat-shaped (Fig. 5.3C) and the nexine is typically laminated (Fig. 5.3A,B). The sexine is either homogenous or has a granular infratectal layer. This pollen type can be found in the Bennettitales, Gnetales (except for *Gnetum*), and Pentoxylon, gymnosperms thought by Doyle and Donoghue (1986b) and Crane (1985) to be closely related to angiosperms.

The lower part of Fig. 5.3 presents the proposed evolutionary sequence of angiosperm pollen development. In this model, the laminated nexine is gone and the sexine goes from homogenous (Fig. 5.3D) to granular (Fig. 5.3E) to tectate/columellate/reticulate (Fig. 5.3F) to columellate/intectate (Fig. 5.3G).

There appears, however, to be a disparity between this theoretical model and the morphology of earliest undoubted angiosperms described in the literature from early Cretaceous horizons. Previous to this study, the oldest undoubted angiosperm pollen grains that have been reported are typically small, globose monosulcates with a reticulate, tectate–columellate sexine. Several workers continue to search for the elusive, boat-shaped magnoliaceous grain in pre-Barremian to Jurassic sediments.

All the samples in this study were deposited in marginal marine to nonmarine environments in the proximity of the early Cretaceous paleoequator. The most important aspects of this study are as follows:

1. It describes angiosperm pollen types in pre-Barremian horizons that are inaperturate and that existed before the monosulcate, tectate–columellate form-genus *Clavatipollenites* (a form-group that most paleobotanists believe represents the oldest recognizable angiosperm pollen type).

2. These data provide information on pre-Barremian, Cretaceous angiosperm pollen morphology which may offer support for some phylogenetic models over others.

3. They may contribute to a new way of interpreting past or future gynmospermous megafossils as a possible clue to the ancestral lines that led to early angiosperms.

4. The pre-Barremian angiosperm pollen described may provide insight as to which extant taxa are closer to the base of flowering plant phylogeny.

5. This study may contribute to a better understanding of the functional morphology of different parts of the exine through knowledge of their temporal inception.

6. It documents that the Helez Formation contains the oldest Cretaceous appearance of pollen with monocotyledonous characteristics.

7. It shows that the late Barremian to Aptian horizons in the Zohar 1 well records the first appearance of several pollen types that are now found associated with some extant magnolialean taxa.

8. This study indicates that samples from the lower Aptian horizon (core 6) in the Zohar 1 well record the earliest appearance of tricolpates (eudicots) in sediments that have now been associated with radiometrically dated volcanics in the field.

9. It may provide information on the geographic origin as well the climatic background of the earliest angiosperms.

GEOLOGIC BACKGROUND

Lower Cretaceous in the Coastal Plain of Israel

The Helez Formation of Israel was deposited in a continental-to-shallow marine environment that transgressed the eroded surface of late Jurassic strata. It is composed of shale, sandstone, limestone, and dolomite that become more shaley to the west, whereas to the east, beds are more sandy. The angiosperm pollen in this study was found in highly carbonaceous shale within the lower and middle sand units. The shale contains numerous carbonized plant fragments, siderite nodules, and fresh to brackish water fauna, and has been interpreted as fluvial and lagoonal sediments formed during a maximum regressive phase of the late Valanginian to late Hauterivian stages (Shenhav, 1970). As in the Negev, the basal Cretaceous in the coastal plain unconformably overlies rocks of Jurassic age. The Helez Formation in the Kokhav 2 well (Fig. 5.1 and 5.2A) was designated as the type section of the unit by Shenhav (1970), who divided it into eight lithostratigraphic members. The first letter of the member is the formation's name and a corresponding number. Palynological material for this study was found in shale layers in cores 2–4 from the middle sand member of unit H4, and in core 5 from lower sand member of unit H1. The shale layers were deposited in marginal marine environments that interbed with continental and nearshore marine sandstones of the middle and lower sand members. The dating of units L.CrIIa (H_m, H_1, H_2, H_3), and L.CrIIb (H_4, H_5) is relevant to this study. Shale layers within the lower sand member (unit H_1), which provided the earliest angiosperm pollen in this study, are dated as late Valanginian to early Hauterivian. This date is based on the presence of the early Hauterivian ammonites *Leupoldia* and *Saynella* in the upper part of L.CrIIa (Kokhav Dolomite), above the lower sand member, and Valanginian foraminifera (Grader and Reiss, 1958) and ostracods (Gerry, 1963) below the lower sand member in L.CrIIa.

A late Hauterivian age is attributed to the angiosperm pollen grains from the middle sand member of the Helez Formation (H_4). This unit overlies unit L.CrIIa, which is early Hauterivian in its upper portion, and underlies unit L.CrIIc, which Grader and Reiss (1958) consider to be of early Barremian age on the basis of its megafossil assemblage (mollusca, annelida, bryozoa, and calcareous algae). Shenhav (1970; Shenhav and Shoresh, 1972) interpreted the sand of unit H_1 as an eolian deposit, whereas a tidal to lagoonal environment was postulated for the sand of unit H_4 (middle sand member).

Lower Cretaceous in Northern Negev

The basal Lower Cretaceous in the northern Negev is younger (late Barremian/Aptian) than the base of the Helez Formation in the coastal plain (Valan-

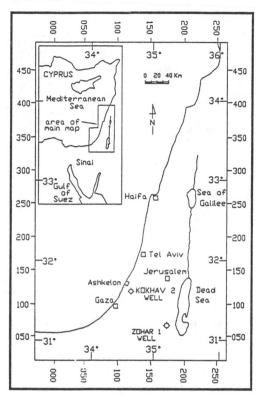

Figure 5.1.   Map of Israel showing well locations.

ginian). The time-transgressive nature of the Lower Cretaceous in this area was demonstrated with palynology by Brenner and Bickoff (1992).

The Lower Cretaceous in the northern Negev was divided by Aharoni (1964) into four formational-divisions within the Kurnub Group (Fig. 5.2B). The type section is a composite from the Zohar 1 and 3 wells. The boundary between the Upper Jurassic Beersheva Formation and the Lower Cretaceous Zeweira Formation can be distinguished by a lithoelectric marker of regional extent. Cores 6 and 7 in the Zohar 1 well (Fig. 5.1) come from the lower part of the Zeweira Formation (Fig. 5.2B). The Zeweira Formation at this level was formed primarily in a fluvial environment with occasional brackish-water inter- vals. Just above core 6, there is a marine intercalation that can be found in outcrop in the Makhtesh Ramon (an eroded anticline in the northern Negev) and near Beirut, Lebanon. In both areas, this marine member of the Zeweira Formation contains ammonites and mollusks of early Aptian age (Dubertret and Vautrin, 1937; Vokes, 1946). Core 6 contains the oldest tricolpates in the Zohar 1 well as well as the first occurrence of *Brenneripollis* and *Afropollis*. This

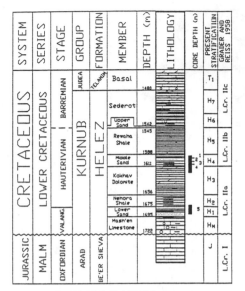

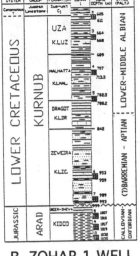

A. KOKHAV 2 WELL

B. ZOHAR 1 WELL

Figure 5.2. (**A**) Stratigraphic columnar-section of the Kokhav 2 well. (**B**) Stratigraphic columnar section of the Zohar 1 well.

assemblage is indicative worldwide of a basal Aptian position. A basalt unit that just overlies this marine intercalation in the Ramon erosional cirque has yielded a ^{40}Ar–^{39}Ar age of 118 ± 1.5 million years ago (Ma; Weissbrod et al., 1990; G. Gvirtzmzn, personal communications 1993). This date is close to the 120 Ma for the Barremian/Aptian boundary suggested by Ozima et al. (1979). Above the Uza Formation, the upper boundary of the Kurnub Group grades into more open-marine-carbonate units of late Albian to early Cenomanian age. The thickness of the Kurnub Group in the type section is 385.5 m. The following are brief descriptions of the four formations of the Kurnub Group starting from the youngest.

Uza Formation: thickness = 77.5 m; consists of limestone, marl, shale, and sandstone with a calcareous matrix. The limestone is argillaceous, white and gray with dark clay inclusions; the marl and shale are medium to dark gray.

Malhatta Formation: thickness = 94.5 m; consists of sandstone, with shale and marl intercalations, and limestone members. The sandstone is fine grained with a calcareous or dolomitic matrix. Glauconite is present in the upper half of the formation. The shale is dark gray with abundant plant remains.

Dragot Formation: thickness = 58 m; alternating limestones, marl, and hematitic shale. There are some sandy lenses at the base of the formation.

Zeweira Formation: thickness = 155.5 m; consists of interbedded fine-grained sand-stone with carbonaceous shale and lignitic layers, as well as some limestone strata ["first marine intercalation" of Bentor and Vroman (1960)]. The base of the Zeweira Formation unconformably overlies the Jurassic Arad Group.

ANGIOSPERM POLLEN FROM THE BASAL CRETACEOUS OF THE COASTAL PLAIN

The angiosperm pollen types described below were found in core samples from the Helez Formation, of Upper Valanginian to Upper Hauterivian age, in the Kokhav 2 borehole from the Coastal Plain of Israel.

Late Valanginian to Early Hauterivian Angiosperm Pollen Types

The angiosperm pollen grains in core 5, from the lower sand member (H_1) of the Helez Formation, are extremely rare, with less than two grains per thousand palynomorphs. Fifteen grains have been examined. All angiosperm pollen grains at this level show tectate–columellate structure; the columellae are closely spaced and have, for the most part, a very fine reticulum (Fig. 5.4A–D), with columellae in most cases connected by tectal bridges.

Figure 5.4D, which is one of the larger grains, shows the tectal connections clearly. One feature of the exine that seems characteristic for these grains is the discontinuous nature of the tectate–columellate condition. On most specimens, the tectum is missing in tiny patches, and the columellae stand free. This condition of the reticulum gives the grains a rather irregular appearance under light microscopy at 1000 times magnification. The grains observed are typically very small, about 15 μm, to as much as 26 μm in diameter, and primarily circular in outline (based on 15 specimens). The nexine or foot-layer on which the columellae rest is less than 1 μm in thickness. The columellae are closely spaced, less than 1 μm apart, except for the few larger grains. None of the grains show any indication of a sulcus, nor does safranin staining indicate any endexinal thickenings.

Late Hauterivian Angiosperm Pollen Types

In the upper Hauterivian Middle Sand Member from cores 2 to 4, inaperturate and weakly monosulcate angiosperm pollen grains appear together. They are still rare but more abundant than those from core 5, being found in amounts equal to about two to four grains per thousand palynomorphs. Patches of endexinal development are commonly found in pollen types with or without apertural development. In inaperturate forms, the endexinal development is

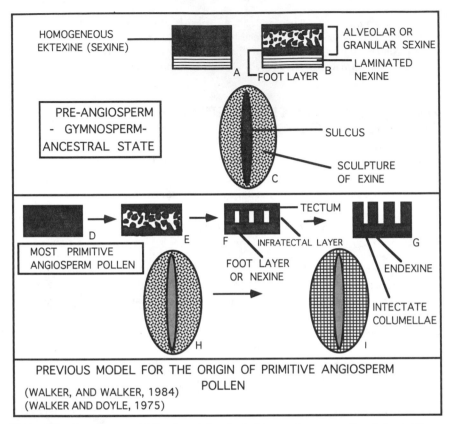

Figure 5.3.   Previous model for the origin of primitive angiosperm pollen (adapted from Walker and Walker, 1984 and Walker and Doyle, 1975.)

found on only one side of the grain. In the sulcate forms, the endexine commonly forms beneath the aperture.

The angiosperm pollen types from this level can be organized into four morphological groups:

1. Pre-*Afropollis* Group (Fig. 5.5A–D). The group is distinguished by the presence of wedge-shaped, sinuous muri that have fluted sides (Fig. 5.5B), a feature common to *Afropollis*. The grains are circular to broadly elliptical in outline, 19-25 μm maximum diameter. Although some grains are weakly monosulcate with ragged apertural margins (Fig. 5.5C,D), most of the grains are inaperturate (Fig. 5.5A,B). No evidence of any opercular structure, as is common in *Afropollis*, was found. In addition to the large columellae that support the tectum, extremely small microcolumellae typically anchor the tectum to the central body (Fig. 5.5B). Columellae are strongly reduced in areas where the sexine separates from nexine and forms a loose envelope around the central body.

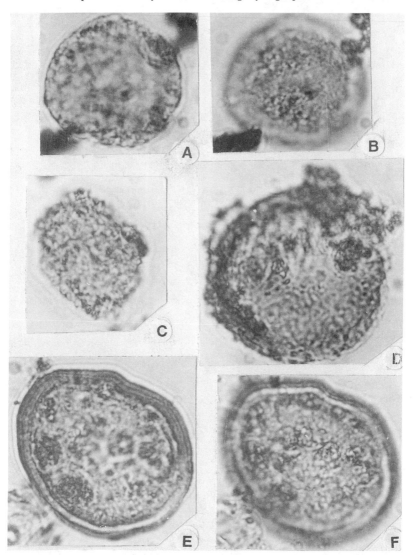

Figure 5.4. Light micrographs of pollen grains from the Helez Formation, lower sand member, Kokhav 2 well, core 5, late Valanginian/early Hauterivian; all specimens are x1820. (**A,B**) Inaperturate, circular form with irregularly placed columellae connected by tectal bridges. (**A**) Medium focus showing columellae along the border. (**B**) Same grain, high focus showing distal ends of columellae. (**C**) Inaperturate, semitectate, broadly elliptical grain, with irregularly arranged columellae. (**D**) Larger, inaperturate grain, circular in outline; large, widely spaced pila with capita connected by tectal bridges. (**E,F**) *Gnetum*-like grain showing dense, closely packed pila with irregular distribution. (**E**) Same grain, medium focus; (**F**) same grain, high focus.

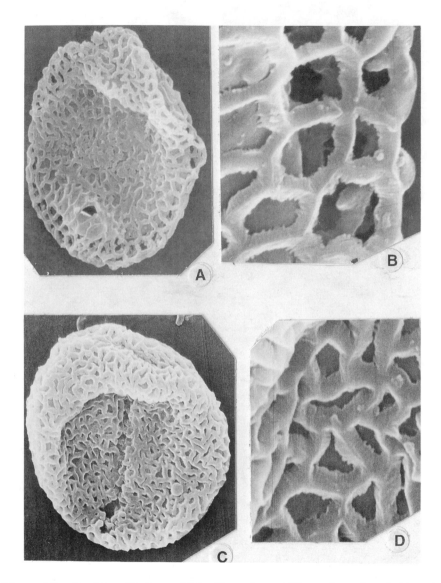

Figure 5.5.   Pre-*Afropollis* Group, Helez Formation, middle sand member, Kokhav 2 well, core 2. (**A,B**) Inaperturate grain. (**A**) SEM view, x2593; (**B**) same grain, SEM, showing wedge-shaped muri and fluted sides. The muri are attached to the foot layer by small infratectal columellae, x15,300. (**C,D**) Sulcate type with weakly developed margins. (**C**) Showing jagged-edged sulcus, SEM, x3124; (**D**) same grain, showing wedge-shaped mural walls and fluted sides underlain by micro-columellae, SEM, x15,300.

2. Spinatus Group (Fig. 5.6A–D). The pollen in this group is all inaperturate, very small (12–20 μm), and distinctly circular to broadly elliptical in outline. The exine is tectate–columellate; the tectum is semitectate–perforate to finely reticulate. In addition to the above morphology, the group is identified by the presence of broadly based supratectal spines, which are straight to curved at their distal ends. The supratectal spines are similar to the ornamentation found on the early Aptian *Peroreticulatus* (*Brenneripollis*) types. On many of the grains, part of the sexine is underlain by a thick development of endexine, which can be seen easily by the denser staining of the endexine with safranine (Fig. 5.6A).

3. *Clavatipollenites* Group (Fig. 5.7A–D). Pollen of the *Clavatipollenites* Group are all weakly monosulcate to inaperturate, very small (12–20 μm), and distinctly circular in outline. The exine is tectate–columellate; the tectum is semitectate-perforate to finely reticulate and covered with supratectal granules that are occasionally spinate. When present, the aperture is expressed by a reduced development of the sexine. The surface of the aperture is distinctly granular to verrucate (Fig. 5.7A,B), in which the grana or verrucae are formed by the reduction of sexine elements. Endexine shows up as dense staining in a small portion of the total exine area. In all cases observed, the endexinal development is positioned beneath any weak expression of an aperture (Fig. 5.7D). Grains in this group resemble the fossil pollen *Clavatipollenites hughesii*, as well as the pollen of the extant *Ascarina difussa* (Walker and Walker, 1984).

4. *Liliacidites* Group (Fig. 5.8A,B). The pollen in this group is monosulcate, larger (20–50 μm) and more elliptical in outline than the above-mentioned groups. The sexine is coarsely reticulate with the lumina of the reticulum quite variable in size. The muri are typically smooth and supported by relatively tall columellae. This sexine structure is similar to what is found in some moncots.

Pollen grains belonging to the *Liliacidites* Group are a very rare component of the angiosperm pollen found in the Helez Formation. They are the oldest grains in this study that show monocot affinities.

ANGIOSPERM POLLEN FROM NORTHERN NEGEV

The Upper Barremian/Lower Aptian Zeweira Formation

In the northern Negev, east of the Kokhav 2 well, the basal Cretaceous Zeweira Formation in the Zohar 1 borehole contains angiosperm pollen of late Barremian to late Aptian age (Brenner and Bickoff, 1992).

Afropollis, *Brenneripollis*, and tricolpate pollen make their first appearance together in core 6 of the Zeweira Formation. The microfloral assemblage in core 6 is, in most respects, identical to that which is found in core 7. Specific spores and monosulcate angiosperm pollen clearly indicate the temporal closeness of

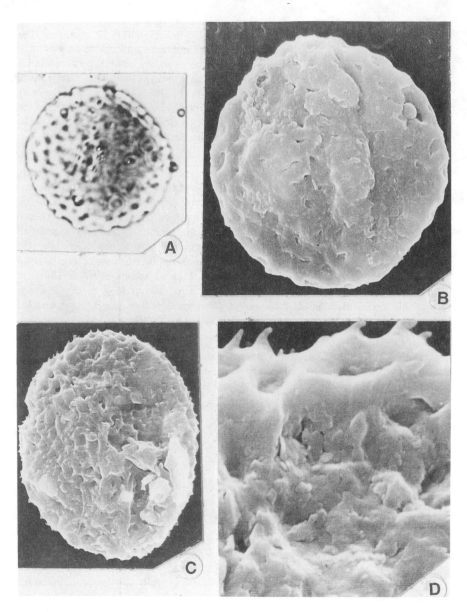

Figure 5.6. Spinatus Group. (**A,B**) Inaperturate, circular form with endexinal staining. (**A**) Light micrograph, shows endexinal staining, indicates locus for potential aperture formation, x2000; (**B**) same grain; short spines can be seen at margin of grain, SEM, x3350. (**C,D**) Inaperturate, slightly elliptical grain. (**C**) SEM, x3015; (**D**) same grain showing well developed supratectal spines, SEM, x17,748.

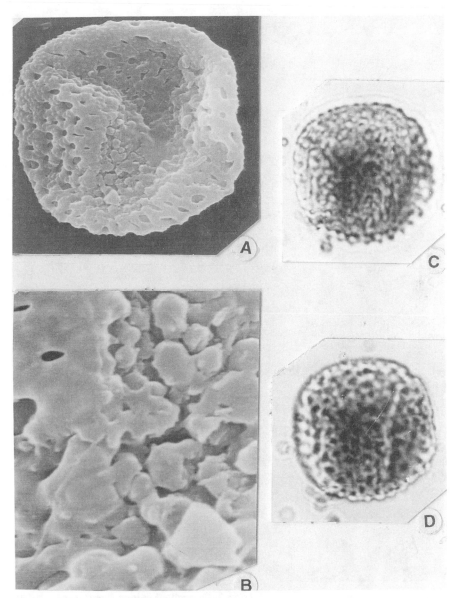

Figure 5.7. *Clavatipollenites* Group, Helez Formation, middle sand member, Kokhav 2 well, core. (**A-D**) Four views of the same grain. (**A**) Circular outline with supra-tectal grana and a weakly developed sulcus, SEM, x3405; (**B**) shows reduction of sexine in area of aperture development, SEM, x15,300; (**C**) light micrograph, mid-focus, showing endexinal staining, x2000; (**D**) high focus, x2000.

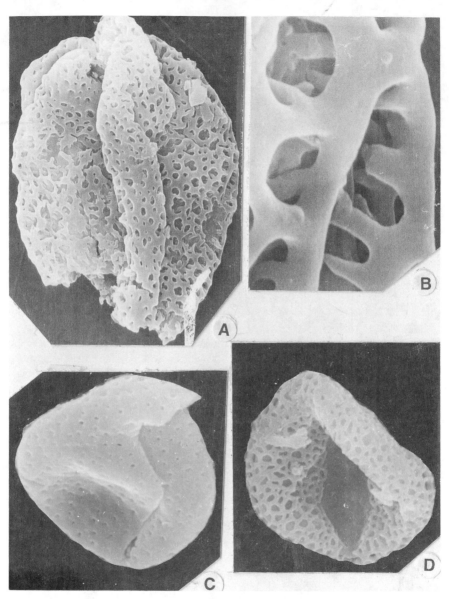

Figure 5.8. *Liliacidites* Group, Helez Formation, middle sand member, Kokhav 2 well, core 2. (**A,B**) same grain, monosulcate type with variable-sized lumina; showing lumina that are slightly finer in the area of the aperture and at the ends of the grain, SEM, x3960; (**B**) showing smooth muri and extremely high columellae, SEM, x17,557. (**C**) *Retimonocolpites* sp. A, Zeweira Formation, Zohar 1 well, core 7; tectate-perforate exine, SEM, x 3350. (**D**) *Retimonocolpites* sp. B, Zeweira Formation, Zohar 1 well, core 7.

cores 6 and 7, and also distinguish this horizon from the much older Helez assemblages in the coastal plain (Brenner and Bickoff, 1992). Although rare occurrences of *Afropollis* have been reported from the late Barremian of England (Penny, 1989) and Morocco (Gübeli et al., 1984), its earliest effective appearance in paleoequatorial sections is basal Aptian. *Brenneripollis* (Fig. 5.9A,B) and tricolpate pollen (Fig. 5.9C) first appear in the early Aptian of Northern Gondwana, with both pollen types becoming more abundant in the Albian. The first appearance of *Brenneripollis* is a recognized palynological marker for the Lower Aptian (Doyle, 1992). The fact that *Afropollis, Brenneripollis*, and the first tricolpate pollen grains appear together for the first time in core 6, but not in core 7, makes it convenient to place the Barremian/Aptian boundary between these two cores.

A date of ~ 120 Ma (the date suggested for the Barremian/Aptian boundary by Ozima et al., 1979), somewhat greater than the radiometric date of 118 ± 1.5 Ma (the date for the basalt directly above the stratigraphic position of cores 6 and 7), is suggested for the first occurrence of tricolpate pollen and *Brenneripollis* in the paleoequatorial region of the Northern Gondwana Province.

Figures 5.9A,B and 5.10A,B show SEM and LM micrographs of several monosulcates that appear for the first time in the Zeweira Formation. In general, the monosulcates in the Barremian–Aptian levels are morphologically much more diverse than those in the Helez Formation. The monosulcates in the Zeweira are generally larger, more elliptical in shape, and have better defined apertural margins than any of the types found in the Valanginian and Hauterivian forms from the Helez Formation. The following types have been found in cores 6 and 7:

1. *Liliacidites* sp. A Brenner and Bickoff 1992 (Fig. 5.9B, cores 6 and 7, Zohar 1 well): large, elliptically shaped tectate–perforate forms with a reticulate sexine. The lumina vary greatly in size, with finer lumina situated along the margin of the sulcus and at the ends of the grain. The mural surface is smooth. This form is a common member of the late Barremian and early Aptian assemblages.

2. *Liliacidites katangataensis* (Fig. 5.9A): large monosulcate pollen (20–40 m), elliptical in outline, with a long and narrow sulcus that extends the full length of the grain. Exine characteristics of this species suggest monocot affinities.

3. *Retimonocolpites* spp:. small, coarsely reticulate *Retimonocolpites* sp. B, (core 7, Fig. 5.9D) and tectate–perforate *Retimonocolpites* sp. A, (Fig. 5.9C) have continuous well-defined apertural margins.

4. *Brenneripollis* spp. (Fig. 5.10A,B, core 6, Lower Aptian): is easily identified by the absence of columellae, the coarse and loosely appending reticulum, and the well-pronounced supratectal spines. The Zohar 1 well contains the first appearance of this Aptian marker, which more abundant in the Albian beds.

More advanced apertural types that appear for the first time in the early Aptian:

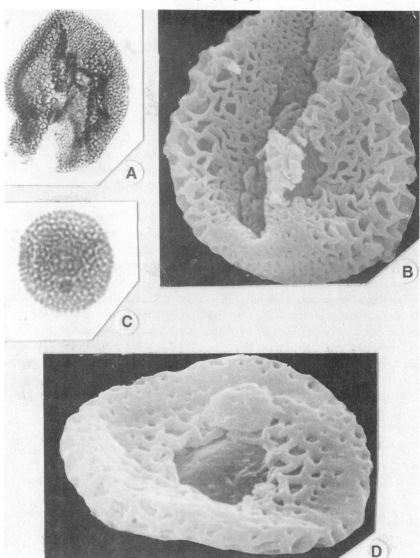

Figure 5.9. (**A**) *Liliacidites katangataensis*, Zeweira Formation, Zohar 1 well, core 7. Muri covered with supra-tectal spines, light micrograph, x950. (**B**) *Liliacidites* sp. C , Zohar 1 well, core 7, SEM, shows finer lumina along the apertural margin and at the ends of the grain as is typical for many extant monocot pollen grains, x1830. (**C**) *Retimonoporites operculatus* Brenner and Bickoff 1962, Zeweira Formation, Zohar 1 well, core 6. Winteraceous monad that shows pore area underlain by densely stained endexine, light micrograph, x1425. (**D**) *Retimonoporites operculatus*, Zohar 1 well, core 6; operculum formed by sexine, SEM, x4070.

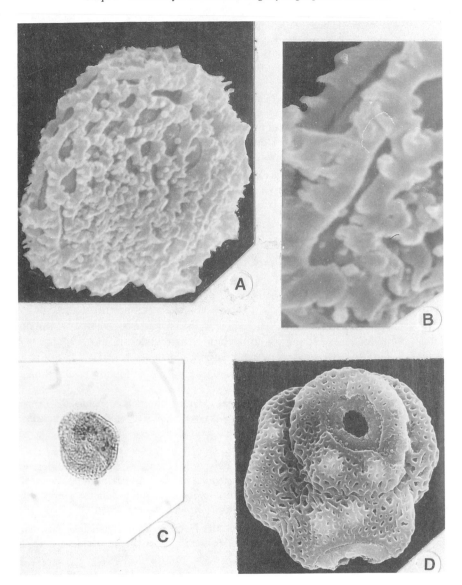

Figure 5.10. (**A,B**) *Brenneripollis* sp., Zeweira Formation, Zohar 1 well, core 6, same grain. (**A**) Absence of columellae shown here is typical of *Brenneripollis,* SEM, x3090; (**B**) showing supratectal spines, SEM, x9500. (**C**) Tectate-reticulate, tricolpate grain from the lower Aptian, Core 6, Zeweira Formation, Zohar 1 well, light micrograph, x1000. (**D**) Winteraceous tetrad from the Dragot Formation, upper Aptian/lower Albian, Zohar 1 well, core 5; SEM, x14,500.

5. Winteracous monad (Fig. 5.9C): *Retimonoporites operculatus* Brenner and Bickoff 1992, ranges from core 6 in the Zeweira Formation (lower Aptian) to core 4 (Albian) of the Malhatta Formation. The grains are very small, 11–15 μm, and conspicuously circular in outline. The sexine is reticulate and forms a circular operculum in the center of one side of the grain. Endexine develops beneath the apertural area (Fig. 5.9C). A winteraceous tetrad (Fig. 5.10D) was found in the overlying Dragot and Malhatta Formations (Walker et al., 1984). The appearance of a monad of winteraceous affinity prior to the appearance of the more typical tetrad condition suggests that the monad condition is more primitive than the tetrad condition.

6. *Tricolpites* spp. Several species of tricolpate pollen occur for the first time in core 6. All grains are globose to oblate in shape. The sexine is tectate–columellate with a well–defined reticulum that varies in arrangement with different species (Fig. 5.10C, core 6, Zohar 1 well).

EVOLUTIONARY IMPLICATIONS OF THE HELEZ ANGIOSPERMS

Almost axiomatic in paleobotanical theory is that the monosulcate condition is the basal type for the earliest angiosperms. Most authors, when constructing cladograms dealing with angiosperm relationships, code the monosulcate condition as the primitive character state. Support by most workers for the basal position of the monosulcate condition is based on the following assumptions:

1. Monosulcate pollen is found in several gymnosperm groups, such as the Bennettitales, Pentoxylales, and Gnetales, that have been picked, on the basis of cladistic analysis, as the most closely related outgroups to angiosperms (Doyle and Donoghue, 1986b).

2. Monosulcate grains are the oldest, commonly found, indisputable, angiosperm pollen-types from several middle Barremian to lower Aptian localities around the world. These pollen types are usually assigned to the generalized form-genus *Clavatipollenites*, which is monosulcate and tectate–columellate.

3. Some botanists claim that the most primitive, extant, magnolialean taxa have large, boat-shaped, monosulcate pollen with a granular interstitium (as in the *Magnoliaceae*; Walker and Walker, 1984).

However, the appearance of inaperturate angiosperm pollen-types in the late Hauterivian that are small, circular, tectate–columellate grains with sulcate-shaped endexinal development together with weakly monosulcate types of similar morphology suggests that the inaperturate condition is at least as old as the reputed primitive monosulcate condition.

Even older, extremely rare grains are found from the lower sand member (core 5), of late Valanginian/early Hauterivian age. All specimens found (15 in

3 slides) were inaperturate, and there is no sign of any endexinal staining or sulcal development. The grains are small tectate–columellate, with the tectum ranging from tectate–perforate to tectate–reticulate. The reticulum is weakly developed in most grains.

The fact that the inaperturate condition, found in the Valanginian, precedes the simultaneous appearance of both inaperturates and monosulcates in the upper Hauterivian of the Helez Formation suggests that the inaperturate condition may have been the initial pollen condition that led to the *Clavatipollenites* group. The absence in the Valanginian of any apertural development or endexinal formation suggests that the ancestral stock that precedes the burst of angiosperm radiation during the Barremian/Aptian was a plant group whose pollen organs contained small, inaperturate, globose pollen with a tectate–columellate and reticulate structure. No endexinal presence can be detected by Safranin staining.

In the upper Hauterivian levels of the Helez Formation, elongate patches of endexinal staining can be found in many of the inaperturate grains; this may indicate the position of potential apertural formation.

Small, circular (globose in the original state) pollen grains, both inaperturate and monosulcate that resemble the pollen found in the late Valanginian and late Hauterivian assemblages are commonly found in several extant pollen forms within the Chloranthaceae, Piperaceae, and Saururaceae. Taylor and Hickey (1992), on the basis of cladistic analysis, suggested that the nonmagno!ialean Chloranthaceae and Piperaceae are nearest to the base of angiosperm evolution.

In the Gnetales, which some botanists consider a sister group to the angiosperms (e.g., Hickey & Taylor, Chapter 8), the pollen of some species of *Gnetum* come closest to the general shape and aperture-condition of these early angiosperm types. The grains of *Gnetum* in the extant, South American Araegnemones, section Gnemonomorphs (Erdtman, 1965), are similar in size and shape to some pollen found in core 5. The *Gnetum*-like grains (Fig. 5.4E,F, same grain) are circular in outline, ~ 25 µm, have a thick sexine that is tectate and composed of closely packed baculum-like elements. Cladistic analyses by several workers have created topologies reflecting the possible relationships of the outgroups to angiosperms (Crane, 1985, 1986; Doyle and Donoghue, 1986b, 1993; Hickey & Taylor, Chapter 8). These studies suggest *Ephedra* as the most basal of the three living genera of the Gnetales (Doyle and Donoghue, 1993), with *Welwitchia* and *Gnetum* having a common ancestry. Of the three extant genera of the Gnetales, *Ephedra* and *Welwitchia* have large, elongate, polyplicate pollen, whereas the pollen of *Gnetum* is small, circular in outline, and inaperturate.

Although nonsaccate ephedroid pollen is known from the late Triassic, it does not make an effective appearance until the early Cretaceous. The absence of any reports of fossil *Gnetum* pollen in the Mesozoic may be due more to the difficulty of recognition than its actual absence. This striking difference between the pollen of *Gnetum* and the other two gnetalean genera may suggest a very different phylogeny for *Gnetum*, one possibly closer to the stem group in angiosperm evolution. Muhammad and Sattler (1982) describe the existence of scalariform vessel structure in *Gnetum gnemon* and *Gnetum montanum*. They suggest that *Gnetum* may be closer to the ancestral stock of angiosperms than has been previously been thought. Erdtman (1965) also notes that there is some resemblance between the South American lianas of *Gnetum* (Araegnemones) and the pollen grains of some species of the Piperaceae, Chloranthaceae, and Saururaceae.

POSTULATED SEQUENCE IN THE MORPHOLOGIC EVOLUTION OF ANGIOSPERM POLLEN MORPHOLOGY AND ITS EVOLUTIONARY IMPLICATIONS

Figure 5.11 illustrates the following exine stages proposed, while Fig. 5.12 illustrates the general pollen (evolutionary) stages.

POLLEN STAGE I: Proto-angiosperms stage (unknown)—Pre-late Valanginian.

POLLEN STAGE II: EXINE STAGE A—Evolution of the early angiosperm basal groups that evolved from a stock that also gave rise to *Gnetum*. Pollen small, circular in outline, tectate-columellate, and without an aperture—**Valanginian** or slightly earlier.

Sporophytic recognition proteins stored in the sexine—of tapetal origin—**Valanginian**.

POLLEN STAGE III: EXINE STAGE B—A possible intine thickening in the site of apertural development, coupled with the development of a thickened endexine above it (incipient sulcus)—**Hauterivian**.

EXINE STAGE C—Evolution of a sulcus in some species for more effective exit of recognition proteins; proteins are of gametophytic origin and are commonly stored in the intine. Monocot and dicot pollen types diverge from basic dicot stock—**Hauterivian**.

POLLEN STAGE IV: EXINE STAGE D—A highly diverse assemblage of monosulcate types dominate the angiosperm pollen assemblage. Rapid diversification and migration of basal angiosperm groups to multiple geographic centers—**Barremian**.

POLLEN STAGE V: Tricolpate pollen evolves in northern Gondwana from either monosulcate or possibly inaperturate forms (see Fig. 5.12). This advance in pollen morphology reflects the evolution of eudicots: by the late Aptian and Albian the eudicot diversification appears at higher latitudes, worldwide—**Lower Aptian**.

The above pollen evolutionary sequence suggests several new possibilities in the evolution of angiosperm pollen and angiosperms in general. The ancestral

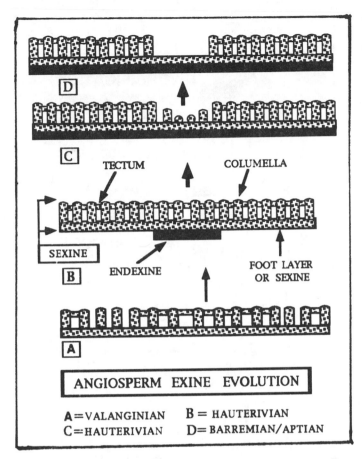

Figure 5.11. Proposed model for the development of the exine during the Early Cretaceous.

pollen type for angiosperms would then be inaperturate, not monosulcate. The formation of a sulcus during the early Cretaceous may have been an adaptation that was a more effective way of releasing recognition proteins involved in pollen-tube development while the later development of the tricolpate condition in the Aptian would be a further extension of this process. The tectate–columellate condition would be the original exine condition. The early existence of this exine structure (although not exclusive to insect pollination) suggests that insect pollination mechanisms had already been established during this early stage of angiosperm diversification.

The endexine in angiosperm pollen would be a new layer that precedes and underlies the locus of apertural development. The formation of endexine before

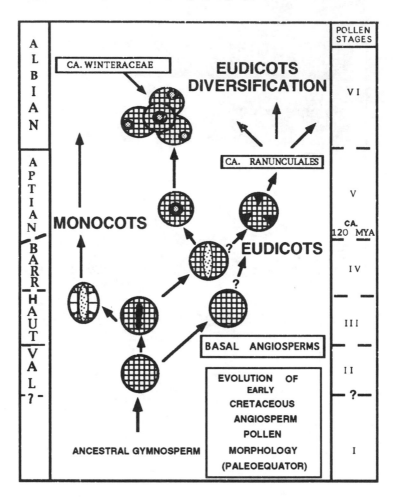

Figure 5.12.   Evolutionary sequence of angiosperm pollen types in the Early Cretaceous from inaperturates to tricolpate pollen.

the development of an aperture may reflect the development of an intine-thickening that is somehow related to the locus of sulcal development. In extant angiosperms, the intine beneath the aperture stores recognition proteins that are formed by the haploid pollen grain and are, therefore, gametophytically controlled (Heslop-Harrison, 1971). It is suggested that the evolution of the aperture from the inaperturate condition gave reproductive advantage to the grains.

The pollen forms found in the earliest stages of angiosperm diversification during late Valanginian time are closer in morphology to the small, circular, inaperturate and monosulcate grains found in the Chloranthaceae, Piperaceae, and Saururaceae than the large, boat-shaped pollen commonly found in the Magnoliaceae and some cycadalean groups. In the Chloranthaceae, the pollen of the primitive genus *Sarcandra* looks very much like some of the inaperturate types found in the Helez formation. The pollen of *Sarcandra*, which has been described in the literature many times as inaperturate, was shown by Endress (1987a) to be polyforate. The exine is reticulate, with a well-developed tectate–columellate sexine. However, with the aid of Astral Blue staining, the pollen shows several circular areas, evenly distributed around the grain, that fail to develop tectal bridges. What Endress then calls apertures are really patches of exine in which the columellae are freestanding. This condition looks very similar to some of the angiosperm pollen from the late Valanginian in core 5 of the Kokhav 2 well. In Fig. 5.4B patches of exine fail to develop a tectum and the columellae are freestanding and frequently missing.

If the angiosperm pollen found in the oldest beds of this study represents the pollen of the most primitive types, then it would suggest the paleoherbs may be the most basal in angiosperm phylogeny, as has been proposed by Taylor and Hickey (1992).

PROPOSED GEOGRAPHIC ORIGIN, PALEOECOLOGY, AND CLIMATIC BACKGROUND OF THE EARLIEST ANGIOSPERMS

The rare occurrence of inaperturate angiosperm pollen during the late Valanginian and Hauterivian reflects what I believe is the early evolutionary stages of those angiosperms that formed the basal stock for all living flowering plants. Monosulcate angiosperm pollen-types diversify rapidly in the Barremian and reached most continental areas by the end of the Barremian Stage, except for the highest latitudes. Tricolpate pollen is first found in the early Aptian at paleoequatorial localities, and then at successively higher latitudes with time (Brenner, 1976). The development of the tricolpate grain reflects the next major stage of angiosperm diversification, the eudicots.

The origin of any taxonomic group can only be resolved by an ongoing search for their earliest occurrence in the fossil record. Pre-Barremian sections present a difficult problem in this respect because few of them were deposited in a terrestrial setting. The presence of inaperturate angiosperm pollen, in nearshore to nonmarine samples from upper Valanginian to lower Hauterivian sediments from the coastal plain of Israel, provides us with additional data on the possible geographic locus of early angiosperm diversification. The Northern Gondwana Province (Brenner, 1976) may then have been the site of the earliest angiosperms. From there, the basal angiosperm taxa that produced the typical monosulcates spread to other areas, forming new centers of diversification. In the Potomac Group of Maryland and Virginia, pollen studies (Brenner, 1963, Doyle, 1992) indicate that the earliest angiosperms did not arrive there until at least late Barremian or early Aptian time; such angiosperm floras are surely not the oldest.

Spore and pollen assemblages from Valanginian to Aptian horizons in Israel are similar to assemblages from the Northern Gondwana Microfloral Province (Brenner, 1976) in that they lack the bisaccate pollen produced by plants of the Podocarpaceae and Pinaceae. However, they differ from the Brazilian microfloras of the same age (personal observations of the author) in that they contain much lower percentages of *Classopollis* and ephedraceous pollen, and much higher diversities of pteridophyte spores. This suggests that the eastern part of Northern Gondwana, including Israel, was humid at this time. Moisture then was readily provided from the east by equatorial trade winds coming from the paleo-Pacific ocean (Parrish, 1982, 1987).

Most of the samples in this study contain low frequencies of dinoflagellate cysts that indicate a marginal-marine environment. Interspersed with these samples from both wells are completely nonmarine layers. If the earliest angiosperm pollen were wind pollinated one would expect them to be more abundant than they are. The palynological recovery of most of the samples is good, so that preservation is not a problem. Thus, the extremely low frequencies of the earliest angiosperm pollen grains in the Helez Formation suggests that they were insect pollinated.

CONCLUSIONS

Almost axiomatic in plant systematics is that the monocolpate condition is the fundamental type for the earliest angiosperm pollen—a feature inherited from their gymnospermous ancestry. An examination of angiosperm pollen in Upper Valanginian to Upper Hauterivian sediments from Israel suggests otherwise.

Late Valanginian pollen grains from lower sand member of the Helez Formation are extremely rare and not diverse. They are all inaperturate, reticulate, with a tectate–columellate sexine, and typically circular in outline and very small. In the late Hauterivian, the sulcus, although poorly developed, makes its first appearance, and both inaperturate and monosulcate forms are present. Endexine, which can be seen with safranin staining, appears for the first time in both the inaperturates as well as in the first monosulcates. Endexinal development in the inaperturates is found on only one side of the grain. The unilateral position of the endexine may somehow be connected to the development of an aperture in subsequent forms. There is a pronounced increase in the diversity of angiosperm pollen types in the late Hauterivian although they are still quite rare in the palynological assemblage. Most of the inaperturates and monosulcate grains are, as in the Valanginian samples, very small and circular in outline; however, a few rare forms are more oval in outline and have an exine morphology that is more typical of pollen from extant monocots (Fig. 5.8A, B). The late Hauterivian types have been grouped into four morphological groups: (1) the Pre-*Afropollis* Group, (2) the Spinatus Group, (3) the *Clavatipollenites* Group, and the (4) the *Liliacidites* Group. These four groups contain a mosaic of characteristics that are commonly found in the more abundant and diverse assemblages from Barremian to Aptian horizons around the world.

The palynological assemblage in the basal Cretaceous of the Zohar 1 well from the northern Negev (Fig. 5.2) is younger than the palynological assemblage from basal Cretaceous of the Kokhav 2 well from the coastal plain. Tricolpates, *Brenneripollis*, *Afropollis*, and a winteraceous monoporate monad, *Retimonoporites operculatus*, appear for the first time in core 6, but are not found in core 7; both cores are in the lower Zeweira Formation. A winteraceous tetrad (Fig. 5.10D) appears for the first time in core 5, of early Albian age. The above palynomorphs place core 6 in the early Aptian (ca. 120 Ma). In outcrop, the lower Zeweira is overlain by the Ramon basalt which has yielded a ^{40}Ar–^{39}Ar date of 118 ± 1.5 Ma.

In conclusion, this study suggests that the ancestral angiosperms, at least those referred to by Doyle and Donoghue (1993) as the "Crown Group," first occurred in the early Cretaceous of the Eastern Gondwana Province. They produced small, globose, inaperturate grains, with a weakly developed, semitectate, reticulate sexine. The sulcus, which first appears in the Hauterivian, would then be a totally new structure, whose origin was somehow connected to the development of a new endexinal layer. By the late Hauterivian, the columellate-reticulate condition, as well as the sulcus, becomes better developed and more diversified, possibly as a response to developments in insect pollination.

The similarity of the inaperturates described in this chapter with pollen types found in some extant members of the Piperales agrees with Taylor and Hickey's (1992) proposal for the herbaceous origin of angiosperms. Within the Gnetales, the pollen of some species of *Gnetum* come closest to the angiosperm grains found in core 5 from the late Valanginian Helez Formation. The possibility then exists that some extant species of *Gnetum* are the remnant of a gymnospermous clade that is the sister group of the angiosperms.

An analysis of the total spore and pollen assemblage in the Helez and Zeweira formations suggests that the climate in which these early angiosperms lived was warm and humid. The Northern Gondwana Province may have been the locus for the initial radiation of angiosperms, which then spread to most other areas by the late Barremian to early Aptian.

The Origin and Evolution of the Angiosperm Carpel

David Winship Taylor and Gretchen Kirchner

The angiosperm carpel is one of the defining characteristics of flowering plants. Carpels are unique to angiosperms and are found in all of its members. Yet, due to the distinctive structure and function of carpels, deducing homologies among carpels and other seed-plant organs has been difficult. Progress in understanding carpel homologies and evolution is being made in a number of directions. Homologies, as well as transformations between them, have been proposed among angiosperms and other seed plants. These include homologies among the reproductive structures of Glossopterids, *Caytonia*, and other fossil taxa (e.g., Thomas, 1925, 1957; Andrews, 1963; Stebbins, 1974; Doyle, 1978; Retallack and Dilcher, 1981a; Crane, 1985; Doyle and Donoghue, 1986b). Recent studies have examined the homologies and transformations among female reproductive organs through outgroup comparison (e.g., Crane, 1985; Doyle and Donoghue, 1986b; Taylor, 1991a) and interpretations of carpel structure within angiosperms (Taylor, 1991a). Developmental, morphogenic, and genetic studies have also provided additional insight on carpel structure and development (e.g., Szymkowiak and Sussex, 1992; Gasser and Robinson-Beers,

1993; Modrusan et al., 1994). Together, data from these diverse sources are leading to a revised view of carpel homology and structure.

Although transformations between the carpel and other seed plant reproductive organs (particularly fossil) have been proposed, difficulties in showing compelling evidence of these homologies has led to hypotheses based on structural interpretations of carpels from living angiosperms. These efforts have generated numerous hypotheses of carpel origin over the last century (see references in Esau, 1965; Cronquist, 1968, 1988; Takhtajan, 1969, 1991). Yet, even though the presence of carpels is a synapomorphy for living angiosperms, carpel morphology is surprising variable within basally placed living angiosperms. Taylor (1991a) developed a new terminology to describe this variation. He also reviewed the carpel characters found in the families of the Magnoliidae as well as the basal families of the Hamamelidae, Caryophyllidae, Dilleniidae, Rosidae, and monocots.

Using Taylor's terminology, the carpel morphology of the Magnoliidae ranges from ascidiate (Fig. 6.1D,E-H,L,M) to ascoplicate (Fig. 6.1I,J) to plicate (Fig. 6.1K,N,O). Ascidiate carpels have the ovules attached proximally to the closure (e.g., Fig. 6.1D,F), plicate ones have them attached along the margins of the closure (e.g., Fig. 6.1K,N) while ascoplicate carpels are intermediate between the two (e.g., Fig. 6.1I,J). The number of vascular traces leading into the ovule(s) ranges from one to many. Although most members have carpels that are free, some are syncarpous.

The placement of the ovules and the placentae is also variable (Taylor, 1991a) and ranges from basal (Fig. 6.1D,E), to lateral [either admedial and facing toward the floral axis (Fig. 6.1F,H), marginal and along the closure (Fig. 6.1K,N), exmedial and facing away from the floral axis, radial and on the lateral walls perpendicular to the floral axis (Fig. 6.1M), or chaotic and along all walls of the carpel (Fig. 6.1L,O)] to apical (Fig. 6.1G) in their placement. The placenta may be situated in the plane extending from the floral axis and bisecting the carpel through the closure (e.g., Fig. 6.1G) or be nonplanar and located on either side of the closure (e.g., Fig. 6.1F,N). Finally, the number and arrangement of ovules and placentae are variable, ranging from single (e.g., Fig. 6.D), double (e.g., Fig. 6.F) to few (e.g., Fig. 6.I) or many in number (e.g., Fig. 6.K), and these may be placed in an alternate (e.g., Fig. 6.N), opposite (e.g., Fig. 6.K), or clustered arrangement.

The last type of variation is found in carpel development (Taylor, 1991a). Some carpels develop from two separate growth areas (Fig. 6.1AI,AII) whose presence may be separated spatially or temporally. The first primordium produced is *U*-shaped or cup-shaped and develops into the gynoecial appendage. The second primordium or growth area is surrounded by the first and is the source of the placentae and ovules. Most carpels initially develop as a cup and have the placenta placed along the admedial portion of the wall (Fig.

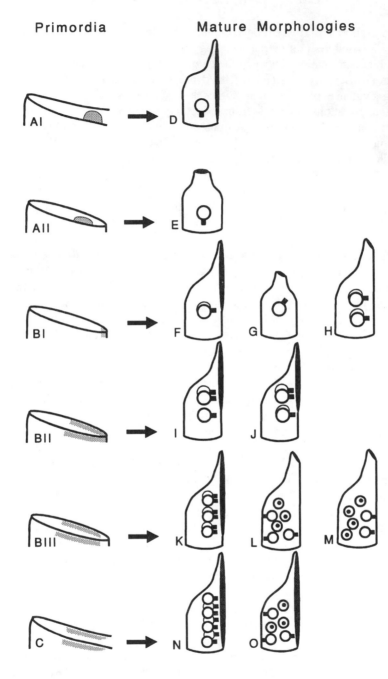

Figure 6.1. Diagrammatic representations of the primordia types: ascidiate (**A**), ascoplicate (**B**) and plicate (**C**), and resulting mature basic carpel types (**D–O**). The shading in the primordia marks the region from which the placenta and ovules develop. The shading in the mature carpels show the closure of the carpel that becomes appressed during development.

118

6.1BI,BII). Finally some have placentae along the margins of the closure (Fig. 6.1BIII,C). These carpels develop from either a cup-shaped or *U*-shaped primordium.

This variation in carpel morphology has led to two major interpretations of carpel evolution, although both agree with interpretations that a portion is leaf derived (Arber, 1937; Troll, 1939b). Although these hypotheses describe the ancestral states, they usually do not address the transformations to the other states that exist in the basal angiosperms. Most hypotheses suggest that seeds are phyllosporous (Lam, 1961) in angiosperms and that the carpel originated from a seed-bearing leaf (e.g., Leinfellner, 1950; Bailey and Swamy, 1951; Cronquist, 1988; Takhtajan, 1991 and references therein). Thus, the gynoecial appendage portion of the carpel is a modified leaf on which the ovules are borne (Sattler, 1974; Sattler and Perlin, 1982). This hypothesis suggests that the carpel is homologous with a megasporophyll.

The phyllosporous-origin or megasporophyll-homology hypothesis has two variants. Most authors suggest that the gynoecial appendage is fundamentally a folded leaf with a proximal attachment of the petiole (e.g., Bailey and Swamy, 1951; Cronquist, 1988; Takhtajan, 1991). Carpels of this type have been termed conduplicate (Bailey and Swamy, 1951) and have many ovules along the margins (or submargins) of the closure (Fig. 6.1N). Others have suggested that the leaf is fundamentally peltate (e.g., Baum, 1949, 1952; Leinfellner, 1950; Baum-Leinfellner, 1953). This is due, in part, to the cup-shaped primordia found in many carpels. Although such ovules are still considered to be marginal (Fig. 6.1I,J), these authors note that ovules can be few in number and placed medially at the base of the closure (at the cross-zone; Fig. 6.1F). In neither variant of the phyllosporous-origin hypothesis have the transformations to all the other carpel types been explained from a developmental, morphological, or evolutionary perspective.

The second major group of hypotheses suggests that the carpel has a fundamentally stachyosporous origin (Lam, 1961). Thus the gynoecial appendage is thought to be homologous to a subtending bract and the placenta is homologous with a shoot bearing distally-placed ovules (e.g., Pankow, 1962; Moeliono, 1970; Sattler and Lacroix, 1988). Again, there has been relatively little discussion of the transformations to other carpel types. Hagerup (1934, 1936, 1938) has addressed the transformations most extensively and suggested that conduplicate carpels also develop from two growth areas (see also Melville, 1962).

Recently, Taylor (1991a) has further elaborated the stachyosporous origin hypothesis of Hagerup (1934, 1936, 1938) and others (e.g., Melville, 1962; Moeliono, 1970; Sattler and Lacroix, 1988) based on outgroup comparison. He has suggested that the carpel is a bract–terminal ovule system and directly homologous to that found in the outgroups including the Gnetales, the closest living sister group. The sterile gynoecial-appendage of the carpel is considered homologous with a subtending bract in the angiosperm sister groups, whereas the placental growth area is homologous with the ovular axis. Based on these homologies, Taylor concludes that ascidiate morphology and low ovule-number (1 or 2) are ancestral (Fig. 6.1D,F,G). He also suggests homologies among carpel types with the basally to admedially placed placenta in a carpel with a sterile gynoecial-appendage (Fig. 6.1AI) as homologous with the lateral placenta of an ascoplicate carpel (Fig. 6.1BI) or marginal placentae of a conduplicate carpel (Fig. 6.1C). The origin of the conduplicate (plicate) and ascoplicate carpel types would be due to the integration of the gynoecial primordia and ovular (placental) growth area. The evolution of these other carpel types may be related to the evolution of carpels with high ovule numbers.

From these different studies of carpel evolution, a group of questions can be posed. First, which type of carpel found in living taxa is most similar to the ancestral form? Second, what is the ancestral number of ovules? Third, why is there variability among carpels in the basal angiosperms? Finally, what are the transformations among carpel types?

POLARIZATION OF THE ANCESTRAL CARPEL CHARACTERS

Due to the lack of clear homologies between reproductive structures of angiosperms and their sister groups, the polarization of the ancestral carpel states has been difficult. Taylor (1991a) examines this problem and presents several lines of evidence. One line is the frequency of carpel types and number of ovules in the basal angiosperm families (Table 6.1). Of the members of the Magnoliidae (sensu Cronquist), two-thirds of the families have members with the ascidiate carpel type and over a half have carpels with one or two ovules. Stebbins (1974) also notes that low ovule number is common and suggests it may be ancestral.

Unlike for carpels, it is possible to polarize many of the ancestral states of the ovule using outgroup comparison (Taylor, 1991a). When compared to the outgroups, the ancestral angiosperm ovule appears to have the chazala opposite the micropyle, an orthoangle, total symmetry, symmetry of the outer integument, a nucellar attachment opposite the micropyle, free integuments, and the micropyle formed from the inner integument [terminology based on Taylor (1991a)]. In conventional terms, the ancestral ovule is bitegmic and orthotropous. This hypothesis is in contrast to most analyses that suggest that the ancestral ovule was bitegmic and anatropous (e.g., Cronquist, 1968, 1988; Takhtajan, 1969, 1991; Bouman, 1974). Interesting, recent mutagenesis analysis of *Arabidopsis* ovules shows that simple mutations can result in a disruption in the development of the normally anatropous ovule so that is does not have the

Table 1.   General Carpel Types and Ovule Types for the Orders of the Magnoliidae (sensu Cronquist).

Order	Carpel Morphology			Ovule number		
	Ascidiate	Ascoplicate	Plicate	1 or 2	Few	Many
Magnoliales	6	1	5	7	6	6
Laurales	7	1	0	8	0	0
Piperales	3	0	1	3	1	1
Aristolochiaceae	1	1	1	0	1	1
Illiciales	2	0	0	2	1	0
Nymphaeales	4	1	0	3	1	2
Ranunculaceae	7	2	3	9	5	5
Total	30	6	10	32	15	15

Note: The number in each column is the number of families in the order which have the characters. If the family is variable, it is recorded under more than one state. Source: Data from Taylor (1991a).

curved anatropous shape (Gasser et al., 1995). This fits with Taylor's (1991a) hypothesis that orthotropous is ancestral and that the anatropous form results from an additional step in the development of the ovule.

Once the ancestral ovule states are polarized, it is possible to examine the association between carpel states and polarized ovules. Taylor (1991a) consistently found that ascidiate morphology and ovule numbers of one or two are associated with the ancestral ovule states, but the correlations are not significant. Taylor did find significant correlation between ascidiate carpels and ovule numbers of one or two with 77% of the ascidiate carpels reported from families having one or two ovules. In contrast, 79% of the plicate and 73% of the ascoplicate carpels are from families with few to many ovules. In addition, 74% of the families that had ascidiate carpels came from families in which it is the only carpel type. In contrast, only 8% of the families with ascoplicate carpels have that type alone and 54% with plicate have plicate alone. Although some of these association data are equivocal, they do suggest that ascidiate carpels with ovule numbers of one or two and plicate carpels with high numbers are real entities in angiosperms, and that the former are likely to be ancestral.

New Evidence

There is now additional evidence that can be used to further understand carpel evolution and origin. First, the growing number of angiosperm phylogenies allow for the assessment of the ancestral carpel states. Second, as the understanding of the identity of angiosperm sister groups and their relation-

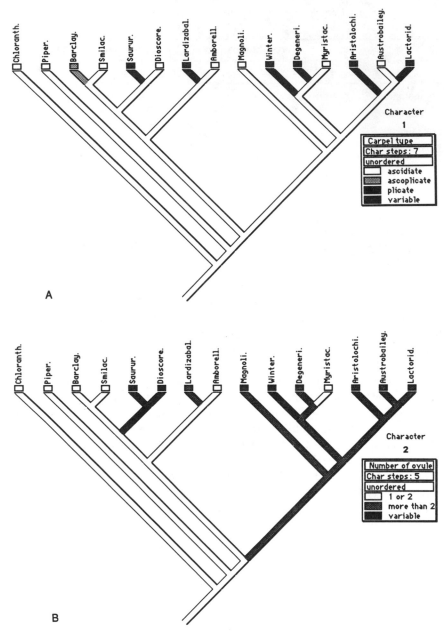

Figure. 6.2. One of three shortest trees from Taylor and Hickey (1992) with the distribution of carpel morphological states (**A**) and ovule number states (**B**). Based on this phylogeny, the ascidiate state and ovule number state of one or two are ancestral.

ships has increased, so has the possibility of using outgroup comparison to determine ancestral states. Finally, the understanding of carpel development and of transformations among types allows tests based on the morphogenesis of carpels. Each of these areas will be discussed in detail below.

Ingroup Phylogeny.   There are a growing number of phylogenetic analyses of angiosperms based on structural (Donoghue and Doyle, 1989a; Loconte and Stevenson, 1991; Taylor and Hickey, 1992; Tucker et al., 1993; Tucker and Douglas, Chapter 7; Loconte, Chapter 10) and DNA sequence [18S rRNA (Zimmer et al., 1989; Hamby and Zimmer, 1992); *rbc*L cpDNA (Chase et al., 1993; Qiu et al., 1993; Sytsma and Baum, Chapter 12)] datasets. These recent analyses support different hypotheses of angiosperm origin suggesting that the basally placed clades are either herbaceous magnoliids (Zimmer et al., 1989; Taylor and Hickey, 1992; Hamby and Zimmer, 1992; Chase et al., 1993; Qiu et al., 1993; Sytsma and Baum, Chapter 12), magnolialeans (Donoghue and Doyle, 1989a) or lauraleans (Loconte and Stevenson, 1991; Loconte, Chapter 10). A method to assess the ancestral carpel states is to place the carpel states on these cladograms and see which states are ancestral based on parsimony.

In the trees from the structural dataset of Donoghue and Doyle (1989a), the woody Magnoliales are basally placed. Although the magnolialeans are variable in carpel morphology, the most parsimonious distribution of states would have the ascidiate morphology as ancestral. The ancestral ovule number in these trees is less clear, as the character is variable within a number of families. If, however, the taxa that are variable are ignored, the ancestral state would then be one or two ovules. The same polarizations (ascidiate and ovule equivocal) are found in the shortest trees based on the 18S RNA dataset (Zimmer et al., 1989; Hamby and Zimmer, 1992). In these trees, the Nymphaeales (sensu stricto) are the clade closest to the base, and the next closest clade is the Piperales (sensu stricto). Both of these groups are herbaceous.

In five other analyses, the most parsimonious distribution of the characters have ascidiate and one or two ovules as ancestral. These include two structural datasets, one by Loconte and Stevenson (1991; also see Loconte, Chapter 10) that has the woody Calycanthaceae and Idiospermaceae as sister groups at the base, followed initially by the Magnoliales, Laurales, and Illiciales. The other by Taylor and Hickey (1992; Fig. 6.2) has the herbaceous families Chloranthaceae and Piperaceae at the base of the tree, with the former most basally placed. The same polarizations are found on the trees of Chase et al. (1993), Qiu et al. (1993), and Sytsma and Baum (Chapter 12) based on *rbc*L cpDNA sequence data. In most of these trees, the basal clades are either the herbaceous magnoliids or that clade in combination with the herbaceous eudicots.

A growing number of phylogenetic analyses support Taylor's (1991a) hypothesis that the ascidiate carpel with one or two ovules is ancestral. It is

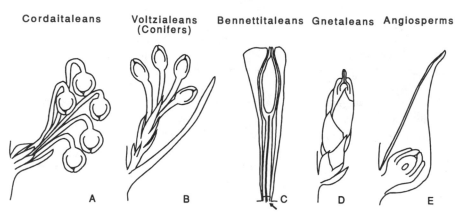

Cordaitaleans   Voltzialeans   Bennettitaleans Gnetaleans Angiosperms
                  (Conifers)

A   B   C   D   E

Figure 6.3.   Comparison of female reproductive structure from the outgroups (**A,B**), sister groups (**C,D**) and angiosperms (**E**). (**A,B**) The main reproductive axis to the right and the female short-shoot subtended by a bract attached to the main axis. (**C**) The bract is obscure and similar to the scales. (**D**) The structure is the entire main reproductive axis with a single, distal, female short-shoot, which is subtended by one of the bracts. (**A–D**) In these taxa the ovules are placed distally and scales proximally on the female short-shoot. Note the general reduction in the number of ovules in a progression from outgroups (**A,B**), to sister groups (**C,D**), to angiosperms (**E**). Our analysis suggests that the bracts are homologous to the gynoecial appendage (carpel wall) in (**E**) and the female short-shoots are homologous to the placenta and ovule in (**E**). Arrow indicates the parenchymatous base (see text).

interesting to note that this inference is not dependent on any single hypothesis of angiosperm origin. Thus, support for the hypothesis is robust even with variable phylogenies.

Outgroup Structure.   If the stachyosporous hypothesis of carpel origin is correct, then it should be possible to polarize angiosperm carpels based on homologies with their outgroups (Fig. 6.3). Recent phylogenetic analyses provide data on the placement of the sister groups and more distantly related taxa (Crane, 1985; Doyle and Donoghue, 1986b, 1992; Loconte and Stevenson, 1990; Rothwell and Serbet, 1994; Nixon et al., 1994). There is a general congruence among most phylogenies with the closest sister groups being the Gnetales, the Bennettitales, and sometimes *Pentoxylon*. We agree that the placement of *Pentoxylon* in this clade is uncertain as there are many loss characters (Crepet et al., 1991) and lack of homologies with other anthophytes (Rothwell and Serbet, 1994). The next closest living seed plant group is the conifers. A potential ancestor to these seed plants are an extinct group called the Cordaitales whose reproductive organs have been suggested to be homologous with those of gnetopsids (Eames, 1952). We review the female reproductive structure of most of these groups below, and propose homologies and transformations among the morphologies. Finally, we discuss the implications for the homology and ancestral state of the angiosperm carpel.

The structure of the gnetopsid female reproductive system appears to be a compound organ (Pearson, 1929; Chamberlain, 1935; Eames, 1952; Martens, 1971; Hickey and Taylor, Chapter 8). In *Ephedra*, a single ovule or pair of ovules each terminate an axis that is distally placed on a short shoot. Proximal to the ovule axes are a series of opposite and decussately arranged bracts (Fig. 6.3D), of which the distal-most subtend the ovule axes and partially encloses them. The ovule axes are composed of terminal ovules and scales with the distal scales enclosing the ovules, except for the micropylar tip. These distal scales form additional integuments. For convenience, we will call the axis with the distally placed ovule(s) and scales a female short-shoot. The entire short-shoot is subtended by a bract or leaf.

In comparison, the structure and homologies of the bennettitalean reproductive structure (Wieland, 1906, 1916; Harris, 1969; Crane, 1985, 1986; Watson and Sincock, 1992) are much more difficult to intepret. In gross morphology, the structure is made up of an enlarged, frequently elongate receptacle. Attached to the receptacle are ovules and sterile organs called interseminal scales, and together these form a dense head (e.g., Wieland, 1906). Most interpretations suggest that the ovules and interseminal scales are attached directly to the receptacle and are scattered across the surface (e.g., Crane, 1985). The interseminal scales have been suggested to be either bract derived or to be sterile ovule homologs.

Taylor (1991b) reexamined specimens of *Cycadeoidea* at Yale Peabody Museum and arrived at a new interpretation based on this material, and Wieland's (1906, 1916) published figures and descriptions. First, a single funiculus may be shared by two ovules. Serial sections show that although this condition is rare, it does occur. In addition, Wieland (1906) notes and shows that the interseminal scales are always attached deeper than the funiculus of the ovules and, thus, are "somewhat shoot-like" (Wieland, 1906, p. 118). The funiculus of the ovule with its single vascular strand joins with several surrounding interseminal scales just above the point where the entire complex is attached to the receptacle (see Wieland, 1906, plates XXVII, 4; XXVIII, 7). Thus, the terminal ovule is attached to a slightly enlarged, apparently parenchymous base, whereas the interseminal scales are attached along the margins of the base (Fig. 6.3C). This is seen clearly in many specimens with female reproductive structures. In these, the ovules and scales have separated from the receptacle, and in this condition, they do not make an even break (see Wieland, 1906, plate XXVIII, 7). Rather they separate in blocks, with each block composed of a central funiculus and several surrounding interseminal scales (see Wieland, 1906, plates XXII, 2; XXIV, XXVII, 4; XXIX, 1, 2). Wieland (1906, p. 118) also notes that the interseminal scales are similar to bracts subtending the entire reproductive axis. Finally, in *Cycadeoidea*, each ovule is usually surrounded by 5 or 7 scales.

Interpretations of this anatomy and morphology must be made with caution. *Cycadeoidea* is a late-occurring taxon and is considered to be derived (Crane, 1985, 1986). In other taxa (e.g., *Vardekloeftia*), ovules are rare in comparison to the number of interseminal scales (Pederson et al., 1989b). Some authors have also described interseminal scales that appear to be structurally intermediate between fertile ovules and sterile, interseminal scales that are suggested to be sterile ovules. Nevertheless, another interpretation is that the ovule–interseminal scale system is directly homologous with the female short-shoot of the Gnetales. Thus, there is a single (sometimes two) terminal ovule(s) subtended by a series of scales (Fig. 6.3C). The fact that there is an odd number of scales might suggest a subtending bract and several pairs of oppositely arranged scales. Further serial sectioning is needed to support this speculation. The club-shape of the interseminal scale could be an adaptation to protect the ovules with a tightly fitting armor and would make them similar in gross structure to ovules by being distally swollen. Similar club-shaped structures have evolved separately in the reproductive organs of many tracheophyte groups (Endress, 1975). As in Gnetales, the interseminal scales in Bennettitales enclose the ovule, except for the micropyle.

Understanding of conifer structure is possible because of the many fossil conifer remains that provide transitional forms to those living today (e.g., Florin, 1939, 1951, 1954; Harris, 1969; Clement-Westerhof, 1988; Mapes and Rothwell, 1991). Conifers are considered to have distally placed ovules on a sometimes modified female short-shoot (Fig. 6.3 B). The ovules may be single, few, or several, whereas the proximal sterile scales may be missing, few, many, or modified into an ovuliferous scale, and the short-shoot is usually subtended by a bract. We consider the female short-shoot of the gnetopsids and the possible short shoot of the bennettitaleans to be homologous with that of the conifers.

The last group that we will describe is the extinct Cordaitales (Florin, 1939, 1951; Rothwell, 1988; Trivett and Rothwell, 1991). This Carboniferous to Permian group had large, strap-like, entire-margined leaves, and variable habit. The female organs are composed of four to six distal funiculi that have a single ovule or that may branch and have two or more ovules (Fig. 6.3 A). Proximally to these are attached many spirally arranged scales. This female short-shoot is placed in the axile of a bract and the bracts has an alternate and distichous arrangement. Comparisons have been made between Cordaitales and gnetopsids (e.g., Eames, 1952; Bold, 1973) and we agree with these homologies between the female short-shoots.

These data for Gnetales, Conifers, and Cordaitales are clear, they all have a female short-shoot composed of distal ovule(s) and proximal sterile scales.

These are subtended by bracts that when grouped together form a larger reproductive organ. Taylor's (1991b) interpretation of these data from *Cycadeoidea* suggest a similar structure for the bennettitaleans. Variable aspects are the number of ovules and the number of sterile scales. If we look at these data from a phylogenetic perspective (Fig. 6.3), we see that the Cordaitales have many scales and ovules. As we look at groups progressively more closely related to angiosperms, the number of scales decreases to only a few and the ovules to one or two. In addition, the distal scales fully enclose the ovules, except for the micropyle to form additional integument(s) (Hickey and Taylor, Chapter 8).

It appears that by outgroup comparison the homology is between the gynoecial appendage of a carpel and the bract subtending the female short-shoot in the outgroups (Fig. 6.3; Hickey and Taylor, Chapter 8). The placenta is then homologous with a female short-shoot with distally placed ovules, each with scales forming the outer integument. Thus, the placenta is an axis and the ovules have a stachyosporous origin. It follows then that the angiosperm carpel is the result of a trend towards the continual reduction in the number of scales and ovules and increased protection of the ovules by the surrounding organs (see also Hickey and Taylor, Chapter 8). Under this interpretation, the ancestral states in the carpel would be ascidiate with basal to slightly lateral placentation and an ovule number of one or two.

Morphogenic Analysis.   The different hypotheses of carpel origin have significant implications for their morphogenesis. One suggests that the carpel is fundamentally a foliar organ, whereas the other suggests that the carpel has a two-parted, foliar–shoot origin. Morphogenic data may be used to further test these hypotheses, including data from studies of mutagenesis, analyses of chimeras, and the manipulation of floral apices or primordia. Each of these areas will be discussed in context of the carpel hypotheses. We realize that developmental data do not always directly reflect the evolution of structure, because selection for the developmental pathway may have occurred independently and for separate functional efficiency (see discussion in Steeves et al., 1991).

The understanding of floral development in *Antirrhinum* and *Arabidopsis* has been greatly enhanced by the use of mutagenesis analysis (e.g., Coen and Meyerowitz, 1991). Many of these mutations affect the placement or the structure of the carpel. These include mutations like *ovulata, pleniflora,* and *deficiens* in *Antirrhinum* (e.g., Carpenter and Coen, 1990), and *apetala, agamous, pistillata,* and pin in *Arabidopsis* (Bowman et al., 1991; Goto et al., 1991). It has been suggested that development of carpels in a whorl is under the control of *peniflora* (*Antirrhinum*; Carpenter and Coen, 1990) and *agamous* (*Arabidopsis*; Bowman et al., 1991). Other genes are also important for carpel number such as *pistillata* (Hill and Lord, 1989). Thus, the control of carpel development is beginning to be understood. Yet the relationship between the gynoecial appendage and the placenta and ovules is less clear.

In several mutants, the resulting organs superficially look like open conduplicate carpels with marginally attached ovules. However, further studies have shown that these carpels are mosaic organs (e.g., Hill and Lord, 1989; Bowman et al., 1989; Irish and Sussex, 1990; Goto et al., 1991; Bowman et al., 1991; Shannon and Meeks-Wagner, 1993). Thus, although portions of these mosaic carpels are frequently due to development of carpel tissues, other parts are due to development of leaf, bract, or sepal tissues. The identification of the different parts of the mosaic carpel is possible due to the existence of cell types specific to a single organ (e.g., Irish and Sussex, 1990). In addition, although the ovules are marginal, the stigma is usually terminal, unlike the marginal stigma proposed to be ancestral in conduplicate carpels. These data suggest that open carpels are a teratology and are not homologous with ancestral conduplicate carpels.

In all, there do not yet appear to be any data that indicate what the ancestral state is. However, some data suggest that carpels may be homologous with short shoots consisting of distally placed ovules on a determinate axis found in the outgroups (see also Hickey and Taylor, Chapter 8). These mutagenesis studies show that most flowers are determinate and that the carpels are placed distally. Flowers would then have a pseudanthial origin (see discussion in Hickey and Taylor, Chapter 8) and be composed of reproductive axes with distally placed female short-shoots, proximally placed male shoots, and sterile bracts basally placed. In addition, carpel and sepal development is initiated by a single gene, unlike the initiation of stamens and petals (Coen and Meyerowitz, 1991). This would suggest that the basic type of flower had bracts (= sepals) and ovules enclosed by bracts (= carpels). The evolution of the bisexual flower and petals was secondary. This occurred with the addition of a whorl of male shoots (stamens) proximal to the carpels, and petals proximal to the stamens and could be due to the evolution of the class B modifying genes (Coen and Meyerowitz, 1991; Doyle, 1994). Finally, some mutagenesis work examining inflorescence genes shows that carpel walls can tightly enclose the floral meristem (Shannon and Meeks-Wagner, 1993), as is found in the scales and bracts of gnetopsids and interseminal scales of bennettitaleans. All these data suggest that at least part of the flower may be homologous with the female short-shoots of the sister groups.

A second area of study uses chimeras to elucidate floral and carpel structure. Dellaporta et al. (1991) used clonal analysis to study the fruit development in *Zea mays*. They used the *Ac* transposon at the *Pericarp* locus as a method for marking the cells in the pericarps (kernels). The gynoecium of *Z. mays* is made up of three carpels in which the single basally placed ovule is developmentally related to the carpel on the germinal face. The carpels are composed of two cell layers, L1 and L2, and the embryo develops from L2. Although these data are interesting from the point of view of the sequestering of cell lineages for reproduction and of the identity of the cell layers from which the embryo develops, they provide little information on the carpel–ovule relationship. Although there can be transposition events that mark the ovule only (Anderson and Brink, 1952; Dellaporta et al., 1991), the examination of kernels does not allow for the identification of long cell-lineages that start before the development of the carpel walls. Therefore, from these studies it is not possible to determine whether the ovule develops from the carpel on the germinal face of the gynoecial appendage or if it develops from a separate growth area.

Interpretation of chimeras and manipulation of floral and ovule meristems have been used to elucidate the floral development of members of the Solanaceae. Satina and Blakeslee (1941, 1943; Satina, 1944, 1945) studied the development of carpels as well as other organs of *Datura* by examining chromosomal chimeras. They found that the floral meristem was composed of three cell layers, L1, L2 and L3. They also found that these three cell layers contributed to all the floral parts of *Datura* in all four whorls (Satina and Blakeslee, 1941, 1943; Satina 1944, 1945).

Datura flowers have bicarpellate ovaries with axile placentation, as do most members of the Solanaceae. This morphology is modified by the existence of a partial false-septum that divides each placenta into two parts. The carpels are initially composed of two arcs of cells that later develop into the carpel walls. Between these two primordia is a separate ridge from which develop the placenta, septum, and false septum. This ridge is thicker in the center and "carries on for a considerable time the functions of the floral apex" (Satina and Blakeslee, 1943, p. 455). As the carpel walls and the ridge develop, the septum and, later, false septum become progressively more connected to the wall.

Chimeral data show that the carpel wall is similar in development to petals and leaves (Satina and Blakeslee, 1943). The difference is that the carpel wall, except for the style, is primarily composed of cells derived from the L3 as opposed to mostly L2 in the petals. Yet, cells derived from the L2 predominate in the marginal regions where the carpel wall meets the septum. These chimeral data indicate that the ridge region is also mostly L3 derived and "that the placental tissue is developing independently of the septum and of the carpel wall" (Satina and Blakeslee, 1943 p. 461). From these observations, the authors concluded that the entire carpel is axile in nature, yet the style has similarities to the petals (Satina, 1944).

These *Datura* data support the hypothesis advanced in this chapter. First, two types of primordia form, the first producing the two gynoecial appendages (carpel walls) and the other the placentae (ridge). Second, we infer that the production of tissues derived from the L2 along the margins of the carpel walls and distally to produce the style shows that the gynoecial appendages are similar to leaf-derived organs. Thus, the gynoecial appendage is a sterile leaf. Finally, the nature of the ridge is similar to a floral meristem and primarily consists of cells derived from the L3, the innermost layer. This would suggest an axile origin of these placentae.

Another member of the Solanaceae in which chimeral plants have been examined is tobacco (*Nicotiana*; Dermen, 1960; Burk et al., 1964; Stewart and Burk, 1970). Tobacco carpels are similar to those found in *Datura* except they do not have a false septum (Hicks and Sussex, 1970; Fig 4A–C). The ovary of tobacco is bicarpellate (Fig. 6.4A) and the septum completely separates the two carpels (Fig. 6.4C). Sandwich chimeras in which the L1 is green, the L2 is white and the L3 is green show that the leaves and sepals are composed of all three layers, whereas the carpel wall appears to be formed only from the L1 and L2, with the megaspores developing only from the L2 (Burk et al., 1964). Similar results have been found in *Pelargonium* (Stewart et al., 1974).

Additional chimeral evidence for carpel development in Solanaceae has been collected from *Nicotiana* by our laboratory (Fig. 6.4D–F). We implemented clonal analysis on the flowers of tobacco, using the *Ac*–GUS reporter system (see Kirchner et al., 1993). To use the GUS bacterial gene as a marker system to trace cell lineages in plants, it is necessary to mark an occasional cell. An ideal method is to use transformed plants containing the GUS construct that has a transposon inserted between the GUS gene and its promoter (Finnegan et al., 1989). This prevents enzyme expression until the transposon randomly excises from the DNA and allows the cell to make the GUS enzyme, which marks the cell and all of its descendants.

In our preliminary work, we have found periclinal and mericlinal clones in the L3. These clones extend from the center of the receptacle (Fig. 6.4D), into the central part of the sepal lobes (Fig. 6.4D, top), and to the base of the petals. In the last whorl, the clones extend into the nectaries, to the base of the carpel walls (Fig. 6.4F), and into portions of the placenta (Fig. 6.4E) and central portion of the septum (Fig. 6.4F). They are not found where the septum connects to the carpel wall nor do they extend into the stigma and style. Our data confirm previous work showing that the sepals have all three layers, whereas the carpel walls only have L1 and L2.

The phyllosporous-origin or megasporophyll-homology hypothesis suggests that the placenta develops from the carpel wall. Because the clonal data shows that the carpel wall is composed of only the L1 and L2, the outermost layers, the placenta would be expected to be composed of the same layers (Fig. 6.4G–I). On the other hand, if the carpel is composed of two growth areas, the carpel wall could be composed of L1 and L2 layers, but the placentae could have L3 in addition (Fig. 6.4J–L). Our data demonstrate that all three layers are present in the placentae (Fig. 6.4E) and the middle of the septum (Fig. 6.4F) but that the L3 is missing from the carpel wall. Again, these data support the

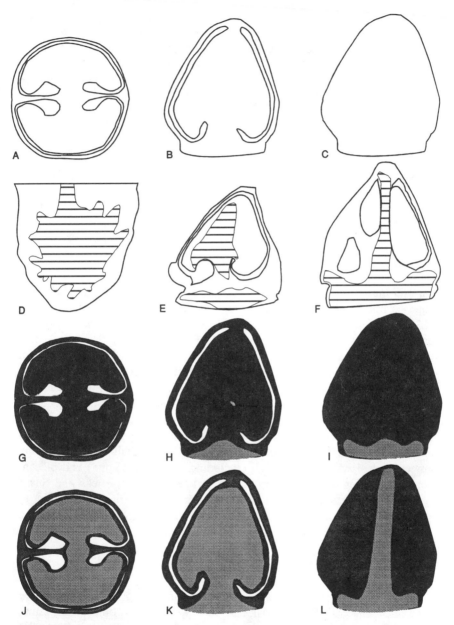

Figure 6.4.   Camera-lucida drawings (**A–F**) and interpretive drawings (**G–L**) of tobacco ovaries (*Nicotiana tabacum*). (**A–C**) Camera-lucida drawings of sections through
tobacco carpels, including a cross section, a longitudinal section perpendicular to septum,
and a longitudinal section through septum, respectively. (**D–F**) Sections of carpels with

marked chimeral tissue in cross-hatching. (**D**) Cross section of a floral receptacle with all organs removed except for a portion of calyx at top. The chimeral tissue is found in the innermost layer, the L3. Note that the clone extends into a sepal lobe at top. (**E**) Slightly tangential, longitudinal section perpendicular to the septum. The marked cells form a mericlinal chimera and in the other half of the section the two marked areas are attached. (**F**) Slightly tangential, longitudinal section through the septum. In places, the locules are exposed, and to the right, the ovules are exposed. The chimera extends into the floral nectaries at the base of the carpel walls. (**G–L**) Interpretations of the L1, L2 and L3 makeup of the carpels based on competing hypotheses of carpel origin. (**G–I**) Interpretation of the carpel wall and placenta layer based on the phyllosporous origin or megasporophyll homology hypothesis. If the placentae are derived from the carpel walls, they would be expected to consist of the same layers (L1 and L2). (**J–L**) Interpretation of the carpel and placenta based on the stachyosporous hypothesis. The placentae would be similar to a shoot with up to three layers and would not be restricted to the two layers found in the carpel wall. The chimeras found (**E, F**) are most similar to this interpretation. Note: The outermost layer, the L1, is black, the middle layer, the L2, is a dark shade, and the innermost layer, the L3, is a light shade.

composite nature of the carpel. The petals and the carpel appear to be mostly composed of L2–derived cells. In contrast, the placentae are mostly composed of L3. As in *Datura*, the megaspore still develops from the L2, since the ovules are usually composed of only L1 and L2 (Bouman, 1984). Thus, the carpel appears to be composed of a sterile gynoecial-appendage that is homologous with a leaf- derived organ and a fertile placenta that is stem derived.

Tobacco flowers have also been used in a number of excision studies. Hicks and Sussex (1970, 1971) cultured floral primordia and specific organ-primordia *in vitro*. They used this system to study organ regeneration after bisection of the flowers at different stages. In some of these experiments, sterile carpels developed after the bisection of the septum. A single half would develop a complete carpel with a carpel wall (including style and stigma) and placenta whereas the other would develop only a style and stigma. Hicks and Sussex suggest that this was due to difficulties in cutting the carpels exactly along the septa. We suggest, that this view is compatible with the two-growth-area hypothesis. When the cut is tangential and parallel to the septum, one of the halves gets both the carpel-wall primordium and placental primordia. The other half gets only the carpel-wall primordium and, thus, develops without producing ovules.

Finally, the importance of the L3 in regulating carpel number has been shown in Solanaceae (Szymkowiak and Sussex, 1992) and other taxa (Tilney-Bassett, 1986). This suggests that the floral meristem is important in controlling carpel development because the L3 is missing from the carpel wall. However, these data do not show whether or not the carpel is composed of two growth areas.

These morphogenic data provide tentative support for the hypothesis that the carpel originated from the integration of two growth areas: the gynoecial appendage and an axis with distally-placed ovules. Yet because there have been few studies, it is not certain that this type of development is found in all angiosperms. Nevertheless, carpels of the Solanaceae appear to have two growth-areas. One develops into the carpel wall, style, and stigma (gynoecial appendage), and the other develops into the placentae and central portion of the septum. These data do not support the interpretation that axil placentation is due to a conduplicate carpel in which the marginal placentae have become fused. In addition, they do not support the peltate phyllosporous hypothesis either, as the placenta develops from the floral apex, not from the cells of the carpel wall. Finally, some of the evidence does suggest that the gynoecial appendage has petaloid characters and, thus, could be bract derived.

EVOLUTION OF THE CARPEL

The evidence presented above strongly supports the two-growth-area model of carpel structure. Yet angiosperm carpels are very diverse in their form. In this section, we will begin with the hypothesis that the ancestral carpel was ascidiate with a basal to slightly lateral placenta from which developed one or two ovules. From this base type we propose an explanation for why there is variability among the carpels from different taxa and what the transformations are among these types.

Evolution of Carpel and Ovule Variability

To provide background for the question of why there is carpel variability, let us review the probable stages of carpel evolution. The first stage is the enclosing of the ovule axis by a subtending bract. Although the ovule(s) are already enclosed by the scales and partially enclosed by bracts in Gnetales and Bennettitales, the micropyle tip still projects past the enclosing organs, and pollen lands directly on the micropyle. The completion of this stage is the enclosing of the entire ovule and the concurrent change to stigmatic deposition and germination of the pollen grains. Thus, enclosure of the ovules would not have been as signficant as the appearance of the stigma. The advantages of stigmatic germination have been discussed by many authors (e.g., Heslop-Harrison and Shivanna, 1977; Lloyd and Wells, 1992). One advantage is the evolution of compatibility systems to make for more efficient pollination systems (e.g., Beach and Kress, 1980; Bernhardt and Thien, 1987). Another advantage to stigmatic germination is the possibility that sperm competition may have increased selection against lethal recessive mutants (Mulcahy, 1979; Mulcahy et al., 1992).

We suggest that the second stage of carpel evolution involved the directing of the micropyle away from the stigmatic region. This would have increased the length of the path that the pollen tube would have had to travel. Thus, the diversity of ovule types may have to do with carpel evolution and stigmatic germination. Others have suggested that curved ovules have evolved to increase contact to the transmission tissue, especially in multiovulate structures (Endress, 1990, 1994a; Lloyd and Wells, 1992). Yet, anatropous ovules curve away from the closure where the stigma or transmission tissue is found, not toward it. A final possibility is that the ovules were directed away from the closure until the fully sealed state or stigma evolved, and later in evolution the ovule became redirected at the transmission tissue.

The third stage would have been the evolution of novel carpel and pistil types and the evolution of high ovule numbers. Numerous functional adaptive possibilities may have opened up to angiosperms once the ovulary axis was enclosed by the gynoecial appendage and there was stigmatic germination. This could have allowed the evolution of morphologies that would have increased the number of ovules in each carpel, the number of carpels in each flower, and the number of ovules in each inflorescence.

Ovule Evolution.   It is striking that outgroups have orthotropous ovules; yet, in angiosperms, orthotropous ovules are relatively rare and only weakly correlated with ascidiate carpels and ovule numbers of one or two (Taylor, 1991a). In addition, orthotropous ovules are variably placed in relation to the gynoecial appendage. If we start with the proposed ancestral carpel (Fig. 6.5A, C) and the ancestral orthotropous ovule (Taylor, 1991a) we can propose three directions in which these organs could have adapted to place the micropyle further from the stigma and transmission tissue. One strategy would be a curved ovule with the micropyle directed basally (Fig. 6.5I). A second would be to move the orthotropous ovule to a more lateral or apical position (Fig. 6.5E). A last strategy would be a terminal stigma (Fig. 6.5B). Thus, selection for increased micropyle–stigma distance would result in new carpel forms as well as other ovule types. These adaptations will be examined in the Magnoliidae (sensu Cronquist) below.

The carpel and ovule morphologies (Taylor, 1991a) provide considerable evidence for adaptations to increased micropyle–stigma distances. We examined these data and looked at the association among carpel types, placentation, and ovule curvature. We note that the proposed ancestral carpel and orthotropous ovule morphologies (Fig. 6.5A) do not exist together in any living angiosperm. There are no living taxa that have ascidiate carpels with a marginal stigma, basal placentation, and orthotropous ovules. The two families with carpels most similar are Amborellaceae and Circaeasteraceae, and they have lateral, admedially placed ovules, although the ovules in the latter family are unitegmic.

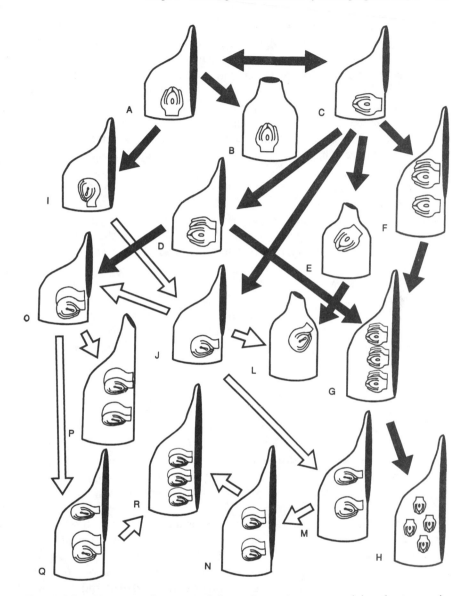

Figure 6.5.   Diagrammatic representations of carpel types containing the two major types of ovules, showing their evolutionary relationships. The proposed ancestral types, ascidiate with basal to slightly lateral placentation, are at top. Note that the carpels with anatropous ovules, or with ascoplicate or plicate morphology, are inferred to be derived. Dark arrows show initial evolution of morphologies having orthotropous ovules. Open arrows show evolution of derived carpels with anatropous ovules.

There are seven families that have members with ascidiate carpels with marginal closures and basal to slightly lateral placentation [Annonaceae, Myristicaceae, Monimiaceae (sometimes apical), Illiciaceae, Schizandraceae, Ranunculaceae (sometimes lateral), and Berberidaceae]. Three of these have members with other carpel types as well (Annonaceae, Monimiaceae, and Ranunculaceae). In all cases, the ovules are curved to place the micropyle away from the stigma and closure. Thus, Annonaceae have anatropous to campylotropous ovules, Berberidaceae have hemianatropous to anatropous ovules, Ranunculaceae have hemianatropous (except when unitegmic), and the remainder are anatropous.

Other families have members with orthotropous ovules but have a variety of carpel types and placements of the placenta. All have either lateral or apical placement of the ovules (Chloranthaceae, Ceratophyllaceae, Saururacae, Lardizabalaceae, and Barclayaceae). Most of these carpels are ascidiate (e.g., Fig. 6.5 C,E) except for some members of Saururaceae and Lardizabalaceae which are plicate (e.g., Fig. 6.5G,H). Taxa of Saururaceae have lateral, admedially (Fig. 6.5C) or marginally placed ovules (Fig. 6.5G), whereas those of Lardizabalaceae can be placed laterally and marginally (Fig. 6.5G) or laterally in a chaotic manner (Fig. 6.5H). Barclayaceae have ovules in a lateral, radial arrangement (Fig. 6.1M). Finally, Chloranthaceae and Ceratophyllaceae (see Endress, 1994b) have the ovules attached at the apex of the carpel (e.g., Fig. 6.5E). Again, in every case, the micropyle is directed away from the stigma.

The last suggested adaptation would be the evolution of a terminal stigma in the cases where there was a basal orthotropous ovule (Fig. 6.5B). This condition is found in some Piperaceae. In fact, the cylindrical shape of the carpel, the terminal stigmatic region, and multiple stigmas in some species have led some to believe that these flowers have a syncarpous ovary. It should be obvious that syncarpy with distal stigmatic-surfaces would also increase the distance between the stigma and the micropyle as is found in Canellaceae, Gomortegaceae, Saururaceae, Nymphaeaceae, Barclayaceae, Sabiaceae, Papaveraceae, and Fumariaceae.

The remaining members from the families of the Magnoliidae have curved ovules and nonbasal placentation. These ovules are found in a variety of lateral and terminal placements (Fig. 6.5J–Q). Most have anatropous ovules with the exception of Canellaceae, Ranunculaceae, Sargentodoxaceae, Menispermaceae, and Sabiaceae. We suggest that in these families both the carpels and the ovules have been under selection, and the derived states have evolved multiple times. It appears that the loss of curvature is rare, although it may have occurred with the change from bitegmic to unitegmic ovules (e.g., Ranunculaceae).

In summary, it appears that adaptations for a greater distance between the micropyle and stigma may have been factors in the evolution of new ovule and carpel characters in angiosperms. This factor may have been sufficiently adap-

Sister Groups   Angiosperms

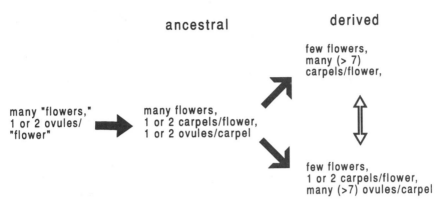

Figure 6.6.   Diagram showing the strategies to increase the number of seeds, based on the evolution of carpel and floral types. The ancestral condition for angiosperms is based on the suggested homologies between the female structures of the sister groups, and angiosperm flowers and inflorescences. Once the angiosperm carpel evolved, other strategies (to right) were available to angiosperms that were not available to the sister groups.

tive that carpels having our hypothesized ancestral carpel and ovule morphologies are no longer found in living angiosperms.

Increased Ovule Numbers.   The number of ovules and, thus, seeds is relatively constrained in the gnetopsids and bennettitaleans (Fig. 6.3). The number of seeds is limited due to the number of ovules at the tips of the female short-shoots and the number of female short-shoots on a primary axis (Fig. 6.6). In the sister group, there are many short shoots, each usually with a single ovule, although the potential for multiple ovules is found in the more distant outgroups (Fig. 6.3). One or two ovules is suggested to be the ancestral condition in angiosperms (Fig. 6.6). The states in the outgroups suggest that the ancestral angiosperms had inflorescences containing many flowers with one or two carpels per flower and only one or two ovules per carpel. Yet the evolution of the carpel with stigmatic germination allows for other adaptations for high ovule and seed numbers. One derived strategy would be to have relatively few flowers with many carpels, each with one or two ovules per flower. A related strategy would be few flowers with few carpels, but each carpel having many ovules. Obviously, combinations of these above strategies might be expected as well.

To test these hypotheses of whether angiosperms have evolved different strategies to increase ovule number, we collected data on the number of carpels per flower (Taylor, 1991a), the inflorescence type, and number of flowers per inflorescence for each family of the Magnoliidae (Cronquist, 1981). We considered the inflorescence to have few flowers if there were one to four per inflorescence and many if five or greater. For the carpel numbers per flower and ovule numbers per carpel, we categorized them as one or two, few (three to six) and many (seven or greater). Because these data were based on family-level comparisons, families with great variability (Winteraceae, Annonaceae, Cannellaceae, Monimiaceae, Aristolochiaceae, Ranunculaceae, and Papaveraceae) were not included because we could not match the combinations from the three characteristics.

Many families from the Magnoliidae have many flowers in racemose, spicate, or cymose inflorescences (Fig. 6.6). In these families, nearly half have flowers with one or two carpels and with one or two ovules per carpel (Myristicaceae, Trimeniaceae, Lauraceae, Hernandiaceae, Chloranthaceae, Piperaceae, Berberidaceae, Sabiaceae, and Fumariaceae). All of these families have ascidiate carpels, except for Berberidaceae (rarely ascoplicate) and Fumariaceae (plicate). About a quarter have members with three to six carpels containing one or two ovules per carpel (Amborellaceae, Gomortegaceae, Piperaceae, Menispermaceae, and Coriariaceae) and carpels with ascidiate morphology. In some cases, the ovaries are syncarpous. The remaining taxa have a variety of carpel types (mostly plicate) and variable carpel and ovule number. Thus, the most common type of female reproductive structure is a many flowered inflorescence with one or two carpels per flower and one or two ovules per carpel. This is the condition that we hypothesized as ancestral by our outgroup comparison (Fig. 6.6).

A smaller number of families have few flowers upon an inflorescence. In the flowers of these families, about half have seven or a greater number of carpels with only one or two ovules (Himantandraceae, Eupomatiaceae, Magnoliaceae, Calycanthaceae, Illiciaceae, Schizandraceae, Nelumbonaceae, Cabombaceae, and Circaeasteraceae). All of these families except for Calycanthaceae have ascidiate carpels. One-seventh of the families had members with one or two carpels containing one or one ovules (Idiospermaceae, Ceratophyllaceae, and Berberidaceae). Except for Berberidaceae, which are occasionally ascoplicate, this group also has ascidiate carpels. Thus, the strategy of low floral-number and high carpel-number with one or two carpels is also found (Fig. 6.6) and, we would suggest, is derived.

Another adaptive strategy for magnoliids with few flowers is fewer carpels per flower with more ovules (Fig. 6.6). This is rarer in basal angiosperms and is found in Degeneriaceae and some members of Winteraceae, Annonaceae, Canellaceae, and Aristolochiaceae. All of the taxa with this strategy have plicate carpels. The remaining families have members with various combinations of carpel types (with the majority being plicate) and numbers of carpels and ovules. A related strategy is the development of syncarpous ovaries, as these can have few carpels that can have many ovules. This strategy of few flowers with few carpels but with high ovule numbers is rare and we suggest it is also derived.

In summary, it appears that angiosperms have used three adaptive strategies to control seed number. The most common strategy is similar to that found in the sister groups with a high number of flowers on an inflorescence, each of which contains a few carpels and ovules. Another strategy is to decrease the number of flowers but increase the number of carpels (with one or two ovules) in each flower. This strategy is most frequently found in families with low numbers of flowers. In both of these strategies, the carpels are usually ascidiate. In other combinations where ovule numbers increase, other carpel morphologies are found, including ascoplicate and plicate. Thus, we suggest that adaptation for ovule number has resulted in multiple originations of these derived carpel types and related lateral placentation types.

Transformations Among Carpel States

The evolution of carpel diversity appears to be due, in part, to selection for positioning the micropyle away from the stigma and for increasing or maintaining high ovule numbers. Both of these adaptations are possible with the advent of stigmatic germination. The transformations among the various carpel types are possible through the integration of the two growth areas that produce the gynoecial appendage and the placenta.

Based on our analyses, the ancestral carpel was ascidiate and had one or two ovules. Furthermore, the outgroup homologies suggest that it had basal to slightly lateral placentation and that the ovule was orthotropous (Fig. 6.5A, C). From this ancestral morphology, selection for increased pollen-tube growth positioned the micropyle away from the stigma. This was accomplished by evolving an anatropous ovule (Fig. 6.5I), with a lateral to apical placement (Fig. 6.5E) or by means of a terminal stigma (Fig. 6.5B). Most of the remaining stages in the evolution of carpel types were probably related to increasing the number of ovules per carpel (e.g., Fig. 6.5, C to F to G, and J to M to N) and formation of terminal stigmas (e.g., Fig. 6.5, J to L). Our model suggests that curved ovules, particularly anatropous ones, have evolved multiple times (e.g., Fig. 6.5, A to I, D to O, E to L, and C to J). Ascoplicate (e.g., Fig. 6.5, C to F, J to M, and O to Q) and plicate carpels (Fig. 6.5, F to G, M to N, and Q to R) also appear to have multiple origins.

CONCLUSIONS

Growing evidence from character analysis, phylogenetic analyses, outgroup comparison, and morphogenic analysis supports the stachyosporous-origin hypothesis. Our analysis indicates that the gynoecial appendage of the carpel is homologous with the bracts subtending the female short-shoot (ovular axis) in the outgroups. The placenta within the carpel is homologous to the female short-shoot itself. Our data consistently support the hypothesis that the ancestral carpel morphology is ascidiate with a marginal stigma and basal to slightly lateral placentation of one or two orthotropous ovules. Transformations to other carpel types are likely to be a result of the integration of the primordia producing the gynoecial appendage and the carpel wall, with the placental growth area. We suggest that the evolution of curved ovules and the placement of the ovules in other positions was to direct the micropyle away from the stigma or pollen-tube transmission-tissue. Based on outgroup comparison, we suggest that reproductive axes with many flowers, few carpels per flower, and few ovules per carpel were ancestral. From these evolved two types of inflorescences: one with few flowers each of which had many carpels and few ovules and the other with few flowers each containing few carpels and many ovules.

ACKNOWLEDGMENTS

The authors thank the reviewers for their insightful comments. We also thank our research assistants, Carla Kinslow, Greg Bloom, and Gail Emmert for their diligent work on the tobacco research. Finally, we thank J. H. Zhou for the transgenic tobacco seeds. This research was supported by grants from the McCullough Fund and the Office of Academic Affairs, Indiana University Southeast.

CHAPTER 7

Floral Structure, Development, and Relationships of Paleoherbs: Saruma, Cabomba, Lactoris, and Selected Piperales

Shirley C. Tucker and Andrew W. Douglas

The current strong interest in the group of taxa known as "paleoherbs" (Donoghue and Doyle, 1989) results from their aggregation at the base of the angiosperms based on morphological evidence (Donoghue and Doyle, 1989a, 1989b) and molecular evidence (Zimmer et al., 1989; Hamby, 1990; Hamby and Zimmer, 1992; Chase et al., 1993). For most authors the paleoherbs included Aristolochiales, Nymphaeales, Piperales, and Liliopsida (monocotyledons) among others. Lactoridaceae are included by most authors mentioned but not by Loconte and Stevenson (1991). Although the paleoherbs are considered monophyletic by Doyle and Donoghue (1989a, 1989b) and Loconte and Stevenson (1991), some phylogenetic analyses suggest that the paleoherbs are not a monophyletic group (Zimmer et al., 1989; Hamby, 1990; Hamby and Zimmer, 1992; Taylor and Hickey, 1992), although the component groups are still viewed by these authors as having been derived early in angiosperm evolution.

Given the current interest in paleoherbs, our goal in this chapter is to use data from comparative floral development to examine phylogenetic relationships among selected taxa via a cladistic analysis. Conclusions about floral form and development will be based on cladistic treatments from the literature, including our analysis of Piperales (Tucker, et al., 1993) as well as new data. The combination of phylogenetic analysis and comparative ontogenies are used with the aim of defining ontogenetic events that result in the existing diversity among paleoherbs.

The representative taxa studied here are *Saruma henryi* Oliver (Aristolochiaceae), *Cabomba caroliniana* Gray (Cabombaceae), *Lactoris fernandeziana* Phil. (Lactoridaceae), and *Gymnotheca chinensis* Decaisne (Saururaceae). These will be compared with each other and with other taxa of Saururaceae (Tucker, 1975, 1976, 1979, 1981, 1985; Liang and Tucker, 1989, 1990), Piperaceae (Tucker, 1980, 1982a, 1982b), *Acorus calamus* (Payer, 1857; Sattler, 1973), *Chloranthus spicatus* (Endress, 1987a), *Illicium floridanum* (Robertson and Tucker, 1979), *Magnolia denudata* (Erbar and Leins, 1981, 1983), selected Liliaceae (Greller and Matzke, 1970; Sattler, 1973), and Ranunculaceae (Tepfer, 1953).

MATERIALS AND METHODS

Provenance

Material of *Saruma henryi* was collected from cultivation (Chinese source) at the U.S. National Arboretum, Washington, DC by Elizabeth Harris. Paula Williamson and Edward Schneider collected material of *Cabomba caroliniana* from the San Marcos River, San Marcos, Hays Co., Texas, where it is native. Tod Stuessy collected *Lactoris fernandeziana* in the Juan Fernandez Islands where it is endemic. *Gymnotheca chinensis* was collected in China by Liang Han-Xing. All material was fixed in formalin–acetic acid–alcohol. Vouchers are in the Louisiana State University herbarium, and liquid collections are retained by the authors.

Scanning Electron Microscopy

Preparation for scanning electron microscopy (SEM) involved dissection of flower buds of various sizes in 95% alcohol with a Wild M5A stereoscopic dissecting microscope, further dehydration in an ethanol–acetone series, critical-point drying with CO_2 in a Denton DCP apparatus, and mounting on aluminum stubs with colloidal graphite. Stubs were coated with gold–palladium in an Edwards S-150 sputter coater, and micrographs were taken with a Cambridge S-260 scanning electron microscope at 25 kV.

Leaf Clearing

Whole leaves or leaf pieces were cleared according to a technique suggested by Foster (1952), using 5% sodium hydroxide, concentrated, aqueous chloral hydrate, safranin staining, dehydration to toluene, and mounting in Permount (Fisher Scientific Co.).

Phylogenetic Analysis

Data were compiled from our original investigations and from the literature for all taxa. Because of the potentially large number of magnoliid taxa, we selected taxa for which published documentation was available on anatomy and floral development. Ingroup taxa and sources of information included *Acorus calamus* L. (Araceae; Payer, 1857; Sattler, 1973), *Anemopsis californica* Hook. and Arn. (Saururaceae; Tucker, 1985), *Cabomba caroliniana* Gray (Cabombaceae s. str., Nymphaeaceae s. l.; Moseley, 1958, 1974; Padmanabhan and Ramji, 1966; Williamson and Schneider, 1993a), *Chloranthus spicatus* (Thunb.) Makino (Chloranthaceae; Endress, 1987a), *Gymnotheca chinensis* Decaisne (Saururaceae; Liang and Tucker, 1989, 1990), *Houttuynia cordata* Thunb. (Saururaceae; Tucker, 1981), *Illicium floridanum* Ellis (Illiciaceae; Robertson and Tucker, 1979), *Lactoris fernandeziana* Phil. (Lactoridaceae; Cronquist, 1981; Engler, 1887; Lammers et al., 1986), *Lilium tigrinum* Ker. (Liliaceae; Greller and Matzke, 1970), *Magnolia denudata* Desr. (Magnoliaceae; Erbar and Leins, 1981, 1983; Skipworth and Philipson, 1967; Skipworth, 1970), *Peperomia metallica* Linden and Rodigas (Piperaceae; Tucker, 1980), *Piper amalago* L. and *P. marginatum* Miq. (Piperaceae; Tucker, 1982a, 1982b), *Saruma henryi* Oliver (Aristolochiaceae), *Saururus cernuus* L. (Saururaceae; Tucker, 1975, 1976, 1979), and *Scilla violacea* Hutch. (Liliaceae; Sattler, 1973). The outgroup taxa used were ranunculids *Ranunculus repens* L. and *Aquilegia formosa* var. *truncata* (F. & M.) Baker (Ranunculaceae; Tepfer, 1953). Ranunculids are an appropriate outgroup, based on the fact that the ranunculid clade appears to be the sister group to the magnoliids (Chase et al., 1993; Qiu et al., 1993; Loconte and Stevenson, 1991; Donoghue and Doyle, 1989a, 1989b). Cronquist's book (1981) was useful throughout.

Parsimony analyses were performed with PAUP 3.1 (Swofford, 1993) using Multiple Parsimony (MULPARS), Delayed Transformation, and the Branch-and-Bound option. All data were unordered and polarized by using Ranunculaceae as the outgroup. Fifty-eight characters were used and are listed in Table 7.1. Bootstrap analysis (200 repetitions) of the data was performed to test the relative frequency of well-supported clades within our data matrix. Decay analysis (Bremer, 1988) was performed similarly to identify branch support in the most parsimonious tree. All trees five steps longer than the most parsimonious tree were obtained using 100 random-starting point heuristic searches. A strict consensus tree was calculated. Then, all trees were filtered one step at a time, and a strict consensus was generated at each

Table 1.   Character States Used in Cladistic Analysis[a]

---

1. Habit: 0, herbaceous; 1, woody
2. Growth form: 0, monopodial; 1, sympodial
3. Vegetative phyllotaxy: 0, alternate or helical; 1, opposite or decussate
4. Chloranthoid tooth at apex: 0, present; 1, not present
5. Stomata type: 0, anomocytic; 1, tetracytic/anisocytic; 2, paracytic
6. Leaf margins: 0, entire; 1, lobed; 2, serrate
7. Fimbriate vein: 0, present; 1, absent
8. Areoles: 0, undefined; 1, incomplete; 2, well defined
9. Veinlet ultimate branching: 0, lacking or incomplete; 1, branched once
10. Leaf shape: 0, elliptic; 1, ovate; 2, oblong; 3, obovate
11. Leaf tip: 0, acute; 1, attenuate; 2, acuminate; 3, rounded; 4, retuse
12. Lamina base symmetry: 0, symmetrical; 1, asymmetrical
13. Lamina base shape: 0, lobate; 1, cuneate; 2, truncate
14. Secretory cells: 0, present; 1, absent
15. Phloem plastids: 0, "*s*-type"; 1, "*p*-type"; 2, protein inclusions
16. Stipules: 0, absent; 1, present
17. Stem primary vasculature: 0, one cylinder; 1, two cylinders; 2, scattered bundles
18. Perforation plates: 0, scalariform; 1, simple
19. Inflorescence position: 0, axillary; 1, terminal
20. Inflorescence phyllotaxis: 0, helical; 1, decussate or distichous
21. Type of inflorescence if present: 0, solitary flower; 1, raceme; 2, spike; 3, cyme
22. Peloria: 0, absent; 1, present
23. Pedicel/peduncle: 0, sessile; 1, pedicellate or pedunculate
24. Flower bract stalk: 0, absent; 1, present
25. Floral bract or subtending leaf: 0, linear, lanceolate or spatulate; 1, ovate; 2, peltate
26. Basal bracts on inflorescence: 0, not present; 1, bracts not showy; 2, showy bracts
27. Common primordia-bract/flower initiation: 0, separate initiation; 1, common primordium
28. Floral symmetry: 0, radial; 1, dorsiventral
29. Floral vasculature: 0, cylinder per flower; 1, single bundle per flower; 2, two bundles per flower
30. Perianth presence and number: 0, lacking; 1, one whorl; 2, two whorls; 3, many, helical or acyclic
31. Stamen connation: 0, free; 1, connate
32. Stamen vasculature: 0, stamen trace distinct from carpel; 1, stamen traces partly fused with carpel trace
33. Stamen primordia: 0, separate initiation; 1, common primordium
34. Stamen sequence of initiation: 0, simultaneous; 1, successive
35. Number of stamen whorls: 0, one; 1, two; 2, three or more; 3, helical or acyclic
36. Median-sagittal stamen initiation: 0, laterals first; 1, median sagittals first; 2, no difference
37. Anther dehiscence: 0, latrorse; 1, apically latrorse; 2, introrse; 3, extrorse

---

[a] For delimitation and justification of states, see text.

Table 7.1. *(cont.)*

38. Filament differentiation: 0, little differentiation; 1, well differentiated
39. Pollen aperture type: 0, monocolpate or monosulcate; 1, tricolpate or triporate; 2, polycolpate or polyporate
40. Carpel number per whorl: 0, helical; 1, five; 2, three; 3, two; 4, one
41. Number of carpel whorls: 0, helical; 1, many; 2, two; 3, one
42. Carpel number: 0, three; 1, one; 2, four; 3, many
43. Position of median sagittal carpel: 0, abaxial; 1, adaxial; 2, both; 3, otherwise
44. Carpel initiation: 0, simultaneous whorled; 1, helical; 2, paired, decussate or successive
45. Origin of carpel cleft: 0, adaxial conduplicate; 1, terminal indentation
46. Ovary type: 0, hypogynous; 1, perigynous; 2, epigynous
47. Carpel fusion: 0, apocarpous; 1, syncarpous
48. Style and stigma number: 0, equal to carpel number; 1, less than carpel number; 2, more than carpel number
49. Style presence: 0, stigma sessile; 1, style present
50. Carpel vasculature: 0, two ventral bundles per carpel; 1, ventral bundle from adjacent carpel; 2, one ventral bundle only
51. Ovule number per carpel: 0, three or more; 1, one or two; 2, less than one on average
52. Ovule orientation: 0, orthotropous; 1, anatropous
53. Placentation: 0, submarginal; 1, parietal; 2, basal; 3, pendulous; 4, axillary
54. Embryo sac: 0, monosporic (Polygonum type); 1, tetrasporic (Fritillaria type)
55. Cotyledon number: 0, one; 1, two
56. Proanthocyanidin: 0, present; 1, absent
57. Endosperm storage: 0, oily; 1, protein and/or hemicellulose; 2, perisperm; 3, starchy
58. Endosperm development: 0, nuclear; 1, cellular; 2, helobial

[a] For delimitation and justification of states, see text.

successively lower step. Strictly supported clades were recorded at each step. In addition, individual character removal (Davis, 1993) and character subset analyses were performed. Because of the computational time involved, we used the PAUP search option (100 random starting-point heuristic searches) with each of the 58 individual character-removal analyses. The data were also divided into subsets based on different floral organs and vegetative characters. A parsimony analysis was performed with each subset to examine clade support in the most parsimonious tree by the different character subsets. In addition, each character subset was independently excluded from an analysis, once again, to examine the support of clades within the most parsimonious tree(s) based on an analysis of all of the data. An advantage to character-subset analysis as performed here is that similar states in each organ-type among taxa can be recognized and compared.

Characters Used in Cladistic Analysis

Of the 58 characters selected (Table 7.1), the following were used and described in a previous article (Tucker et al., 1993): 1, 16–19, 21–25, 27–29, 31–34, 36, 37, 42, 43, and 46–54. The remaining characters will be described below, with references as necessary.

2. Stem architecture. Data came from personal observations of live plants and herbarium specimens.

3. Vegetative phyllotaxy. Data were obtained from personal observations. Phyllotaxy is prevailingly alternate among Piperales (with minor divergences in *Peperomia* and some other taxa; Tucker et al., 1993), as well as in *Saruma* and *Lactoris* (Cronquist, 1981; Lammers et al., 1986). *Cabomba* has decussate phyllotaxy (Moseley et al., 1984), the only exception among those taxa included.

4. Chloranthoid teeth. Tooth types were described and illustrated by Hickey and Wolfe (1975). They define a chloranthoid tooth as having "a medial vein braced by two prominent lateral [veins]." This type of marginal tooth is found in Piperaceae, *Saururus* and *Gymnotheca* (but not in *Houttuynia* or *Anemopsis*; Saururaceae), *Cabomba*, *Saruma*, *Lactoris*, *Chloranthus*, Liliaceae, and Ranunculaceae. Although common among magnoliids and paleoherbs in particular, chloranthoid teeth are absent in *Magnolia* and *Illicium*. Data were obtained by personal observation of live plants, fixed material, and herbarium specimens.

5. Stomata type. The principal reference used was Metcalfe and Chalk (1950). Although states are not ordered here, the anomocytic state may be considered most generalized (and other types derived), based on ontogeny. The least number of cell divisions in the protoderm results in the anomocytic state, whereas all other types require increased numbers of cell divisions. These data were obtained from personal observation.

6–13. Leaf characters. References by Hickey (1973) and Hickey and Wolfe (1975) were consulted. Some data came from personal observations of live plants and herbarium specimens. For characters 7–9, 12, and 13, data are original from personal observations of cleared leaves and herbarium specimens.

14. Secretory cells. This character is useful to distinguish among the paleoherbs, since it characterizes all Piperales it was not considered usable in the work of Tucker et al. (1993). Although used here simply in terms of presence or absence, oil cells are heterogeneous and differ in both late development and in contents (West, 1969).

15. Phloem sieve-tube plastids. Behnke and Dahlgren (1976) and Cronquist (1981) compiled the information on presence/absence of types of sieve-tube plastids.

20. Phyllotaxis of inflorescence. Inflorescence phyllotaxy, or that of the distal portion of a mixed branch in taxa where flowers are solitary, is categorized. Ontogenetic homology is evident between the two. Each may be either helical or decussate among the taxa included. Data are based on personal observations.

26. Basal inflorescence bracts. In some Saururaceae, the flower-subtending bracts are of two types: showy or nonshowy. The showy ones are much larger than the others and are restricted to the base of the inflorescence. Many other taxa have basal bracts, but these are neither showy nor as large.

30. Perianth presence/whorl number. Because taxa of Piperales uniformly lack a perianth, that character was not useful in the work of Tucker et al. (1993). Here, several taxa have a perianth. A whorl may form simultaneously (as in the pairs in flowers of Piperales) or successively (as in the perianth of *Magnolia denudata*; Erbar and Leins, 1981, 1983).

35. Number of stamen whorls. The character states included here are one, two, three, or a multiple/helical condition. A single whorl may include different numbers of stamens: two in *Peperomia* or six in *Cabomba*. *Chloranthus* is coded as having a single stamen whorl; the three stamens all arise from the same common primordium (Endress, 1987a). Rutishauser (1993) mentions two monomerous whorls (one stamen, one carpel) in flowers of *Sarcandra*, another member of Chloranthaceae. Having two successive stamen whorls is the most common state among paleoherbs studied, although three or more are found in *Saruma* and *Aquilegia*. The peculiar orders of initiation in Piperales are coded as three or more whorls (three or four initiation events), except in *Peperomia* with only two stamens (one initiation event) and *Houttuynia* with three (two initiation events). The helical or acyclic order [see Rutishauser (1993) for terminology] in *Magnolia*, *Ranunculus*, and *Illicium* is not whorled and generally results in numerous stamens per flower.

37. Anther dehiscence. Although used in the work of Tucker et al. (1993), a few additional comments are needed. References used include Baillon (1871), Cronquist (1981), Dickison (1992), Endress (1987a), Endress and Hufford (1989), Leins and Erbar (1985), Leins et al. (1988), Moseley (1958), Moseley et al. (1984), Sattler (1973), and Tepfer (1953). *Magnolia* species have latrorse dehiscence (Endress and Hufford, 1989).

38. Filament/connective differentiation. Endress and Hufford (1989) described differences in level of anatomical differentiation in stamens for anther, filament, and connective. Taxa of Magnoliaceae generally show a low level of differentiation.

39. Pollen aperture type. Data on pollen were obtained from Cronquist (1981), Dickison (1992), Walker (1974, 1976a, 1976b) and Walker and Doyle (1975).

40. Carpel number. We divide carpel number into two characters in this chapter, comparable with our treatment of the androecium characters. The total number of carpels among the paleoherbs sampled includes these states: one or two, three, four, or numerous. Although the presence of numerous carpels in magnoliaceous flowers has been considered a generalized or primitive state since the time of Bessey (1897, 1915), recent investigations suggest that paleoherbs, with their oligomerous flower structure, may be an additional or alternative early group of angiosperms (Doyle and Hickey, 1976).

41. Carpel whorl number. Number of whorls of carpels (distinct ontogenetic events) may be one, two, three, four, five, or alternate helical, among the taxa studied.

44. Carpel initiation. Order of carpel initiation differs among the taxa studied, with states including simultaneous whorled, helical, and paired/decussate. Flowers with few carpels tend to show simultaneous whorled initiation, whereas those with numerous carpels show a helical order of initiation. The Piperales show paired/decussate or simultaneous-whorled order (Liang and Tucker, 1989; Tucker, 1976, 1980, 1981, 1982b, 1985). Where there is only one carpel as in *Peperomia*, it is arbitrarily termed a single-member (monomerous) whorl.

45. Origin of carpel cleft. In the majority of the taxa studied, a crease or cleft develops adaxially on the carpel primordium and eventually becomes the locule of the ovary within which the seeds form. *Chloranthus* (Endress, 1987a) and *Cabomba* differ in having a terminal indentation on the carpel primordium as the first indication of the cleft.

51. Ovule number per carpel. Although this character was used in the work of Tucker et al. (1993), one state ("less than one") may need to be explained here as well. In a syncarpous ovary of several carpels, the ovule number may be fewer than the carpel number. As an example, a tricarpellate, syncarpous ovary in *Piper* species has a single ovule in a shared locule. The remaining paleoherb taxa have either two ovules per carpel or many (here coded as three or more).

52. Ovule orientation. The order of character states does not imply a judgment about the primitive state in angiosperms. Tucker et al. (1993) cited prevailing opinion that the orthotropous state is derived.

55. Cotyledon number. One obvious difference between taxa of dicots and monocotyledons is the number of cotyledons. There is an undocumented reference by Cronquist (1981) to an extra cotyledon rudiment in Araceae, based perhaps on Eames' (1961, p. 343) assertion that embryos of *Arum* and *Arisaema* have two cotyledons; *Acorus* has a single cotyledon.

56. Proanthocyanidin. The information on presence of this compound was obtained from Cronquist (1981), Giannasi (1988), and Loconte and Stevenson (1991).

57. Endosperm storage products. Oil, starch, protein, and hemicellulose in different combinations are found as storage products among the taxa studied. It is common for more than one to be mentioned for a taxon (Cronquist, 1981).

58. Endosperm development. The types of endosperm development include nuclear, cellular, and helobial. Cronquist's (1981) book was used as a reference.

Terminology

The terms zygomorphic and dorsiventral are synonymous for the type of symmetry in which only one plane bisects the flower into two mirror-image halves. Both terms are used here, because "dorsiventral," although preferred by Europeans, is less familiar than "zygomorphic" to American readers. "Actinomorphy" means radial symmetry. The term "whorl" has different meanings, depending on whether defined descriptively or developmentally. Here, a whorl is a set of similar organs arising simultaneously or, less commonly, successively, as in the trimerous whorls of *Magnolia denudata* described by Erbar and

Leins (1981, 1983). In *Chloranthus*, there are monomerous whorls with only one organ. In Piperales, a simultaneous pair or a single organ initiation constitutes a whorl, a single developmental event. The terms "helical" and "acyclic" are used synonymously; the latter is preferred by some European workers (Rutishauser, 1993). There is a problem, however, with the use of "acyclic" having two meanings. The term is used by Rutishauser for both helical development and for absence of order in organogeny. Although the latter condition is rare, examples are found in stamen initiation in *Begonia* (Merxmüller and Leins, 1971) and *Achlys* (Endress, 1989b). Another set of terms used here applies to the longitudinal planes of section in bilateral and zygomorphic flowers. The sagittal median plane bisects a zygomorphic flower and its subtending bract, if present. The frontal plane bisects a bilateral structure, such as a floral apex with paired bracteoles. Sagittal and frontal planes are perpendicular to one another. Both planes are "median longitudinal," but sagittal and frontal indicate precisely which median plane is intended.

RESULTS

Floral Development of Saruma henryi

Flowers are borne singly per axil, attached on the petiole of the vegetative leaves, which are arranged alternately. Each flower (Figs. 7.1A and 7.9M) has 3 sepals, 3 petals, 12 stamens, and 6 carpels, and is perigynous. The dome-shaped floral meristem is attached just above the base of the subtending leaf primordium (Fig. 7.2A).

Three sepal primordia are initiated simultaneously and at the same level (Fig. 7.2B,C). The floral apex is slightly concave after sepal initiation (Fig. 7.2D). Next, three petal primordia are initiated simultaneously and equidistantly (Fig. 7.2D) in positions alternating with the sepals. The floral apex is now more deeply concave (Fig. 7.2E). The petal primordia enlarge as the floral apical diameter increases (Fig. 7.2E–G). Six stamen primordia are then initiated simultaneously in pairs alternating with the petals (Fig. 7.2F–G, arrows). A whorl of three additional stamens forms next ("3" in Fig. 7.2H), each stamen situated between the two members ("2") of one of the earlier pairs. A third whorl of three stamens next is initiated simultaneously in antepetalous positions ("4" in Fig. 7.2H) on the flanks of the concave floral apex. The six carpel-primordia form as two closely successive and alternating whorls of three (C in Fig. 7.3A) on the sides of the concave meristem below the stamen primordia. Later stages show subtle evidence of two whorls, with the outer having slightly larger carpels, the styles of which meet apically before the later three (Fig. 7.3D–F).

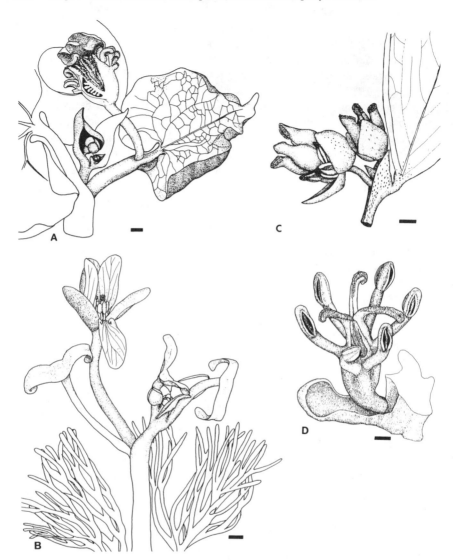

Figure 7.1.   Flowers of four paleoherbs. (**A**) *Saruma henryi* (Aristolochiaceae). Each flower is solitary and has 3 sepals, 3 petals, 12 stamens, and 6 carpels and is perigynous. (**B**) *Cabomba caroliniana* (Cabombaceae). The flower is solitary and has 3 sepals, 3 petals, 6 stamens and 3 carpels and is hypogynous. (**C**) *Lactoris fernandeziana* (Lactoridaceae). A perfect flower is shown. It has 3 tepals, 6 stamens in 2 whorls, 3 carpels and is hypogynous. (**D**) *Gymnotheca chinensis* (Saururaceae). Flowers are in a spike. Each is bract-subtended and lacks a perianth. Each has 6 stamens and 4 carpels and is epigynous. Scale = 2 mm in (**A**) and (**B**); scale = 1 mm in (**C**) and (**D**).

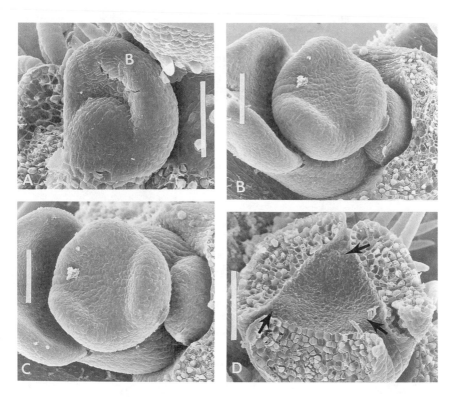

Figure 7.2. Organogenesis in *Saruma henryi*. Sepals have been removed in (**D–H**). (**A**) Floral apex and subtending bract primordium (B). (**B, C**) Two views of a floral apex with three young sepal primordia. The two views show that the sepals are synchronous and equal in size. (**D**) Simultaneous initiation of the three petals (at arrows). (**E**) Flower with three petal-primordia (P), sepals (S), and a concave floral apex. (**F**) Initiation of six stamen-primordia in three pairs (polar view; one pair at arrows); each pair lies between two petal-primordia. (**G**) Oblique view of paired stamen-primordia, one pair at arrows. (**H**) Flower (polar view) with two additional trimerous whorls of stamen primordia. One whorl of three is antesepalous (3) with each stamen initiating between the two members of an earlier pair (2). The last trimerous whorl of stamens (4) is antepetalous. Scale = 100 μm.

The floral organs enlarge in the order of their initiation. The sepals enlarge early, become valvate, and enclose the rest of the flower. In older buds, the enclosing sepals are densely covered with erect hairs (Fig. 7.3G). The petals are slow to enlarge (Fig. 7.3B–C) but close to the time of anthesis, they have enlarged, have an abaxial crease, and appear crumpled in the bud (Fig. 7.3F). The stamens all differentiate concurrently, with arcuate tetrasporangiate anthers and short dorsifixed filaments. The carpels are initially peglike (Fig. 7.3B) but

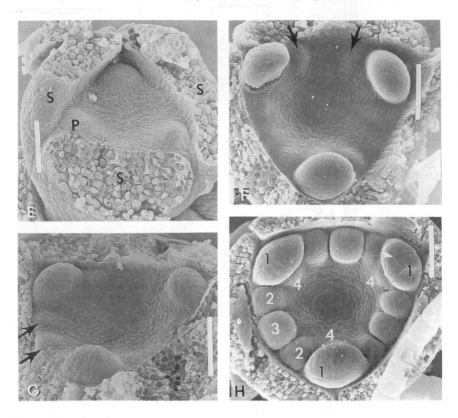

Figure 7.2. (*cont.*)

later enlarge abaxially at the base (Fig. 7.3E) as the ovules are formed submarginally. The six carpels synchronously develop clefts and remain open for a time (Fig. 7.3C) before the margins of each fuse. The carpels remain distinct through a height of 1 mm but eventually become fused basally (Fig. 7.3G,H). Stigmatic hairs line the suture the full length of the carpel (Fig. 7.3H).

Floral Development of Cabomba caroliniana

Each flower is solitary in a bract axil and about 2.5 cm in diameter. Each has two perianth whorls of three organs (Figs. 7.1B and 7.9L). The members of the two perianth whorls (sepals and petals) are white, with pinkish or purplish margins, with the petals having basal yellow patches bearing nectaries (Schneider and Jeter, 1982). There are six stamens and three carpels; the flower is hypogynous. Carpel number in the genus varies from one to six (Moseley, 1974) although *C. aquatica* has only two (Raciborski, 1894).

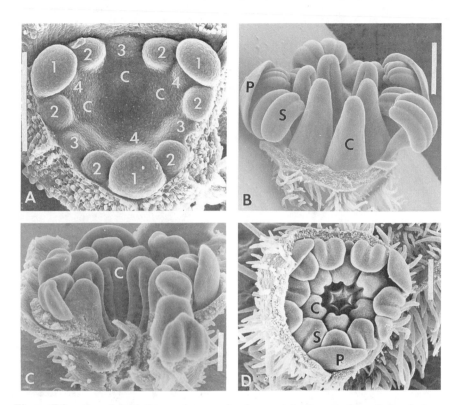

Figure 7.3. Carpel initiation and organ development in *Saruma henryi*. Sepals have been removed in all figures, as well as some other near-side organs in (**B**), (**C**), (**E**), (**G**), and (**H**). (**A**) Initiation of six carpel-primordia (three labeled, C) on the sides of the concave floral apex. Petals and stamen primordia are numbered in the order of initiation. (**B**) Lateral view of differentiated anthers (S) and peg-shaped carpel primordia (C). P = petal. (**C**) Lateral view of carpel primordia (C) with ventral clefts. (**D**) Polar view of flower with stamen anthers (S) curving inward, and the six carpel-primordia (C). P = petal. (**E**) Lateral view of carpel primordia (C), each with a dorsal enlargement. Ovules (at arrow) are present. (**F**) Polar view of flower near maturity, with folded petals (P), incurved stamens, and six stigmas (C). (**G**) Longitudinal section of perigynous flower showing ovules (at arrows) in locules. (**H**). Carpels near time of anthesis, with ovules exposed (at arrow). Scale = 200 μm in (**A**), (**C**), and (**D**); scale = 500 μm in (**B**) and (**E**); scale = 1 mm in (**F**)–(**H**).

The dome-like floral apical meristem (Fig. 7.4A) of *C. caroliniana* produces three sepal primordia simultaneously in a whorl (Fig. 7.4B). Three petal primordia are initiated next, alternately with the sepals and at the same level, on the convex floral apex (Fig. 7.4B-D). Next, six stamen primordia are initiated above and alternate with the sepal and petal primordia (Fig. 7.5A,B). The floral

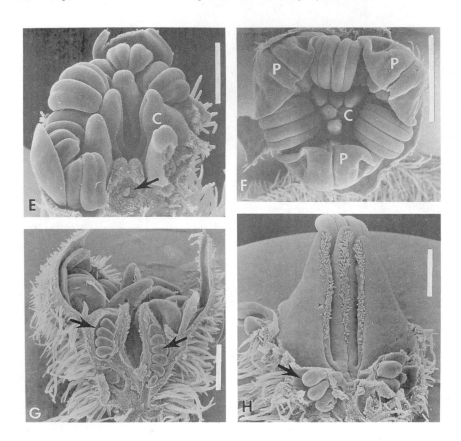

Figure 7.3. (*cont.*)

apex remains convex throughout (Fig. 7.5A,C,D). The stamen primordia enlarge equally around the floral apex (Fig. 7.5C,D). Three carpel primordia
eventually partition the remaining floral meristem to form a whorl (Fig. 7.5E).

The sepal primordia enlarge as ovate structures that do not overlap. The
petal primordia remain small until the late bud stage (Fig. 7.5G); at anthesis
they are flared (Fig. 7.1B). Each stamen becomes bifid (Fig. 7.5E,F) and
later becomes tetrasporangiate, dorsifixed to the filament (Fig. 7.5H) and with
extrorse dehiscence. Each carpel primordium grows as a flat-topped cylinder
and then develops a short cleft or depression on the upper side (Fig. 7.5F,G).
The carpels remain separate; each later becomes flaskshaped (Fig. 7.5H) with
pearl glands covering the ovary. Each has a short style, a globose, papillate
stigma, and two or three ovules, the position of which is highly variable
(Moseley et al., 1984).

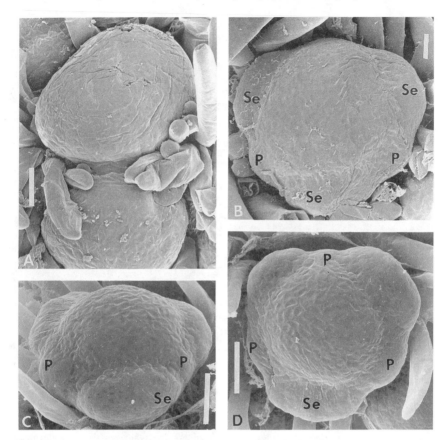

Figure 7.4. Sepal and petal organogenesis in *Cabomba caroliniana*. (**A**) Two floral apices lacking organs. (**B–D**) Three sepals (Se) have been initiated synchronously on the convex floral apex and have subsequently enlarged. Three petals (P) have just been initiated between the sepals. Scale = 50 μm.

Floral Development of Lactoris fernandeziana

Flowers may be either perfect or functionally carpellate. Each perfect flower (Figs. 7.1C and 7.9N) has one perianth whorl of three tepals, two whorls of three stamens each, and three carpels, and is hypogynous. Female flowers differ in that the six stamen primordia are suppressed after initiation (Fig. 7.6H,I). Each flower is solitary in the axil of a bract. The floral meristem is convex (Fig. 7.6A). Three sepal primordia are initiated simultaneously and equidistantly (Fig. 7.6B). The floral apex remains convex as the sepal primordia enlarge and grow upward (Fig. 7.6C–E) and eventually enclose the rest of the flower (Fig. 7.1C and 7.7B). Three stamen primordia next are initiated equidistantly and

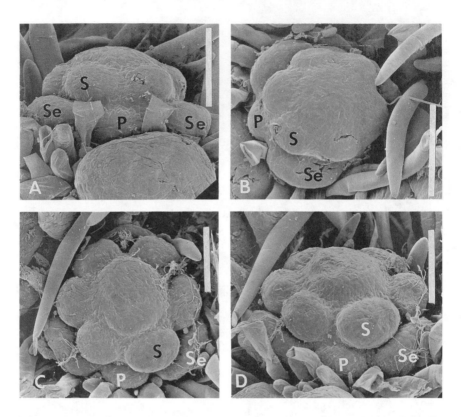

Figure 7.5. Stamen and carpel organogenesis and development in *Cabomba caroliniana*. (**A,B**). Initiation of whorl of six stamens above the sepals (Se) and petals (P) (lateral and polar views). (**C,D**) Stamens have enlarged around the convex floral apex (polar and lateral views). (**E**) A whorl of three carpel-primordia (C) have initiated and used up the remainder of the floral apex (polar view). Each has a concavity, which will become the cleft. Stamens (S) show the beginning of anther formation. (**F,G**) The carpels (C) have slit-like clefts on their summits. Stamens (S) have formed anthers. In (**G**), sepals and one petal (P) have been removed. F, young floral apex. (**H**) Carpels (C) have differentiated, with capitate stigmas and short hairs on the ovaries. Stamens (S) have filaments. Sepals and petals have been removed. Scale = 100 μm.

simultaneously (Fig. 7.6E) in alternisepalous positions. The three stamens of the second whorl are initiated simultaneously (Fig. 7.6F, G) and in positions alternating with those of the first three. In functionally female flowers, such as that shown in Fig. 7.6F,G, stamens are initiated but are thereafter suppressed. In perfect flowers, the stamens continue to enlarge and form anthers (Fig. 7.7A). Dehisced stamens are seen in Fig. 7.7C.

Carpel initiation occurs as a simultaneous equidistant whorl of three (C in

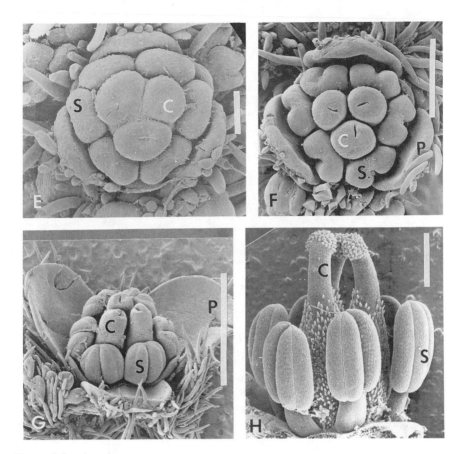

Figure 7.5. (*cont.*)

Fig. 7.6F,G). The carpels each enlarge (Fig. 7.6H,I), remaining distinct. An ovary and style differentiate; the styles become twisted in the bud (Fig. 7.7A). The carpels at anthesis (Fig. 7.7D) each have a short style, a capitate stigma, and a suture running their length. The fruits are follicles that split along the suture (Fig. 7.7E).

Floral Development of Liliopsida

Araceae. *Acorus calamus* (Fig. 7.9I) has been viewed as out of place in Araceae (Cronquist, 1981) although no better affinity has been offered. The taxon has been shown by Chase et al. (1993), Duvall, Clegg et al. (1993), and Qiu et al. (1993) as the basal branch of monocotyledons. Sattler's (1973)

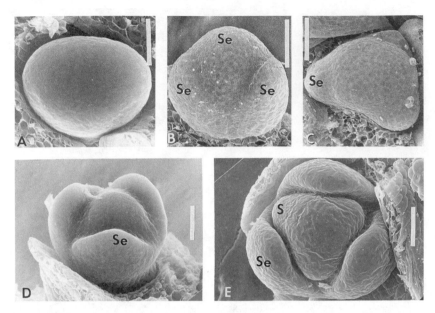

Figure 7.6.   Floral organogenesis in *Lactoris fernandeziana*. (**A**) Floral apex before organs form (oblique frontal view). (**B**) Synchronous initiation of three sepals (Se). (**C,D**) Three sepal primordia (Se) enlarging around convex floral apex. (**E**. Simultaneous initiation of three stamens (S), alternate with sepals (Se). (**F,G**) Simultaneous initiation of three carpels (C) around low-convex floral apex. Two trimerous whorls of stamen primordia (S) are seen. (**H**) Near-polar view of three carpel-primordia as they become arcuate around locules (at arrows). (**I**) The carpel primordia tips (C) are tapered and convergent. Three of the aborted stamens (S) are visible at base. Scale = 50 μm.

account of floral development differs from that of Payer (1857). According to Payer, the members of each whorl arise simultaneously. Sattler described a regular alternation of a single initial, followed by a simultaneous-initiating pair. Each "whorl" of three is formed in this fashion, with alternating solitary and paired initiations.

Liliaceae.   The floral development of members of three monocotyledonous groups has been described in the literature: Liliaceae, Helobiae (Alismatidae), and Arecaceae. In Liliaceae, two examples will be described briefly, both with whorls, but differing in order of initiation between simultaneous whorls and helical whorls. *Lilium tigrinum* (Fig. 7.9J; Greller and Matzke, 1970) has three sepals, three petals, six stamens in two whorls of three, and three carpels. The floral organs are in trimerous whorls. Each whorl is helical; the direction of the phyllotactic spiral alternates at each new whorl. Later equalization of organs

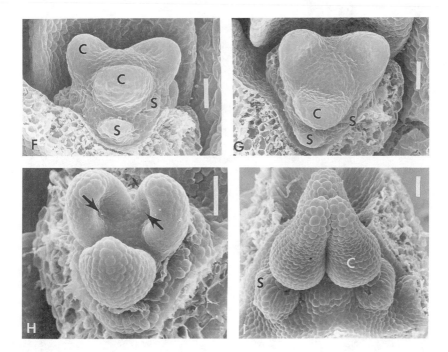

Figure 7.6. (*cont.*)

prevents the order of initiation from being detectable in the open flower. Another member of the family, *Scilla violacea* (Fig. 7.9K; Sattler, 1973), has the same number and arrangement of organs as *Lilium*. The three organs of each whorl are initiated simultaneously in this species.

Helobiae (Alismatidae). Extensive work on floral development in this group has been done by Sattler, Singh, Posluszny, Charlton, Leins, Kaul, and others; the results are summarized in a recent article by Posluszny and Charlton (1993). Floral organogeny usually consists of simultaneous or helical whorls, with a rare unidirectional exception in *Ranalisma humile.*

Arecaceae. Uhl (1988) has summarized the results of her considerable work on floral structure and development in palms. The simplest palm flowers are exemplified by *Nannorhops* (Uhl, 1969, 1988) and *Phoenix* (De Mason and Stolte, 1982). Sepals and petals are initiated in either simultaneous or helical, trimerous whorls. In the androecium, the basic pattern has two whorls of three, showing the same order (simultaneous or helical whorl) as in the perianth of the particular taxon. Polyandry is common in palm taxa and occurs in a diversity of developmental sequences. The three carpels are initiated in an irregular whorl.

Figure 7.7. Organ development in *Lactoris fernandeziana*. (**A**) Side view of young flower with differentiated stamens, and carpels (C) with twisted styles. Sepals and petals removed. (**B**) Undissected mature flower enclosed by sepals. (**C**) Open flower with dehisced stamens (S) and developing follicular fruits (C). (**D**) Carpels with capitate stigmas. Sepals, petals, and stamens removed. (**E**) Follicles with dehiscing sutures, polar view. Scale = 500 μm.

Summarizing for Liliopsida, simultaneous or helical initiation of trimerous whorls is the common pattern of floral organogeny. The exception is in *Acorus calamus*, with initiation of single or paired organs.

Floral Development of Gymnotheca chinensis

As representative of Piperales, floral initiation will be described in *Gymnotheca chinensis*. Its complete organogeny and development have been described previously (Liang and Tucker, 1989).

About 50–70 flowers are arranged in a short spike. Each flower (Figs. 7.1D and 7.9B) is subtended by a bract and lacks a perianth. It has six stamens and four carpels and is epigynous. The floral apex is tangentially broad before organogenesis (Fig. 7.8A) and when the first of six stamens is initiated in an adaxial-median position in the sagittal plane (Fig. 7.8B, "S_1"). The next two stamens are lateral, on either side of the first (Fig. 7.8C, "S_2"). The fourth and fifth stamens are next to be initiated; they are also paired and lateral, on the abaxial side of the flower (Fig. 7.8D, "S_3"). The sixth and last stamen to initiate forms abaxially and in the median sagittal-plane (Fig. 7.8D, "S_4"). Stamen size becomes equalized soon thereafter (Fig. 7.8E). At anthesis (Fig. 7.1D), stamens have long filaments and basifixed anthers with latrorse dehiscence.

The four carpels are initiated in two pairs, the lateral or frontal pair arising first and simultaneously (Fig. 7.8F), and the sagittal pair last and successively (Fig. 7.8G, "C_2"). In older stages (Fig. 7.8H), the carpels become similar in size due to equalization after initiation. The gynoecium is compound and inferior with short styles and four elongate, reflexed stigmas (Fig. 7.1D).

Order of floral organ initiation varies greatly among taxa of Saururaceae and Piperaceae. Diagrams indicating the order among stamens and carpels are shown in Fig. 7.9A–G. All, however, show initiation as pairs or solitary organs, never trimerous whorls as suggested by early authors (Schmitz, 1872; Eichler, 1875–1878).

Chloranthaceae

Chloranthus (Fig. 7.9H) appears to be a member of the paleoherbs according to the analyses of Chase et al. (1993) and Qiu et al. (1993). Chloranthaceae have been included in Piperales by Cronquist (1981) and Burger (1982). Endress (1987a) refutes this placement on the basis of numerous morphological character differences and asserts that Chloranthaceae have the closest relationship to Trimeniaceae in the Laurales.

Phylogenetic Analysis

The matrix (Fig. 7.10) shows distribution of character states, which are listed in Table 7.1. A single most parsimonious tree (Fig. 7.11) was obtained from

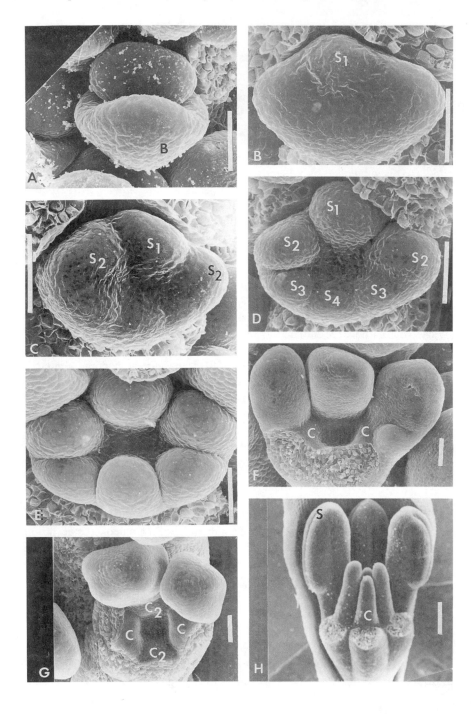

Figure 7.8. Organogenesis in *Gymnotheca chinensis*. (**A**) Floral apex subtended by a bract (B). (**B**) Floral apex at initiation of first stamen (S_1) medially on adaxial side. (**C**) Initiation of two lateral stamens (S_2) on either side of the first stamen (S_1). The floral apex is now concave. (**D**) A second lateral pair of stamens (S_3) have been initiated, and finally an abaxial stamen (S_4) in the sagittal median plane. (**E**) Size equalization of the six stamen primordia obscures their order of initiation. (**F**) Initiation of a pair of opposite carpels (C) laterally around the concave floral apex. Near-side stamens have been removed. (**G**) The second pair of carpel primordia (C_2) is being initiated in the median sagittal-plane. All but the adaxial median and one lateral stamen have been removed. (**H**) Flower with partly differentiated stamens (S) and with the four carpels (C) fused basally and having elongated styles. The three near-side stamens have been removed. Scale = 50 μm in all except (**H**), in which scale = 200 μm.

parsimony analysis, with 214 steps (c.i. = 0.51, r.c. = 0.33, r.i. = 0.65). In the tree, the paleoherbs appear monophyletic. The tree supports a monophyletic Piperales with monophyletic Saururaceae and Piperaceae, a monophyletic monocotyledon line, a monophyletic clade of *Cabomba*, *Saruma*, and *Lactoris* basal to the monocotyledon/Piperales, and a monophyletic clade of woody magnoliids (*Chloranthus/Magnolia/Illicium*) which is basal to all other paleoherbs.

There is little support for a stable topology of the taxa based on decay indices and bootstrap analysis, except for Piperales, monocotyledons, and woody magnoliids. Decay indices are illustrated above the nodes in regular print, and bootstrap values are underlined beneath the nodes in Fig. 7.11.

Using the character substitution method of Davis (1993), we could identify the robustness of clade support in our most parsimonious tree. The deletion of any one character from an analysis generally resulted in the same tree. There were, however, several characters that, if excluded, resulted in a total loss of paleoherb/magnoliid clade structure. The most unstable clade from the character-substitution method is the *Cabomba/Saruma/Lactoris* clade (marked *** in Fig. 7.11); that is, Piperales and monocotyledons remained monophyletic but were part of a polytomy including *Cabomba*, *Lactoris*, and *Saruma*. Other clades of our most parsimonious tree that are unstable via single-character substitution are similarly coded * in Fig. 7.11. Clades within our most parsimonious tree that became unresolved following the deletion of one character had a clade stability index of 0.017 (Davis, 1993). The characters that resulted in loss of structure were characters 3, 9, 19–21, 25, 40, 43, 57, and 58. Interestingly, the single deletion of inflorescence characters (19, inflorescence position; 20, inflorescence phyllotaxy; 21, type of inflorescence, and 25, floral bract) consistently resulted in a loss of structure among *Saruma*, *Lactoris*, and *Cabomba*.

Subset analysis was used to determine cladistic resolution of certain morphological features, as well as to examine certain apparently homoplasious

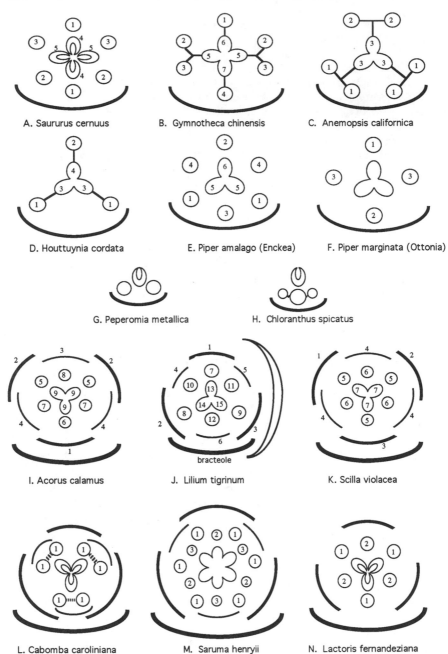

Figure 7.9. Floral diagrams of selected paleoherbs. The numbers on organs indicate the order of organ initiation in each flower.

	1-5	6-10	11-15	16-20	21-25	26-30	31-35	36-40	41-45	46-50	51-55	56-58
Ranunculus repens	00000	10011	00000	01111	30100	00002	00013	22110	03310	00010	-1211	100
Aquilegia formosa	00000	10011	00000	0011-	30100	00002	00012	22111	13300	00010	01011	100
Piper amalago-Enckea	01001	00201	21010	11110	20002	10110	00013	01103	20110	01002	20211	020
Piper marginata-Ottonia	01001	00201	21010	11110	20002	10110	00012	11103	20110	01002	20211	020
Peperomia metallica	01101	00201	20010	02110	20002	10110	00000	01104	31000	0-202	10211	021
Saururus cernuus	01002	00201	21010	10110	20111	11110	01012	10103	22220	00000	10001	021
Gymnotheca chinensis	01002	00201	21010	10110	20110	10110	11112	10103	22220	21001	001-1	021
Houttuynia cordata	01012	00211	31010	10110	21000	21110	11111	00103	20120	11001	00101	021
Anemopsis californica	01012	01112	21010	10110	21000	20110	11112	00103	30100	21001	00101	021
Cabomba caroliniana	00110	11001	30010	00-01	00102	00002	00000	20102	30001	00010	01001	122
Saruma henryii	01100	00201	20011	01101	00111	00002	0-012	23102	23200	1100-	01101	101
Lactoris fernandeziana	11100	00003	40110	11101	00101	00001	00011	23102	30000	00010	01001	120
Chloranthus spicatus	10000	20210	00110	11001	10000	10100	10110	22024	31001	00002	20301	101
												1
Magnolia denudata	10012	00213	20110	1101-	00101	00023	00013	20000	03310	00000	11001	001
				1								
Illicium floridanum	10010	00210	00110	0-000	00101	000-3	0-013	20010	03310	00001	21211	001
Acorus calamus	01012	01002	10201	02010	20000	00002	0-011	20102	30010	0110-	003-0	001
						1						32
Lilium tigrinum	00010	01202	10201	02011	10100	00002	00011	20102	30100	01110	01400	000
				1								12
Scilla violacea	00010	01202	10201	02010	10100	00002	00011	20102	30100	01110	01400	000
				1								12

Figure 7.10.   Matrix for tree in Fig. 7.11. A dash in the matrix represents an unknown.

conditions. We divided the matrix into subsets of characters. Using PAUP, we ran 25 random starting-point heuristic searches of each subset. The subsets included vegetative characters (1–18, 55–58), floral characters (19–54), anther characters (31–39), and carpel characters (40–54). Similarly, we excluded the carpel character subset in one analysis and the anther character subset in another analysis. Analyses of character subsets allows one to examine character congruence among different plant organs and to identify additional support for nodes within the most parsimonious cladogram based on all of the data. In addition, the variable rates of evolutionary changes among different organs are identifiable, as well as cases of convergence or lack of divergence.

In most subset analyses, numerous most parsimonious trees were obtained. A strict consensus was calculated to examine and compare clades supported in our most parsimonious cladogram (Fig. 7.11). An Adams consensus was also calculated. Although an Adams tree does not necessarily identify monophyletic groups (Swofford, 1993) and did not do so here, we found the technique useful to examine the nesting patterns of the taxa, in spite of the small number of characters used in each analysis.

An analysis of the vegetative character subset resulted in one tree (not

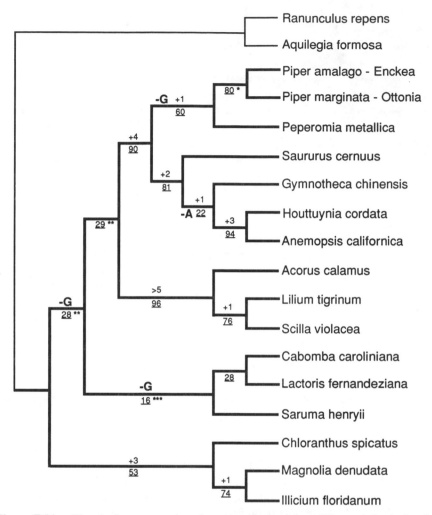

Figure 7.11. The single most parsimonious tree obtained from 58 morphological and ontogenetic characters (Table 7.1). The cladogram requires 214 steps and has a C.I. = 0.51, a R.C. = 0.33, and a R.I. = 0.65. Both families of Piperales are monophyletic and sister taxa to each other. *Chloranthus* appears to be a component of the woody magnoliid clade and not a piperalean taxon. The paleoherbs appear monophyletic, although there is little support for the positions of *Cabomba*, *Lactoris*, and *Saruma* based on bootstrap values (underlined), decay indices (above each node) and single character substitution (***) (see text for explanation of methods). −G represents clades that are not supported when the gynoecium characters (40–54) were excluded from a phylogenetic analysis. −A represents the clade that is not supported when the anther character subset (31–39) was excluded from a phylogenetic analysis.

shown) in which the clades of monocotyledons and woody magnoliids are maintained. Piperales form a grade, however, that is basal to the monocotyledon and woody magnoliid clades. *Cabomba, Lactoris,* and *Saruma* form a successive grade of taxa basal to the Piperales grade and the magnoliid/monocotyledon clade.

The analysis of the subset of floral characters, on the other hand, resulted in 24 trees. In the strict consensus, Saururaceae are monophyletic within a polychotomy of paleoherbs and *Chloranthus. Magnolia* and *Illicium* form a clade basal to the polychotomy. *Saruma* and *Lactoris* form a clade as well. In the Adams consensus, Piperales are grouped together. The monocotyledon taxa examined remain unresolved, although they nest together next to the Piperales group. Basal to the monocotyledon/Piperales group, *Cabomba, Lactoris, Saruma,* and *Chloranthus* form a polychotomy which is a node above the other woody magnoliids.

The carpel character subset (40–54) resulted in 24 trees. In the strict consensus, the paleoherbs form a polychotomy including *Chloranthus.* The families Saururaceae and Liliaceae (representing monocotyledons) are strictly supported. Piperaceae are also supported, although *Chloranthus* and *Acorus* form a grade basal to the Piperales. In the Adams consensus, liliaceous taxa nest with Piperaceae, next to *Acorus.* In the analysis of these data that excluded the carpel character subset, a single most parsimonious tree resulted. Compared to our cladogram (Fig. 7.11), there was loss of clade resolution among Piperaceae (polyphyletic), *Lactoris, Saruma,* and *Cabomba* as well as at the node that supports a monophyletic paleoherb group (marked–G in Fig. 7.11).

The anther character subset (31–39) resulted in 834 trees. A strict consensus demonstrates that the anther characters support monophyly of Saururaceae and of the monocotyledons. All other taxa and the nodes of the two monophyletic groups are unresolved. In the Adams consensus, *Piper marginata* grouped with Saururaceae. From the analysis of the data lacking the anther character subset, two most parsimonious trees were found. The topology of the two trees was identical to our most parsimonious tree (Fig. 7.11) except that in one of the two, *Gymnotheca* and *Saururus* form a monophyletic dichotomy in the Saururaceae (marked–A in Fig. 7.11).

DISCUSSION

Shared Characters of Paleoherbs

Paleoherbs share certain characters: monocolpate or monosulcate pollen, acropetal initiation of floral organs, and herbaceous, upright habit (woody exceptions in *Piper* pro parte, Aristolochiaceae pro parte, and *Lactoris*). Most paleoherbs have alternate phyllotaxy and pinnate leaf venation. Distribution of other characters and character states (Table 7.1) can be seen in the matrix (Fig. 7.10).

Three other character states (perianth presence, trimery, and radial floral symmetry) are sometimes attributed to paleoherbs (Taylor and Hickey, 1992), but Piperales have contrasting states for all three. It appears that there is not one, but two groups of paleoherbs: a radially symmetrical group with trimerous perianth (Aristolochiales, Nymphaeales, Liliopsida, and Lactoridaceae) and a dorsiventrally symmetrical group (clade) lacking a perianth and lacking trimery (Piperales). Differences in order of floral organ initiation also are associated with each of the two groups. The radially symmetrical group has simultaneous whorls or helically initiated whorls, whereas the dorsiventral group has initiation of solitary or paired organs in the flower.

Canalization in Paleoherbs. A common shared character of paleoherbs examined here is that, in most, the flowers have relatively stable numbers and positions of floral organs in each whorl, as well as distinct, differentiated whorls. Exceptions can be seen in relatives of *Cabomba*; flowers of *Brasenia* (Cabombaceae) and *Nymphaea* (Nymphaeaceae) have higher numbers of organs and more variation in number than in *Cabomba*. Canalization of floral organ numbers and positions appears to have occurred several times in angiosperm history. For example, it has occurred once among Chloranthaceae and again among higher angiosperms. From a developmental perspective, the diversity of these constrained floral features, in the taxa examined, makes it difficult to determine homologous patterns. The events leading to the different patterns should become better understood when developmental analyses of additional paleoherb taxa become available.

Unshared Characters

Symmetry. Two types of floral symmetry and underlying organogenesis occur among the paleoherb taxa examined. Dorsiventral symmetry and initiation of organs singly or in pairs prevail in Piperales; symmetry is established at the start of organogeny in all taxa of Saururaceae (Liang and Tucker, 1989, 1990; Tucker, 1975, 1981, 1985) and Piperaceae (Tucker, 1980, 1982a, 1982b). Radial symmetry and organogenesis in simultanous or helical whorls prevail in the grade including all other groups of paleoherbs sampled (Aristolochiales, Nymphaeales, Lactoridaceae, and Liliopsida, including Liliaceae, Helobiae, and Arecaceae). Kubitzki's (1987) assertion that trimery, found in all the radially symmetrical groups sampled, arose separately in monocotyledons and dicotyledons is not upheld by the current work. Dahlgren (1983) speculated that trimery preceded separation of monocotyledons from dicotyledons. These two strikingly different patterns of floral development (actinomorphic and zygomorphic) suggest that there are two major lines of paleoherbs. Phylogenetic analysis of paleoherb taxa using morphological and developmental characters supports this assertion.

Floral Diversity Among Paleoherbs. Among the paleoherbs sampled, other than Piperales, floral organization is trimerous and actinomorphic. There is considerable variation, however, in organs present (no petals in *Lactoris*), in paired organs occupying a single organ site (first stamen whorl of six in *Saruma*), and in type of gynoecium (hypogynous in most, but perigynous in *Saruma*). Also, the stamen number varies: 6 in *Cabomba, Lactoris, Lilium,* and *Scilla,* but 12 in *Saruma.* The order of organ initiation within the whorl also varies among paleoherbs, either helical or simultaneous. Both orders occur in Liliaceae: helical in *Lilium tigrinum* (Greller and Matzke, 1970) and simultaneous whorled in *Scilla violacea* (Sattler, 1973). Somewhat unusual is the simultaneous initiation of six organs, as in the stamens in *Cabomba* and the first whorl of six in *Saruma.*

Background on Taxa of Paleoherbs Studied

Aristolochiaceae. *Saruma henryi* is considered the most primitive member of Aristolochiaceae based on its actinomorphic flowers, perigyny, corolla, and follicle fruit type (Dickison, 1992; Gregory, 1956). Its organogeny has not been described in detail previously, although the floral development of several other aristolochiaceous taxa has been studied, including *Asarum* and *Aristolochia* (Leins and Erbar, 1985) and *Thottea* (Leins et al., 1988). *Saruma* and *Asarum caudatum* have the same number of organs and the same order of organogeny except for the last two whorls of stamens, which initiate in reverse order in the two. *Aristolochia gigantea, Thottea tomentosa,* and *T. siliquosa* lack the petal whorls and the last two stamen whorls found in *Saruma.* Stamen number and order vary in *Thottea* species; the latter can be simultaneous or in two successive whorls of three. Flowers of *Saruma* and *Asarum* are radially symmetrical, whereas those of *Aristolochia* and *Thottea* become dorsiventral. Interestingly, Huber (1977) and Leins and Erbar (1985) concluded that Aristolochiaceae form part of a natural group ("Vermittlergruppe") of angiosperms: Nymphaeaceae/ Piperaceae/Annonaceae, of which the Monocotyledoneae and Magnoliidae form two branches.

Cabombaceae. Baillon (1871), Padmanabhan and Ramji (1966), Raciborski (1894), Schneider and Jeter (1982), and Williamson and Schneider (1993a) described some aspects of floral organogenesis in species of *Cabomba.* Raciborski (1894) dealt with *C. aquatica,* which can be compared with *C. caroliniana,* described here. The organogeny is the same, except that the former has two carpels, the latter three. The corolla spurs and peculiar basal nectaries on the petals of *C. caroliniana* were described by Schneider and Jeter (1982).

Cabomba and *Brasenia* are generally considered the least specialized of Nymphaeaceae s. l. However, a recent phylogenetic analysis of Nymphaeaceae (Moseley et al., 1993), using vascular anatomical characters, showed *Cabomba* and *Brasenia* as a clade, which is part of a polychotomy involving several other genera of the family. *Cabomba* and *Brasenia* in this analysis do not appear to be more basal than other taxa with more complex flower structure.

Phylogenetic Analysis

Our tree (Fig. 7.11 and 7.12G) is compared to several published phylogenetic analyses of Magnoliidae using various data sets: those of Donoghue and Doyle (1989a, 1989b; Fig. 7.12A), Qiu et al. (1993; Fig. 7.12B,C), Hamby (1990), and Hamby and Zimmer (1992; Fig. 7.12D), Loconte and Stevenson (1991; Fig. 7.12E), and Dahlgren and Bremer (1985; Fig. 7.12F). Only the parts of the original trees that include the paleoherbs sampled here are shown in Fig. 7.12A–H; all trees were rerooted at Ranunculaceae. In a sense, the trees in Fig. 7.12A–F represent taxonomic nesting patterns. Figure 7.12H is the result of an Adams consensus and a partial combinable components analysis of the redrawn published trees, illustrating the levels of nesting and apparent taxonomic congruence among the different analyses.

This comparison shows a number of interesting congruencies. First, the Piperales are monophyletic in all analyses. There is also partial support for Aristolochiaceae and Lactoridaceae as sister taxa to the Piperales (Fig. 7.12A–C, F). There appears to be a majority of taxonomic congruence for the Aristolochiales/ Piperales clade although our analysis for congruency (Adams consensus, Fig. 7.12H; Adams, 1972) should be used only as a comparative measure. The other paleoherbs (Nymphaeales, monocotyledons), however, do not appear to hold taxonomically stable positions among all analyses. *Lactoris* and *Saruma* appear to be sister taxa in Chase et al. (1993), Qiu et al. (1993), and Donoghue and Doyle (1989a, 1989b). They appear as sister taxa in our analysis. In all of the trees examined, Chloranthaceae does not appear to be closely related to Piperales. Chloranthaceae seems to have common ancestry with the woody magnoliids (Fig. 7.12H). From our developmental investigations, Piperales are distinct from other paleoherbs based on synapomorphic floral characters associated with zygomorphy. The actinomorphic paleoherbs, on the other hand, represent a grade of variations, one of unstable topology. Although the flowers of Chloranthaceae are superficially similar to those of Piperaceae (specifically *Peperomia*; compare Fig. 7.9G and 7.9H), the development of the flower, anther, and inflorescence as well as leaf and ovule characters are dissimilar (Tucker, 1980; Tucker et al., 1993; Endress, 1987a).

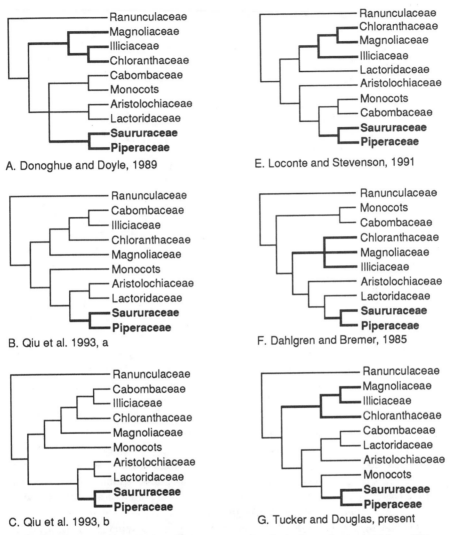

Figure 7.12. Seven redrawn, published trees of paleoherbs and selected Magnoliidae, obtained from different datasets. The trees included other taxa; only those parts of the trees including the same taxa as in Fig. 7.11 were retained. The trees were also rerooted with Ranunculaceae. (**A**) The strict consensus of morphological characters by Donoghue and Doyle (1989a). (**B,C**) Two of the trees by Qiu et al. (1993) using *rbc*L sequence data. (**B**) No weighting was done for this tree. (**C**) Weighting was used for this tree. (**D**) The most parsimonious tree by Hamby and Zimmer (1992) using 18S rRNA sequence data. This analysis did not include *Lactoris*. (**E**) Loconte and Stevenson's (1991) tree using morphological data. (**F**) One of numerous trees obtained by Dahlgren and Bremer (1985) in their morphological cladistic analysis. (**G**) Tree by Tucker and

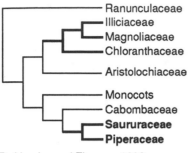

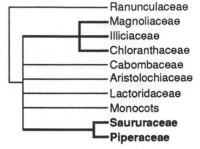

D. Hamby and Zimmer, 1992   H. Adams consensus of 7 trees

Figure 7.12.   (*cont.*)
Douglas (this chapter). (**H**) An Adams consensus of trees in (**A–G**) representing an attempt to identify taxonomic congruence and nesting patterns of various paleoherb taxa. In all trees, Saururaceae and Piperaceae are monophyletic (in bold). Woody magnoliids including Chloranthaceae are also monophyletic (in bold). Other analyses infer Aristolochiaceae and Lactoridaceae as sister taxa to Piperales.

Therefore, the floral similarities between *Peperomia* and Chloranthaceae can be interpreted as a result of different reduction patterns [see Tucker et al. (1993) and Endress (1987a)].

Incongruencies among the other trees primarily involve position of the paleoherbs in general. For example, in some analyses, monocotyledons (Fig. 7.11) appear as the sister taxon to Piperales [present analysis (Fig. 7.12G), Hamby (1990) and Hamby and Zimmer (1992; Fig. 7.12D), Donoghue and Doyle (1989a, 1989b; Fig. 7.12A, unresolved), Loconte and Stevenson (1991; Fig. 7.12E)]. The monocotyledons appear to be basal to Aristolochiales, Lactoridaceae and Piperales in Qiu et al. (1993; Fig.7.12B) and Dahlgren and Bremer (1985; Fig. 7.12F). From a comparison of the different trees, it becomes evident that Magnoliidae, particularly the paleoherbs, should undergo more detailed analyses of floral and vegetative form as well as additional data sets, perhaps from genes other than those in current use.

Our conclusion that paleoherbs are monophyletic is tentative until a larger number of related taxa are examined and can be included in a future analysis. From the separate data analyses (bootstrap, character substitution, decay indices, character subsets, and taxonomic-congruence comparison of other published phylogenies) there is little support for monophyly of paleoherbs (sensu Donoghue and Doyle, 1989a). The relationships revealed among the paleoherbs (particularly the position of the monocotyledons, Nymphaeales, represented by *Cabomba*, and Chloranthaceae, represented by *Chloranthus*) should be considered tentative, due to possibly inadequate sampling, for example in Nymphaeales. Convergences might be present but not be apparent; homologous characters now support clades that may actually be homoplasies or merely the result of plesiomorphies.

The character subset analyses that we performed suggest different rates of evolutionary divergence among different vegetative and floral organs as well as convergences in floral form. For example, taxa lacking a perianth (such as Piperales and Chloranthaceae) converge, based on carpel characters. A more thorough analysis of characters as well as an increase in taxon sampling among magnoliids should clarify relationships among these basal taxa.

In our analyses, we attempted to survey characters from various parts of the plant as well as to clarify interpretation of floral characters via study of their inception and development. The lack of phylogenetic support for the actinomorphic paleoherbs sampled could reflect a bias in sampling, a taxon sampling error, or the effect of using an actinomorphic outgroup, Ranunculaceae.

Radially Symmetrical Paleoherbs. In the majority of analyses, Lactoridaceae and Aristolochiaceae appear to share ancestry with Piperales. A feature common to all four families, as well as to aroids including *Acorus*, is the sympodial pattern of vegetative growth (monopodial in some *Peperomia*, *Macropiper*, and *Pothomorphe*). Cabombaceae includes both sympodial and monopodial states (Williamson and Schneider, 1993a): sympodial in *Cabomba. aquatica* (Raciborski, 1894), monopodial in *C. caroliniana* (personal observation), as well as *Nymphaea* and *Nuphar* (Cutter, 1957, 1959).

Piperales. From all analyses, the Piperales sensu stricto (Piperaceae and Saururaceae) appear monophyletic. All have flowers with dorsiventral symmetry and all lack a perianth. Although the members differ in number of floral organs and in the order of their inception, there are some shared features: initiation singly or in pairs; and lateral organs always are initiated in pairs on either side of the median sagittal-plane.

In 1977, Burger remarked that Piperales are very similar to the monocotyledon groups Arales and some Najadales, and hypothesized that the former order (together with Nymphaeales) is more closely related to monocotyledons than to other dicotyledons. He later proposed (Burger, 1981) that "the earliest angiosperms were small monocotyledon-like plants" with a simple primary body, clasping leaves, and scattered vascular bundles. The 1977 article unfortunately derived the trimerous actinomorphic flower of monocotyledons and *Lactoris* from an aggregation of *Chloranthus*-like flowers, a developmentally unlikely scenario. Also, reduction in chloranthaceous flowers is generally viewed as a specialization, whereas trimery and radial symmetry are viewed as plesiomorphies.

CONCLUSIONS

Floral development is herein used to elucidate relationships among representative taxa of "paleoherbs" (Aristolochiales, Piperales, Nymphaeales, and Liliopsida) and to help determine whether the group is monophyletic. Floral development was compared in *Saruma henryi* (Aristolochiaceae), *Cabomba caroliniana* (Cabombaceae s. str., or Nymphaeaceae, s. l.), *Lactoris fernandeziana* (Lactoridaceae), *Gymnotheca chinensis* (Saururaceae), and piperaceous and monocotyledonous taxa from the literature. The first three taxa have a perianth of one or two whorls, while *Gymnotheca* and other Piperales lack a perianth. Organs are initiated in simultaneous whorls, usually trimerous, in *Saruma*, *Cabomba*, and *Lactoris*, all of which are radially symmetrical. Monocotyledonous flowers sampled also show actinomorphy and either simultaneous whorled initiation or spiral initiation in each whorl, with the exception of *Acorus* which has paired and solitary initiations of floral organs [according to Sattler (1973)]. In contrast to the simultaneous whorls or helical order of most monocotyledons, flowers of Saururaceae (represented by *Gymnotheca chinensis*) and Piperaceae all show dorsiventral (zygomorphic) symmetry and initiation of solitary or paired organs. In *Gymnotheca*, order of stamen initiation is unidirectional from the adaxial side. Other taxa of Piperales vary in order of floral organ initiation, but it is always solitary (in the median sagittal plane) or paired (in lateral positions).

Are the paleoherbs monophyletic? Although they appear monophyletic in our tree, they probably are not, based on our analyses of morphological and floral developmental data as well as comparisons with other published datasets. Piperales (with dorsiventral/zygomorphic flowers lacking a perianth) and Monocotyledoneae (with actinomorphic, trimerous flowers having a perianth) are each monophyletic and are sister groups. The floral developmental patterns of these two groups differ markedly. More extensive analyses of monocotyledons (Duvall, Clegg et al., 1993) and Piperales (Tucker et al., 1993) can be consulted. A poorly supported clade of actinomorphic, trimerous paleoherbs (*Saruma/ Cabomba/Lactoris*) is basal to Piperales/Monocotyledoneae. *Chloranthus/Magnolia/Illicium* form a sister clade to the other taxa examined. Hence, radial symmetry, the presence of a perianth, and trimery appear to be general (possibly plesiomorphic) characters, rather than diagnostic features of a basal clade.

Note added in press: A new book by Endress (1994a) presents abundant information on flower structure, function, and evolution that is relevant to the subject of paleoherbs, as well as other flowering plants. He speculates that the earliest angiosperms had small oligomerous flowers with only stamens and carpels; a simple perianth developed somewhat later in angiosperm evolution.

Other relevant subjects covered by Endress include carpel and ovule evolution, and floral diversity among primitive woody magnoliids.

ACKNOWLEDGMENTS

This work was supported in part by National Science Foundation grant DEB-9207671 to S.C.T. The authors thank Elizabeth Harris, Edward Schneider, Tod Stuessy, and Paula Williamson for collecting some of the material used in the project. Permission was given to republish certain micrographs from *American Journal of Botany*.

CHAPTER 8

Origin of the Angiosperm Flower

Leo J. Hickey and David Winship Taylor

No topic in plant evolution has been more controversial or generated a larger literature than the origin of the angiosperm flower and its components. From the time of Charles Darwin, the failure of either the fossil record or modern comparative morphology to provide plausible intermediate stages between the angiosperm flower and somewhat comparable fertile structures in the gymnosperms has produced a number of contradictory hypotheses and interpretations, none of which have received anything like universal acceptance. Certainly, the leading contender at the present time would be the Magnolialean hypothesis which regards the base of the flowering plants as lying closest to the modern Order Magnoliales (Tahktajan 1953, 1969, 1980), whose flowers are regarded as reduced simple shoots with attached floral organs that are all interpreted as leaf homologs (Bessey, 1897; Bailey and Swamy, 1951; Tahktajan, 1991).

However, the past two decades have seen unprecedented progress in the emergence of new information and techniques for investigation that, we feel, have brought this historic puzzle to the verge of a solution. We note in particular the codification of rigorous methods of phylogenetic analysis, the emergence of molecular techniques, and a renewed interest in the developmental pathways followed during the growth of plant organs.

Twenty-one years ago G. Ledyard Stebbins (1974, p.198) reckoned that:

...fossil remains of any plants that might provide plausible connecting links between angiosperms and any other groups of vascular plants are still completely lacking. We must, therefore, rely entirely upon deduction and inference.

There has now been a nearly exponential increase in the amount of information from the fossil record of the early angiosperms and their potential precursors. Botanists and paleobotanists have also made great progress in understanding the origin of the flowering plants and identifying their basal or primitive members. As a part of this trend, the studies of Crane (1985), Doyle and Donoghue (1986a, 1992), Doyle et al. (1994); Doyle (1994) have resulted in a growing consensus that a group of gymnosperms known as the gnetopsids may be the closest living relatives of the flowering plants. Additionally, among the angiosperms, morphological studies by Burger (1977, 1981) and Taylor and Hickey (1992, Chapter 9) as well as molecular analyses by Zimmer et al. (1989), Hamby and Zimmer (1992), Chase et al. (1993), and Doyle (1994) have raised the possibility that the base of the clade may lie among a group of mainly herbaceous plants with simple flowers that fall within the Chloranthaceae, Piperales, or possibly the Nymphaeales, rather than among the magnolialeans. Parallel fossil studies over the last 20 years have also tended to emphasize the importance of non-magnolialean dicotyledons in the early angiosperm record (Kuprianova, 1967; Doyle and Hickey, 1976; Crane, 1989) and to draw attention to fossils with a number of gnetopsid features as somehow relevant to the origin of the flowering plant lineage (Cornet 1986, 1989; Crane, 1988a).

Starting from the twin hypotheses that the gnetopsids are the sister group of the angiosperms and that the base of the flowering plants lies with the Chloranthaceae/piperalean group, it will be our objective in this chapter to see if, by the process of reciprocal illumination, we can adduce some plausible homologies and character-transition sequences to bridge the gap between the flowers and inflorescences of the basal angiosperms and those of the gnetopsids. Having compared the fertile structures of one potential group of basal angiosperms to that of the gnetopsids, we will then examine the implications of these homologies for angiosperm evolution in general. Our decision to treat the Chloranthaceae/Piperales as basal for this discussion is bound to be somewhat controversial in view of some recent anaylses that have placed the Nymphaeales, especially Cabombaceae and Ceratophyllaceae, in that position (Zimmer et al., 1989; Hamby and Zimmer, 1992; Chase et al., 1993; Doyle et al., 1994). However, this choice of basal taxa is based on the results of our recent analysis

(Taylor and Hickey, 1992; also see Loconte, Chapter 10) and on what we regard as more derived features in the aquatic-adapted Nymphaeales.

THEORIES ON THE ORIGIN OF THE ANGIOSPERM FLOWER

With no clear transitional series between the angiosperm flower and gymnosperm reproductive structures, a number of hypotheses have been developed to bridge the gap between what were regarded as primitive flowering plants and their presumed ancestors. A complete summary of the various ideas regarding the transition from gymnosperms to the angiosperms would occupy a full chapter here. Instead, we give only a brief synopsis and refer the reader to the summaries found in Stebbins (Chapter 10 of 1974), Weberling (1989), Takhtajan (1991), and Friis and Endress (1990) that form the partial basis for our argument. Basically, the most influential of these scenarios fall into two major groups, the Euanthial or Strobilar hypothesis and the Pseudanthial hypothesis, whose basic outlines were formulated during the early twentieth century. Both hypotheses were strongly conditioned by assumptions concerning the identity and morphology of the basal angiosperms and their presumed ancestors, and both, in turn, have greatly influenced the search for basal angiosperms and their immediate precursors.

Euanthial or Strobilar Hypothesis

The Euanthial or Strobilar hypothesis, formulated by Arber and Parkin in 1907 (see also 1908), has been demonstrably the most influential theory of origin in the development of angiosperm systematics, with important elements from it having been subsumed into the classification systems of Bessey (1897, 1915), Hutchinson (1926, 1934, 1969), Takhtajan (1953, 1969, 1980), Cronquist (1981, 1988), and Thorne (1968; Chapter 11). Essentially, this hypothesis views the angiosperm flower as having developed from a simple gymnosperm strobilus such as characterizes the Cycadales and the Bennettitales (Fig. 8.1a,b). It is made up of a central axis with spirally attached fertile organs and sterile parts, consisting of bracts and sporophylls. A rather straightforward transition was then inferred between the strobili of these groups and the superficially strobilar flowers of the living members of the Order Magnoliales which, since the time of Bessey (1897, 1915), had come to be regarded as the most primitive of the living angiosperms. The picture of the primitive angiosperm flower was of a relatively large, radially symmetrical, uniaxial structure with numerous, free, spirally arranged parts of foliar derivation, laminar stamens, multiovular leaf-like carpels, and large seeds developing from bitegmic, anatropous, cras-

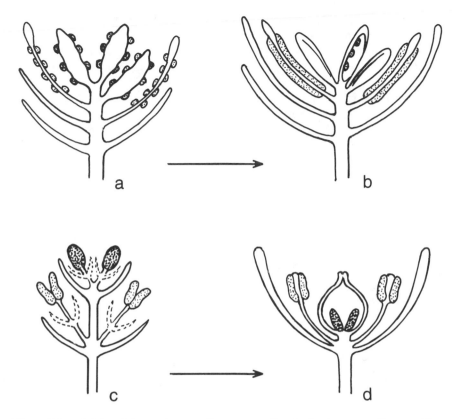

Figure 8.1. Origin of the angiosperm flower according to the two currently prevailing hypotheses. In the Euanthial hypothesis of Arber and Parkin (1907) a uniaxial strobilus (**a**) gives rise to the flower (**b**), whereas according to the Pseudanthial Hypothesis of Von Wettstein (1901, 1907) a pluriaxial cone (**c**) was the source of the flower (**d**). Pollen sacs are lightly stippled and the ovules are heavily stippled. Redrafted after Weberling (1989).

sinucellar ovules through fertilization by insect-borne pollen (Bailey and Swamy, 1951; Barnard, 1961; Takhtajan, 1969, 1980, 1991; Cronquist, Chapter 5 of 1988). As Eyde (1975) emphasized, a consequence of the Euanthial hypothesis was that all of the organs of the flower that were directly attached to the receptacle were interpreted as modified leaves. This has had a profound effect on their morphological interpretation and even on the basic terminology for these organs that persists to the present.

As the competing Pseudanthial hypothesis fell increasingly out of favor, the Euanthial reconstruction became the majority opinion of angiosperm phylogenists, although often with some minor modifications. One of these involved a shift to a model for the primitive angiosperm flower with a somewhat smaller size and fewer parts. This occurred after the demonstration of the derived features in the flower of Magnoliaceae [summarized by Stebbins (1974, p. 213 and ff.)], the realization that Winteraceae occupied a more basal position than Magnoliaceae (Cronquist, 1988; Takhtajan, 1969), and the accumulating fossil evidence for small flowers in the Early Cretaceous (Friis et al., 1986; Taylor and Hickey, 1990a). Gradually, as it became clear that cycads and bennettitaleans were only distantly related to each other (Thomas and Bancroft, 1913; Harris, 1932; Delevoryas, 1968; Crane 1985, 1986; Doyle and Donoghue, 1986a, 1992, 1993), the bennettitaleans tended to be placed closer than the cycads to the main axis of angiosperm derivation.

Pseudanthial Hypothesis

The basic formulation of this hypothesis was by Wettstein (1901, 1935). According to this idea, the angiosperm flower developed from a compound gymnosperm strobilus with a central axis to which were attached numerous secondary axes subtended by bracts (Fig. 8.1 c,d). Both sets of axes may bear both sterile and fertile appendages. Homologous structures are the cones of conifers, the gnetopsids, and the Cordaitales (Wettstein, 1907b; Eames, 1952). In effect, what may appear to be a simple angiosperm flower has developed by the aggregation and condensation of a structure that is really an inflorescence. Under this view, the primitive angiosperm flower would have consisted of small, simple, bilaterally symmetrical, unisexual units built up from structures on several orders of axes, a single whorl of small, bract-like perianth structures, and small, simple carpels with one to few ovules. In its original formulation, this hypothesis fit the ideas of Engler (Engler and Prantl, 1897-1909) that the flowers of the amentiferous-hamamelid orders Casuarinales, Fagales, Myricales, and Juglandales were primitive. Wettstein (1907b) later added the families Chloranthaceae and Piperaceae with their small simple, mostly unisexual flowers to this inferred basal group. However, the presence of tricolpate pollen is considered to represent a derived state in the amentiferous hamamelids (Doyle, 1969; Muller, 1970; Cronquist, 1988), and accumulating fossil evidence for the late appearance of Engler's "primitive" families cast serious doubt on this theory (Doyle and Hickey, 1976; Hickey and Doyle, 1977; Muller, 1981, 1984).

Other theories

A number of other theories of floral origin represent variants or at least hybrids of the Pseudanthial hypothesis and are mostly of historical interest [described in the works of Friis and Endress (1990), Stebbins (1974), and Weberling (1989)]. Only two of these are important in the discussion that follows because they are based on an interpretation of the fossil record and have some similarity to the ideas proposed here. The first of these, the so-called "Gonophyll theory" of Melville (1962, 1963, 1969a) postulates a polyphyletic origin for the flower of the angiosperm from a fertile, branching axis—the gonophyll—that arises from a leaf, in the fashion of the glossopterids. In the second hypothesis, the "Anthocorm theory" of Neumayer (1924) and Meeuse (1963, 1990), the angiosperm flower ("functional reproductive unit") has several separate origins. Those of most Magnoliidae and their dicotyledonous derivatives are modified pluriaxial systems (holanthocorms) that originated from the gnetopsids via the Piperales, whereas the modification of an originally uniaxial system (the gonoclad or anthoid) gave rise to the flowers of Chloranthaceae. Finally, Meeuse (1963) postulates a separate origin for the monocotyledons from the fossil order Pentoxylales through the monocotyledonous order Pandanales.

It goes without saying that any attempt to discover the stages of evolution that led from gymnosperms to the first flowering plants will be strongly dependent on what group of angiosperms is regarded as basal. Certainly, the prevailing concept that the primitive angiosperm was a shrubby to arborescent member of the Order Magnoliales with large, multiparted, relatively complex, somewhat strobilar flowers has strongly influenced the acceptance of the Euanthial hypothesis. Recently, however, Taylor and Hickey (1992) postulated that the earliest angiosperm was actually an herb with a diminutive stature, a rhizomatous to scrambling habit, and small, simple flowers. This hypothesis was based on our phylogenetic analysis, accumulating molecular evidence (Zimmer et al., 1989; Hamby and Zimmer, 1992), and earlier suggestions by William Burger (1977, 1981). A comprehensive description of the hypothetical earliest flowering plant appears in Taylor and Hickey (1992) and in Chapter 9 of this volume. In the sections that follow, we shall reexamine the possibility that the angiosperm flower had a pseudanthial origin.

THE EOANGIOSPERMS

For want of a better name at the time, we used the designation "paleoherb" for the ancestral angiosperm (Taylor and Hickey, 1992). However, we now feel that our expansion of this term to include the most basal angiosperms is not useful because it changes the original concept as first given by Donoghue and Doyle (1989a). These authors considered that the group was made up of derivative, not ancestral forms within the Magnoliidae, and that it was characterized by "anomocytic stomata, two perianth cycles, and trimery in both the perianth and the androecium (except for the loss of one or both perianths in *Lactoris* and the Piperales and secondary multiplication of parts in the Nymphaeaceae)." Donoghue and Doyle included the Aristolochiaceae, Cabombaceae, and monocots in the paleoherbs, in addition to the two taxa previously mentioned. On the other hand, Taylor and Hickey (1992) regarded the group as basal, changed its taxonomic boundaries to include the Chloranthaceae and to exclude the rhizomatous but nonherbaceous Lactoridaceae as well as the monocots, included a wider range of stomatal types, and saw its lack of two cycles of perianth as ancestral, rather than as derived.

We, therefore, propose the designation of "eoangiosperm" for this group of predominantly herbaceous plants with tectate–columellate, acolpate to monosulcate pollen, mostly apocarpous gynoecia, and small, simple floral units or their immediate derivatives, and which lies below the origin of the monocots, the tricolpates, and the magnolialeans on our 1992 cladogram. As we define them, the eoangiosperms consist of the following families: Chloranthaceae, Saururaceae, Piperaceae, Aristolochiaceae, Barclayaceae, Cabombaceae, Nymphaeaceae, and Ceratophyllaceae. The eoangiosperms, as thus conceived, are an admittedly paraphyletic and highly plesiomorphic group in relation to the rest of the flowering plants, but, as will be shown, critical scrutiny of their characters can produce new insights into the origin of the group.

Assuming that the base of the angiosperms lies within this group of simple-flowered, largely herbaceous forms, or eoangiosperms, we should now find it very instructive to ascertain what the general characteristics of the group are and to determine which among these features might have been inherited from their angiosperm precursors. Of the eoangiosperms, the Family Chloranthaceae seems to retain the largest number of basal character states (Taylor and Hickey, 1992; Nixon et al., 1994; Loconte, Chapter 10), so that we will begin by examining the features of this family, especially those of its fertile structures, and then move to a more general consideration of other eoangiosperm taxa. Although we regard the Chloranthaceae as basal in this study, some of its features, such as expanded anther filaments and pinnate leaf venation, almost certainly represent advances over the primitive state; thus, we cannot regard any living member of the family as the complete archetype. In addition, only 2 out of 29 characters in the analysis of Taylor and Hickey (1992) support a basal placement of Chloranthaceae, whereas other eoangiosperm families occupy a position below that of Chloranthaceae in reports by Donoghue and Doyle (1989a), Zimmer et al. (1989), Hamby and Zimmer (1992), Chase et al. (1993), Qiu et al. (1993), and Sytsma and Baum (Chapter 12). The basal placement of *Ceratophyllum* in the molecular tree of Chase et al. (1993) has attracted a good deal of interest, especially in view of previous morphological and fossil evidence for this placement (Les, 1988; Les et al., 1991). However, we feel that this aquatic plant, with a highly reduced vegetative body and pollen wall, tenuinucellate, unitegmic ovules, and problematical early fossil record is a poor candidate for the basal-most position. *Ceratophyllum* is also highly distant from its node, and this portion of the tree by Chase et al. (1993) is only weakly supported by the *rbc*L data, with a different position for the genus being equally parsimonious in an unrooted alternative topology of the flowering plants (Qiu et al., 1993; Baum, 1994; Sytsma and Baum, Chapter 12). Similarly, Cabombaceae, with a number of aquatic modifications, has a leaf-type that only appears in the middle Albian Stage of the Cretaceous (Doyle and Hickey, 1976). Nevertheless, even if another eoangiosperm proves to be more basally placed than the Chloranthaceae (see Doyle, 1994), we feel that many of the conclusions concerning the transition from angiosperms to gymnosperms that are derived from the analysis below remain valid.

The Chloranthaceae and the Eoangiosperms

The Chloranthaceae consist of 5 genera and about 75 species with a highly scattered, mainly tropical distribution. Considering its importance in the Early Cretaceous fossil record (Kuprianova, 1967; Doyle, 1969; Doyle and Hickey, 1976; Hickey and Doyle, 1977; Muller, 1970, 1981, 1984; Walker and Walker, 1984; Crane, 1989; Doyle and Donoghue, 1993) and longstanding arguments over the origin and phylogenetic significance of its apparently simple floral structures, investigations of it are surprisingly few. Much of the current discussion is drawn from Swamy and Bailey (1950), Swamy (1953d), Humbert and Capuron (1955), Burger (1977, 1981), Cronquist (1981), Crane (1989), and especially from an important series of articles by Endress (1983a, 1985, 1987a) and Friis and Endress (1990).

Two genera within the family, *Sarcandra* and *Chloranthus*, appear to be basal by reference to characters of their wood and of their bisexual flowers, whereas *Hedyosmum* and *Ascarina* and the monotypic Madagascan genus *Ascarinopsis* are regarded as more advanced. We base this conclusion on the alternative phylogenetic hypotheses of Crane (1989, Fig. 3) in which the discovery of vessels in *Sarcandra* (Carlquist, 1987) forces *Sarcandra* and *Chloranthus* together at the base of either of Crane's alternatives.

Sarcandra and *Chloranthus* are weedy herbs or small shrubs, with the majority of species in *Chloranthus* being perennial, or even annual herbs. Only two species of *Chloranthus, C. officinalis* and *C. spicatus,* have somewhat suffructescent stems capable of 3 or 4 years of growth. *Sarcandra* and *Chloranthus* are also insect pollinated. On the other hand, the more derived genera, *Ascarina* and *Hedyosmum,* are both generally large shrubs, although they range in stature from suffructescent shrubs to medium-sized trees, whereas *Ascarinopsis* is a small tree. *Ascarina* and *Hedyosmum* are wind pollinated and have what are inferred to be the most reduced floral structures of the group (Swamy, 1953d; Burger, 1977; Endress, 1987a), whereas *Ascarinopsis,* with its close similarity to its namesake genus, is inferred to be wind pollinated as well. The

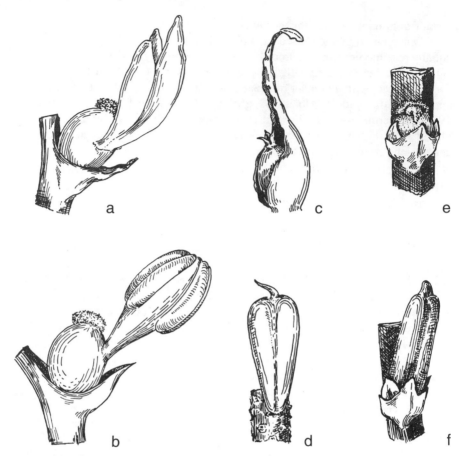

Figure 8.2. Flowers of the Chloranthaceae: (**a**) *Chloranthus Henryi*, bisexual flower consisting of a subtending bract, a trimerous stamen and a single pistil with a tufted stigma. (**b**) *Sarcandra glabra*, bisexual flower, (**c**) female flower and (**d**) male flower of *Hedyosmum orientale*, (**e**) female flower and (**f**) male flower of *Ascarina lanceolata*. Note the paired bracteoles in (**e**) and (**f**). (From Swamy, 1953, by permission.)

leaves of this family are pinnately veined, with semicraspedodromous secondary venation, marginal chloranthoid teeth, sheathing leaf bases and a decussately opposite phyllotaxy (Hickey and Wolfe, 1975; Taylor and Hickey, 1990a, 1992; Todzia and Keating, 1991).

Chloranthaceae have simple, inconspicuous flowers that are usually arranged in decussately opposite pairs basally that often become spiral apically. Flower-bearing axes are often organized into still larger-scale units by decussately opposite attachment to a main inflorescence axis (Endress, 1987a, 1987b; Friis and Endress, 1990). These inflorescences are properly classified as indeterminate single or double thryses, consisting of up to three orders of spikes or racemes, or they may be reduced thryses [see Weberling (1989) for inflorescence terminology], but have also been termed spikes, panicles, or even heads by some workers (Swamy, 1953d; Burger, 1977; Cronquist, 1981).

Basically, flowers of Chloranthaceae are small, with a single, dorsal floral bract subtending the fertile parts (Fig. 8.2). Sometimes this bract is lacking, as in the male flowers of *Hedyosmum* (Fig. 8.2d) and in some species of *Ascarina*, whereas in other species a pair of opposite bracteoles occur laterally above the bract (Swamy, 1953d; Endress, 1987a). These bracts and bracteoles are the only sterile parts of the flower and, except for the possibility raised by the trilobate scales on the stigma of *Hedyosmum* (Fig. 8.2c), no sign of a perianth or its early ontogenetic stages has ever been reported. With Burger (1977), we regard these scales as androecial remnants, especially in light of observations by Endress (1987a) that they are epigynous and lie in the anterior and lateral positions, as in the stamens of *Sarcandra* and *Chloranthus*, but not in the posterior position. Endress (1987a) also noted strong early developmental similarities of these structures to the stamens of *Hedyosmum*, as shown in his Figs. 164 and 171. In defense of the hypothesis for a perianthial origin of these scales, Endress (1987a, p. 206) pointed to their vascularization and inferred that they represented the retention or the reacquisition of what he considered to be the primitive character-states for the family.

Flowers of the family are bisexual in the case of *Sarcandra* and *Chloranthus*, generally unisexual in *Ascarina*, always so in *Hedyosmum*, and probably so in *Ascarinopsis*. The androecium consists of a single stamen or, occasionally, two, three, or sometimes five in *Ascarina*. In *Sarcandra* and *Chloranthus*, the stamen shows various degrees of inflation and occurs attached about midway up on the dorsal side of the ovary wall (Figs. 8.2a,b and Fig. 8.3). Anthers are quadrilocular except for the unusual, trimerous, eight-locular anthers of *Chloranthus*, which will be treated further in the discussion that follows. The gynoecium consists of a single ascidiate, styleless carpel containing a solitary, orthotropous, bitegmic ovule that is pendant from the roof of the locule (Edwards, 1920; Swamy, 1953d; Burger, 1977; Cronquist, 1981; Endress, 1986a, 1987a; Taylor, 1991a; Taylor and Hickey, 1990a).

Accepting the ground plan of the chloranthoid flower as basal, we developed a search image for characters of the primitive angiosperm. Thus, we inferred that the prototypical angiosperm inflorescence and flower had the following set of characters: The inflorescence was a branched, indeterminate thryse made up of racemes and sometimes spikes attached basally in an opposite-decussate arrangement, becoming spiral distally. Flowers were arranged in decussately opposite pairs, possibly becoming spiral distally. They were bisexual, small, and simple, with bilateral symmetry, lacking a perianth but with a subtending bract and decussately opposite bracteoles, followed by from one to three stamens

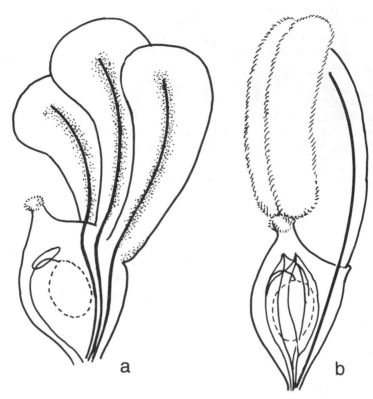

Figure 8.3.   Flowers of *Chloranthus Oldhamii* (**a**) and *C. multistachys* (**b**) showing how the attachment of the stamen to the carpel wall can be explained as a case of adnation. Note especially the origin of the separate vascular traces from the base of the flower. (Redrawn after Swamy, 1953).

arranged in an opposite-decussate fashion in the dorsal and lateral planes, and with a single, terminal carpel. The stamens had anthers with four locules and two thecae, and a filament that projected beyond the anther attachment, making an apiculate tip. The carpel was ascidiate. It contained a single, somewhat laterally attached, orthotropous, bitegmic, crassinucellate ovule (Crane, 1985; Doyle and Donoghue, 1986a). The ovule also underwent onagroid embryogenesis (Loconte and Stevenson, 1991). The seeds had two cotyledons, double fertilization, and endosperm.

If, as we infer, the flowers of the Chloranthaceae closely represent the basic groundplan of the ancestral angiosperm flower, this pattern seems to have been lost rapidly in the other eoangiosperms. Thus, in the Saururaceae, only *Saururus* (Fig. 8.4a) has apocarpous carpels. Nevertheless, the family as a whole possesses bitegmic, orthotropous ovules (Cronquist, 1981; Taylor, 1991a) and

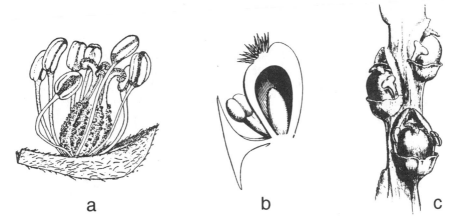

a   b   c

Figure 8.4.   Some other eoangiosperm flowers: (**a**) *Saururus cernuus* (from Raju, 1961, by permission), (**b**) *Peperomia blanda*, and (**c**) *Piper nigrum* (from Engler, 1894).

retains a tendency for the stamens to appear in pairs on the opposite side of the gynoecium, rather than in spirals (Tucker, 1975; Tucker and Douglas, Chapter 7). The closely related Piperaceae (Fig. 8.4b,c) are the only eoangiosperms to possess erect, orthotropous ovules. Among the Piperaceae, *Piper* appears to have become syncarpous, whereas *Peperomia* is apocarpous with a unitegmic ovule (Burger, 1977; Cronquist, 1981; Taylor, 1991a; Taylor and Hickey, Chapter 9). In the Aristolochiaceae, only *Saruma* retains monosulcate pollen, four to six carpels that are distinct above the base, and a whorled arrangement of its perianth and stamens, but its ovules are anatropous (Burger, 1977; Cronquist, 1981). Finally, among the Nymphaeales, although Barclayaceae and the unitegmic Ceratophyllaceae retain orthotropous ovules and simple flowers, which in the case of Ceratophyllaceae are probably drastically reduced, the rest of the order has anatropous ovules and Nymphaeaceae have larger and considerably more complex flowers (Cronquist, 1981; Taylor, 1991a).

Returning to the chloranthoid flower, the apparent simplicity of these "ugly things" (Verdecourt, 1985) poses a considerable challenge to the study of angiosperm evolution because they do not fit the image of the ancestral flower. However we, like Endress (1987a), feel that they have considerable potential to produce insights into early angiosperm evolution, especially when evaluated in light of new fossil material and modern methods of phylogenetic analysis. We will, therefore, be developing further aspects of the chloranthoid inflorescence and flower in later sections of this chapter, but first we will turn to an examination of the inflorescences and "flowers" of the gnetopsids and related gymnospermous lines.

THE GNETOPSIDS

The Class Gnetopsida is one of the most problematical and controversial of any group of living plants. It shares many undoubted characters with the gymnosperms but has no obvious close relatives among them. However, gnetopsids also possess a substantial number of characters that are usually thought of as belonging to the angiosperms. Because of this apparently transitional status, they have been proposed as a separate class, the Chlamydospermopsida, meaning "vestured seed," in reference to the outer integuments or integumentary analogs that surround their ovules (Pulle, 1938) (see Fig. 8.5).

The gnetopsids comprise three separate genera, each of which belongs to its own family and order in the classification used here. Although it has no more than 70 species, no other group of gymnosperms contains such a variety of habits and statures (Pearson, 1929; Chamberlain, 1935; Bierhorst, 1971; Martens, 1971; Gifford and Foster, 1989; Kubitzki, 1990). The genus *Ephedra* consists of about 35 species of shrubs, sometimes weakly rhizomatous, and a few small trees that grow in warm, xeric regions, dry mountains, and rocky areas on every continent except Australia and Antarctica. *Gnetum* is mostly a liana, although the best known species, *G. gnemon*, is a small tree. It is found in tropical rain forests in northern South America, west Africa, Indo-Malaysia, and several of the Pacific Islands north of Australia. The third genus, *Welwitschia*, consists of only one species, found in the Namib Desert of southwest Africa, and looks like a giant woody tuber (Gifford and Foster, 1989). Among the three genera, there is a general consensus that *Ephedra* is isolated from the rest and belongs at the base of the lineage and that *Gnetum* and *Welwitschia* are advanced (Fig. 8.6) (Crane, 1988a; Doyle and Donoghue 1992, 1993; Hasebe et al., 1992a; Chase et al., 1993).

Despite the diversity in vegetative habit displayed by the gnetopsids, a few generalizations can be made. One of the most conspicuous features of the group is the decussately opposite to whorled arrangement of the leaves and other appendages in all genera. The leaves have sheathing bases and exhibit a great variety of form, from scale-like to needle-like with from two to three veins in *Ephedra*, to broad and strap-like in *Welwitschia*, and broad and petiolate in *Gnetum*. *Welwitschia* has parallelodromous primary venation with chevron cross-veins whereas *Gnetum* has pinnate leaves with sheathing bases, brochidodromous secondary veins, and reticulate intercostal venation that closely resemble those of angiosperms. Vegetative similarities to the angiosperms include two cotyledons in the embryo; syndetocheilic stomata in *Gnetum* and *Welwitschia* (characters shared with the bennettitaleans); an apical meristem with a tunica and corpus, as in angiosperms (Gifford, 1943; Esau, 1965; Martens, 1971; Bold, 1973; Gifford and Foster, 1989; Kubitzki, 1990); vessels,

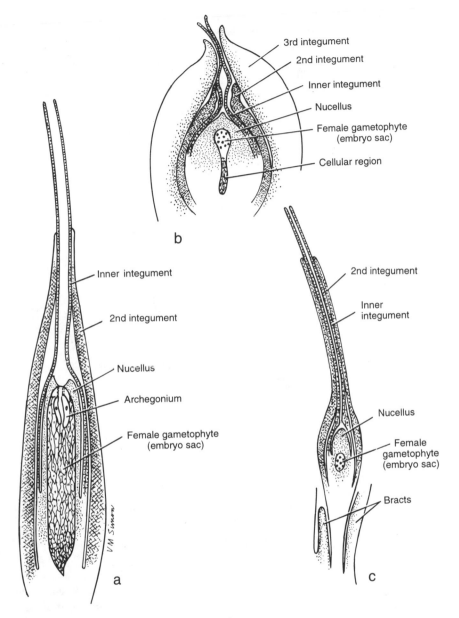

Figure 8.5. Ovules of *Ephedra* (**a**), *Gnetum* (**b**), and *Welwitschia* (**c**) showing the nucellus, the inner integument with its micropylar tube, the outer integumentary envelopes, and the female gametophyte. Homologous structures have the same stipple pattern. (**a**) after Chamberlain, 1935; (**b**) after Martens, 1971.

Female Strobili

Male Strobili

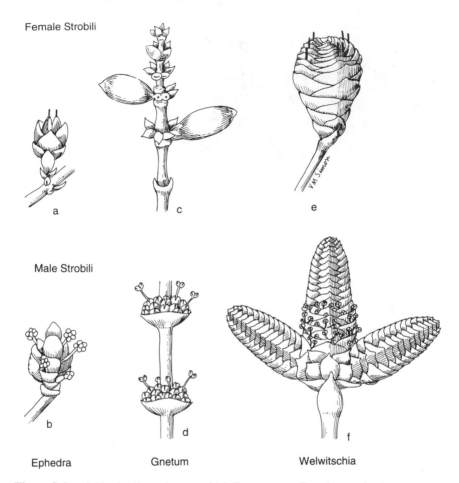

Ephedra   Gnetum   Welwitschia

Figure 8.6.   Habit sketches of gnetopsid inflorescences. Females are in the upper row and males in the lower: (**a, b**) *Ephedra nevadensis,* (**c**) *Gnetum gnemon,* (**d**) *G. africanum,* and (**e, f**) *Welwitschia mirabilis.* Scale is approximately x1.3 for *Ephedra* and x1.0 for the others. Redrawn after Gifford and Foster, 1989 (**a, b, c**), Kubitski, 1990 (**d**), and Chamberlain, 1935 (**e, f**).

although with foraminate pores; astrosclerids and silica bodies in the leaves; lignin with the Maule reaction; a weak development of the rhizomatous habit in *Ephedra* (Martens, 1971); and small asymmetrical chromosomes (Ehrendorfer, 1976). A fuller discussion of the vegetative similarities between the gnetopsids and angiosperms will be delayed until a later section of this chapter.

In the treatment of the fertile organs of the gnetopsids that follows, even the choice of terms to describe them reflects the controversial status of the group and a lack of precision in defining equivalent units. Two of the alternative systems are based either on a conifer-oriented terminology (Chamberlain, 1935; Eames, 1952) with its "strobilus" or "cone" or an angiosperm-oriented system employing terms such as "inflorescence" and "flower" (Crane, 1985). A third set provides a noncommittal but cumbersome and imprecise nomenclature with terms such as "primary" or "secondary reproductive unit". Nevertheless, if we are to discern homologies in the inflorescences of the gnetopsids, it is necessary to specify the precise position of what are relatively simple organs in a complex reiterative hierarchy. The current systems fail to do so and are often highly equivocal. A good example of this is seen in Fig. 8.6 where the imbricately bracteate and, as we shall see, patently nonequivalent inflorescence units of *Ephedra* and *Welwitschia* as well as the axillary inflorecences of *Gnetum* have all been subsumed under the term "strobilus". Therefore, despite our genuine reluctance to proliferate an already abundant terminology, we now do so, in order to introduce an element of rigor and precision into the analysis that is essential to understanding the evolutionary context in which we infer that the gnetopsids have played so important a role.

Some Definitions

Because the detailed analysis of fertile morphology that follows both introduces some new terms and depends on the precise usage of some existing terms that may be unfamiliar to the reader, we offer the following list of terms and definitions. This list, together with our figures and Table 8.1, will serve as a ready reference when they appear in the text. Our use of the terms is modified after Weberling (1989), except for those marked with an asterisk, which represent either new terms or new usages. Our modification of Weberling's system involves the understanding that his terminology applies to the configuration of the fertile system whether consisting of flowers or not. We have thus added the stipulation "with fertile organs" to many of his definitions:

General terms:
  Fertile Truss*—the stereotypic configuration of axes, scales, bracts, perianth, leaves, and sporangia that together form the smallest integral, reiterative reproductive unit of the plant and that is attached to its vegetative system (Figs. 8.7 and 8.18). Its organization may be as follows:
    simple—with its sporangia borne directly on the fertile axis or scales attached to it, (Fig. 8.1a,b)
    compound—composed of a fertile axis bearing secondary axes that carry the sporangia (Fig. 8.7a).
  Fundamental axis*—the highest-order axis to which the reproductive organs or their axes are attached that can continue indeterminate apical growth and that produces normal foliage leaves or phaerophylls at some point along its length (Fig. 8.15).

Phaerophyll—an unmodified foliage leaf that subtends an axillary shoot (Fig. 8.15).

Proanthophyll*—the bract or prophyll that subtends a primary anthion or its homolog (Fig. 8.7).

Prophyll—a modified leaf that subtends a fertile axis (Fig. 8.15).

Some inflorescence terms:

Determinate—main axis of the inflorescence ending in a flower or fertile organ.

Indeterminate—main axis of the inflorescence produces only lateral flowers or fertile organs (Fig. 8.7a).

Inflorescence axis*—as used here, any of several orders of axes lying between the base of the primary anthion and the fundamental axis and which does not produce vegetative leaves (phaerophylls) anywhere along its length (Fig. 8.15).

Raceme—a simple, indeterminate inflorescence that bears stalked flowers or fertile structures in the axils of its bracts.

Spike—a simple, indeterminate inflorescence that bears sessile flowers or fertile structures in the axils of its bracts (Fig. 8.4c).

Homothetic—a compound inflorescence consisting only of lateral simple or partial inflorescences without the main axis itself ending in a simple inflorescence (Fig 8.7a).

Heterothetic—a compound inflorescence consisting of both terminal and lateral simple or partial inflorescences (Fig. 8.15).

Panicle—an indeterminate compound inflorescence where the main axis appears to be terminated by a flower or fertile structure that is actually pseudoterminal (Fig. 8.6e).

Thryse—An inflorescence with cymose partial inflorescences. Cymose refers to the property of branching exclusively from the axils of the bracts or prophylls (Fig. 8.7).

Terms applying to the anthion and its derivatives:

Anthion* (Gr. *anthion*, n. dim. floret)—a homothetic or heterothetic, doubly compound raceme extending from its distal megasporangiate organs to its junction with the axis from which it originates. The axis from which the anthion originates is the fundamental axis or an inflorescence axis, in the case of the primary anthion, or the anthial axis, in the case of second-order anthions (Fig. 8.7b).

Anthial axis*—the main (primary) axis of the anthion to which all of its appendages and subunits are attached (Fig. 8.7).

Anthial bracts*—The bracts attached to the anthial axis. These can be *sterile* if they do not subtend a fertile, higher-order axis; *andranthial* if the higher order axis they subtend is male; or *gynanthial* if the subtended axis is female.

Paraclades—The lateral (secondary) axes of the anthion that serve as the attachment for the *paracladal bracts* and the tertiary axes of the anthion (Fig. 8.7a). Paraclades repeat only a portion of the anthial structure and are divided into the following groups:

Androcladal axes*—paraclades to which the microsporangial axes or microsporangiophores as well as the *androcladal bracts* are attached.

Gynocladal axes*—paraclades to which the megasporangial or ovular axes as well as the *gynocladal bracts** are attached.

THE INFLORESCENCE OF THE GNETOPSIDS

In building the picture of the basic floral morphology of the gnetopsids, we will use the inflorescence of *Ephedra*, as that is the most basal genus of the class and its inflorescence appears to have undergone the least modification in comparison to that of the cordaites, which has been a plausible outgroup for the gnetopsids since Bertrand's suggestion in 1898 and detailed comparisons by Florin (1950) and Eames (1952). Later, we shall discuss the transitional states of the fertile structures seen in the Bennettitales which, from our analysis (below) and those of others (Doyle and Donoghue, 1986a, 1986b; Crane, 1988a), appears to be interpolated between the gnetopsids and the cordaites.

Despite its apparent complexity, the inflorescence of *Ephedra* is built of alternating reiterations of indeterminate, terete units—the axes—that arise in the axils of determinate, bifacial structures—the bracts (Figs. 8.6 and 8.7). We shall call the fundamental reproductive unit of the *Ephedra* inflorescence the *anthion,* which is the approximate equivalent of the term *strobilus* as used for *Ephedra* by Pearson (1929) and Chamberlain (1935). The primitive anthion is a bisexual, indeterminate, homothetic, doubly compound raceme that includes an axis, sterile bracts, fertile bracts and associated lateral microsporangiate and megasporangiate axes. All of these lateral appendages are attached in a decussately opposite, or rarely whorled, arrangement. The *anthial axis* of first-order anthions arises from the fundamental axis or from an inflorescence axis in the axil of the *proanthophyll* and terminates at the apex of the anthion. In *Ephedra* second-order anthions can arise from the first-order anthial axis (Fig. 8.7a). The organs of an anthion are organized in a definite sequence distally, with *sterile anthial bracts* occurring proximally on the anthial axis, followed by fertile anthial bracts from whose axils arise the *androcladal axes,* which, in turn, have paired androcladal bracts proximally and a pair of microsproangiophores distally (Fig. 8.8). Finally, the one or two pairs of fertile anthial bracts at the distal end of the anthial axis subtend *gynocladal axes* that bear the *gynocladal bracts* on their lower part and, at their distal end, ovular axes with terminal ovules and their subtending ovular bracts (Fig. 8.7 and Table 8.1). This aggregation of anthions and their fundamental branches has been called the inflorescence, but this term should not be used when referring to a single anthion, as has frequently been the case when the anthions of *Ephedra* have been discussed in the literature.

The characterization of the anthion as a bisexual structure disagrees with the frequent statement that the gnetopsids have monoecious inflorescences (Cronquist, 1988). In *Ephedra*, the anthions of most species are monoecious, but

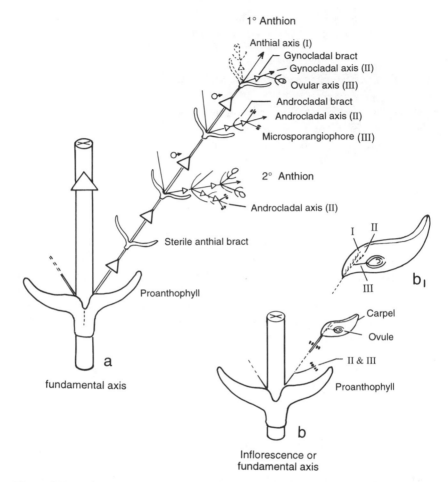

Figure 8.7.   Lateral plan-view of an idealized primitive anthion of *Ephedra* (**a**) compared with that of an idealized primitive anthion of the an eoangiosperm (**b**). Nodes that emerge in the vertical plane are shown as open triangles and only the right-hand anthions are shown. Arrowheads indicate indeterminate axes. Homologous axes are labeled as (I) anthial axis, (II) paracladal (gynocladal or androcladal) axes, and (III) ovular or microsporangiophore axes. Note the secondary anthion five nodes below the top of the primary anthion of *Ephedra* in **a**. Fusion of the anthial and gynocladal axes to the tissue of the carpel is indicated by the cross-hatching in inset **b**₁. Dashed lines in **a** indicate failure of a structure to develop. For further explanation see Figs. 8.8 and 8.18 and Table 8.1.

functionally bisexual anthions have been reported in at least five species (Pearson, 1929; Mehra, 1950). Furthermore, both *Gnetum* and *Welwitschia* exhibit the condition known as pseudomonoecy, where nonfunctional ovules are present in otherwise male anthions. This persistence of bisexuality in the Gnetopsids has led many authors to conclude that it represents the ancestral condition for the class (Chamberlain, 1935; Martens, 1971; Gifford and Foster, 1989). All Gnetopsids have an inner integument that is prolonged and exerted as a so-called micropylar tube (Fig. 8.5), a structure that they share with the Bennettitales. An additional and highly controversial feature is a second envelope surrounding the ovule that has been equated with the outer integument of the angiosperms (Martens, 1971; Gifford and Foster, 1989) and which is the source of the alternative name, chlamydosperms, for this group.

The androcladal axes arise from the middle nodes in the anthion of *Ephedra*. In most species, the androcladal axis appears to consist of an axis or filament that bears from eight to as few as two microsporangia on its distal margin, subtended by a pair or a whorl of small bracts. However, Eames (1952) showed that in primitive species of *Ephedra*, the microsporangiate structure is considerably more complex. Here it consists of a second-order axis (the androcladal axis), arising in the axil of a bract (andranthial bract) on the anthial axis. A pair of filaments, with four distally attached microsporangia, is inserted just below the apex of the secondary (androcladal) axis. In more advanced species of *Ephedra*, these separate microsporangiophores fuse over the apex of the androcladal axis to assume a pseudoterminal position and undergo a reduction in the number of their spore sacs (Fig. 8.9a–d; Table 8.1). Thus, under the Eames hypothesis, the microsporangiophores within *Ephedra* would not be a strictly homologous structure because the single terminal microsporangiophore of the advanced species would actually represent a pair of originally lateral microsporangiophores and the axis to which they were attached.

Continuing distally, the main or anthial axis in bisexual and female anthions is terminated either by a pair of ovules on very short axes, with the abortive apex of the anthial axis lying between (Fig. 8.14), or by a single ovule in the pseudoterminal position. Most often, the lateral ovules are in a single pair, oriented in the lateral plane, as reported by Eames (1952) and illustrated in Fig. 8.9e–i. Sometimes, however, this pair will lie in the medial plane or there will be a whorl of three ovulate structures (Mehra, 1950). The short axis bearing the ovule is termed the gynocladal axis, and this arises from the anthial axis in the axil of the gynocladal bract. We infer that, by comparison to the androcladal units of the anthion, the gynocladal axis bears the ovules laterally on what were once ovular axes that, in turn, bore ovular bracts (Table 8.1). This is effectively the reconstruction envisioned by Eames (1952) in our Fig. 8.9e, except that the axes terminated by the ovules (the ovular axes in the present analysis) would have had at least one pair of ovular bracts attached to it.

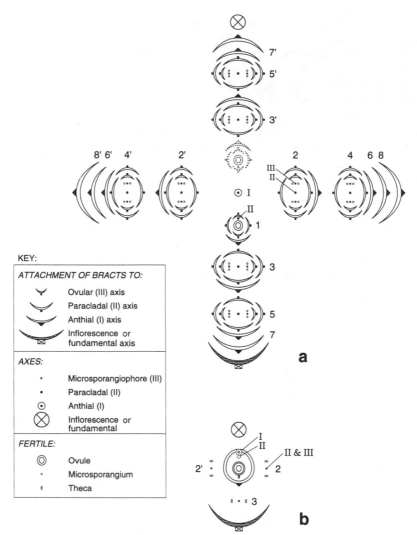

Figure 8.8. Floral diagrams comparing an idealized primary anthion of *Ephedra* (**a**) with that of an idealized primitive eoangiosperm (**b**). Nodes are indicated with arabic numbers, starting from the apex of the anthion and with the ventral, or left member of the pair designated with a prime mark (^{1}). Roman numbers indicate the order of axes in nodes 1 and 2. The ovular axis (III) does not appear in node 1 because it is beneath the ovule itself. The axis to which each bract is attached is indicated by the symbol on its keel. Dashed structures in **a** indicate failure to develop, whereas the open, dashed circles in **b** indicate the fusion of axes I and II to the inner wall of the carpel. Finally, in **b** the position of the ventral suture is indicated even though it occurs at a higher level than the plane of this cross section. See Figs. 8.7 and 8.18 and Table 8.1 for further explanation.

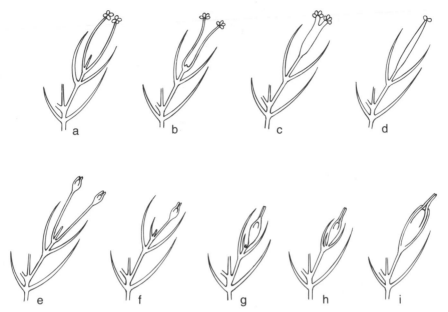

Figure 8.9.   Diagrams of the higher-order inflorescences (paraclades) of *Ephedra* showing the stages of evolutionary modification proposed by Eames (1952): (**a–d**), modification of the microsporangiophores to form a single, terminal structure; (**e–i**), formation of a single, terminal ovule through the loss of one of the ovule pairs and the shifting of the remaining one to a terminal position. The anthial axis is vertical. In these diagrams, all appendages are shown in the medial plane. (Redrafted from Eames, 1952.)

The ovule in the gnetopsids is subtended by from one to three pairs of bracts or, in the case of *Gnetum* and *Welwitschia*, ring-like structures that have been equated with the outer integument of the angiosperms. In our model, these bracts are of mixed origin, with the inner two arising from the ovular axis, and the outer one from the gynocladal axis in a pseudoterminal position. The gynocladal axis arises, in turn, from the anthial axis and is subtended by a gynanthial bract (see Figs. 8.7a and 8.8a and Table 8.1). An important characteristic of the primitive gnetopsid anthion is that primary (anthial) and secondary (paracladal) axes were indeterminate with at least the theoretical capability for continued apical growth, unless an ovule or a fused pair of microsporangiophores has assumed a pseudoterminal position on the axis (Fig. 8.9e–i).

Inflorescence Structure

The configuration of the inflorescences of *Gnetum* and *Welwitschia* appears to be a modification of the primitive anthion and inflorescence structure seen

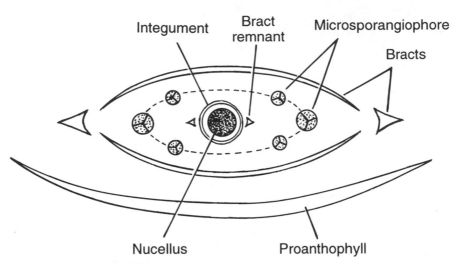

Figure 8.10.   Floral diagram of a pseudomonoecious male anthion of *Welwitschia*. The nonfunctional ovule in the center is surrounded by a whorl of trisporangiate microsporangiophores. Note, however, that the pair in the lateralmost position are slightly larger than the pair that lies medially from them. (Redrawn from Chamberlain, 1935, Fig. 371 with modifications from Bierhorst, 1971).

in *Ephedra* (Table 8.1). However, in both of these genera, a set of axes that lack normal vegetative leaves appears to have been interposed between the base of the anthion and the fundamental axis. Following the general system of Weberling (1989), we shall call these the *inflorescence axes*. In *Gnetum*, the inflorescence axes are indeterminate and occur in up to two orders. The so-called strobilus of *Gnetum* (Fig. 8.6c–d) is inferred to be built of either female or pseudomonoecious male anthions that have become condensed into a whorl and appressed against the indeterminate inflorescence axis at each of its nodes (Bierhorst, 1971). Secondary inflorescence axes that repeat this pattern of anthial attachment may arise at these nodes as well. In *Gnetum*, the anthion is inferred to have been reduced to a simple, determinate, unifacial panicle—a *drepanium* in Weberling's terminology—where all lateral appendages are dorsally placed and the ventral side has become fused to the inflorescence axis (Fig. 8.18g). *Gnetum* anthions appear to lack the sterile and andranthial types of anthial bracts, but the male axes are elongate and have one pair of bracts, inferred to be androcladal bracts, surrounding each of the bisporangiate micro-sporangiophores. In most species, male anthions are terminated by a single abortive ovule, sometimes accompanied by the vestiges of what may be anthial bracts (Table 8.1). Female anthions are characterized by a single ovule that appears to have assumed a pseudoterminal position, with three fully enclosing envelopes. The middle one of these is inferred to result from the fusion of a pair of ovular bracts or scales, whereas the outer one may be homologous with the gynanthial bracts, based on the presence of an abortive node in the vasculature above them.

In *Welwitschia*, the inflorescence axes are determinate and occur in up to three orders to form a determinate, heterothetic, triply compound thyrse. The inflorescence axes have an opposite-decussate arrangement and each is terminated by the precocious development of "strobili" (Fig. 8.6e,f). The "strobili" in *Welwitschia* are inferred to be composite structures of either female or pseudomonoecious male anthions, each subtended by a proanthophyll and arranged in whorls or decussately opposite pairs on an inflorescence axis. The anthions are determinate, simple panicles that are either female or pseudomonoecious male with an abortive ovule. In the male anthions, a whorl of six, basally connected, trilocular microsporangiophores occur around an abortive ovule. However, the two microsporangiophores occurring at 9 and 3 o'clock are somewhat larger than the pairs occurring at 10:30 and 7:30 and at 4:30 and 1:30 respectively, indicating that what appears to be a single whorl may be a composite structure (Fig. 8.10). The whorl of sporangiophores is subtended by four bracts in the lateral and medial positions that are inferred to represent sterile anthial bracts. Andranthial bracts appear to be missing in *Welwitschia* as they are in *Gnetum* and in the angiosperms (Table 8.1). Distal to the whorl of sporangiophores is a single, abortive, pseudoterminal ovule with a single integument. The female anthions are similar, yet they lack the whorl of sporangiophores and their ovules have two integuments.

The presumed homologies from the preceding analysis are outlined in Table 8.1. Summarizing these results, several evolutionary trends can now be inferred within the gnetopsids. The first of these is the reduction of the anthion from an indeterminate, doubly compound raceme in *Ephedra* to a determinate, simple panicle in the two advanced genera. This anthial reduction is accompanied by the loss of the paracladal bracts and the reduction of the anthial bracts as well. The second is the insertion of a system of inflorescence axes into the fertile system that has the effect of "forcing" the anthial units distally out on to more ephemeral and easily replaced branches of the plant. These two apparently contradictory trends toward the elaboration of the inflorescence accompanied by the reduction of its anthial units may be related to optimizing the number of microsporangia and ovules. Thus, as the male and female paraclades within the anthion became reduced, the number of microsporangia and ovules in the

Table 8.1. Comparison of the Inferred Homologies for Each of the Systems of the Inflorescence of the Gnetopsids and of the Angiosperms

Systems	Taxon				
	Gnetopsids			Angiosperms	
	Ephedra	*Welwitschia*	*Gnetum*	*Eoangiosperms*	*Higher angiosperms*
Fundamental	Axis		Axis	Axis	Axis [receptacle]
	Proanthophyll		—	[Proanthophyll]	[Perianth]
	—	Phaerophyll	Phaerophyll	Phaerophyll	—
Inflorescence	None	1–2 orders of axes	0–1 orders of axes	0–3 orders of axes	Receptacle or 1–2 orders of axes
		Proanthophylls	Proanthophylls	Proanthophylls	
		Bracts	Bracts	Bracts	Bracts/[perianth]
Anthial					
Primary	Anthial axis	Anthial axis	Anthial axis	*Anthial (floral)* axis & part of "chair"	Part of "chair"
	Anthial bracts:	*Anthial bracts:*	*Anthial bracts:*	*Anthial bracts:*	*Anthial bracts:*
	Sterile	Sterile	—	—	—
	Andranthial	—	—	—	—
	Gynanthial	Gynanthial	3rd integument	Foliar part of carpel	Foliar part of carpel
Secondary	*Paracladal axes:*	—	*Paracladal axes:*	*Paracladal axes:*	*Paracladal axes:*
	Androcladal	—	—	Part of filament & apical tip	Part of filament & apical tip
	Gynocladal	—	*Gynocladal*	Placenta & part of "chair"	Placenta & part of "chair"

Table 8.1. (*cont.*)

Systems	Taxon				
	Gnetopsidis			**Angiosperms**	
	Ephedra	*Welwitschia*	*Gnetum*	*Eoangiosperms*	*Higher angiosperm*
	Paracladal bracts: Sterile Androcladal Gynocladal	— — —	— — —	—	— — —
Tertiary	Axes: Microsporangiophore axis Ovular axis Bracts: Ovular = 2nd integument	Axes: Microsporangiophore axis — Bracts: Ovular = 2nd integument	Axes: microsporangiophore axis — Bracts: Ovular = 2nd integument	Axes: Part of filament and apical tip Funiculus Bracts: Ovular = 2nd integument	Axes: Part of filament and apical tip Funiculus Bracts: Ovular = 2nd integument
Quaternary	Microsporangia Ovule	Microsporangia Ovule	Microsporangia Ovule	Anther theca Ovule	Anther theca Ovule

Vestigal organs are indicated by italics, whereas a square bracket indicates that an organ may not always be present.

inflorescence would have been reduced. The number could have been recovered by increasing the number of branches in the inflorescence, thus increasing the total number of anthions. The third trend is the condensation and fusion of the anthions into larger units that produces the "strobili" of *Gnetum* and *Welwitschia* (Fig. 8.6). In these genera, the number of ovular paraclades has been reduced to one and this has assumed a terminal position in the anthions. Finally, we note that the more or less terete anthion of *Ephedra* has become dorsiventrally flattened in both *Gnetum* and *Welwitschia* as a result of its appression against an inflorescence axis.

Rules of Organization for the Gnetopsid Flower

Before going on to review the inflorescence structure of other groups of seed plants, including the eoangiosperms, we will summarize some of the empirical rules by which the fertile structures of the gnetopsids seem to be organized:

1. Inflorescence and anthion structure is primitively indeterminate and apical in development.
2. Fertile organs are all appendicular, that is, attached to an appendage of the anthial axis, rather than cauline, or attached to the axis itself (cf. Eames, 1952).
3. The gnetopsid inflorescence is made of repeating units, the anthions, and these can be either compound or simple.
4. The type of organ that develops in the anthion is related to its position on the anthial axis. The sequence proceeds distally from sterile appendages (bracts), to fertile appendages subtending male (androcladal) axes to ovular (gynocladal) axes, in the most distal position. Both sets of fertile paracladal axes are subtended by anthial bracts (see Table 8.1 and Figs. 8.7a and 8.8a).
5. Insertion of all appendages of the inflorescence and the anthion is decussately opposite with transitions to the whorled condition.
6. In the medial plane, development on the anterior (12 o'clock) side of the anthial axis is often inhibited toward its apex, especially when it undergoes condensation (foreshortening). Conversely, development in the posterior (6 o'clock) position is often favored, possibly as a result of proximity to the vascular trunk of the subtending bract.

THE INFLORESCENCE IN CORDAITES

The final group that we regard as having an important bearing on the question of angiosperm origins is the cordaites. Over 40 years ago, Eames (1952) suggested that the closest affinity of the gnetopsids was with the cordaites, an extinct group lying near the base of the seed plants and believed to be directly ancestral to the conifers as well (Florin, 1950; Stewart and Rothwell, 1993).

Thus, an understanding of the fertile organs of the cordaites could provide important insights for inferring homologies between the fertile organs of gnetopsids and angiosperms.

Like the inflorescence of the gnetopsids, the inflorescence of the cordaites is a compound structure that could be categorized as a homothetic, doubly compound raceme (Weberling, 1989). However, unlike the anthions of the gnetopsids, only two orders of fertile axes are present (Fig. 8.11a) and the structures are unisexual and, thus, outside of the definition of anthion as used in the gnetopsids. The primary fertile axes arise in the axils of the spirally arranged leaves on the main, or fundamental, axis; and the secondary fertile axes are arranged in a four-ranked (distichous) spiral on the primary axis, with each one subtended by a bract (Fig. 8.11b,c). These secondary axes are, in turn, covered by spirally arranged "scales," the distal ones of which bear either microsporangia or ovules on their distal margins. We infer that these secondary axes represent the homologs of the paracladal axes of the gnetopsid anthion, even though their scaly appendages are spirally arranged and do not subtend any higher-order axes (Fig. 8.18). Unfortunately, nothing is known of the positioning of the unisexual inflorescences of the cordaites, even as to whether the two sexes occurred on the same plant. Eames (1952), by analogy to their derivatives, the conifers, argued that the fertile truss of cordaites was bisporangiate, with the megasporangiate structures distal to the microsporangiate ones. We have tentatively accepted this suggestion based on the appearance of this condition in the Bennettitales and the gnetopsids. If true, then the axis of the bisporangiate fertile truss was also an important vegetative axis of the cordaite plant; in other words, the truss occupied an axial position in the vegetative system, rather than being lateral as in the gnetopsids (Fig. 8.18). (Much of this and the discussion that follows is based on material found in the works of Crane [1985, 1988], Rothwell [1986, 1988], Taylor and Taylor [1993], and Stewart and Rothwell [1993].

The fertile structures of the cordaite inflorescence have been considered to be leaves and have thus been termed "microsporophylls" or "megasporophylls." In many cases, these are dorsiventrally flattened structures that are nearly identical to the more proximal sterile scales of the inflorescence (Fig. 8.11b,c) (Stewart and Rothwell, 1993). However, their terete vascular traces and the attachment of sporangia to their distal margins are unusual foliar characteristics. Unfortunately, there has been a strong inclination in comparative morphological studies to categorize vegetative organs as either axes or leaves (Eyde, 1975). In this case, because of the mixed set of characters for these structures, rather than committing ourselves to implicitly recognizing them as leaves, we have chosen the noncommittal term "scale" for these structures in the discussions that follow.

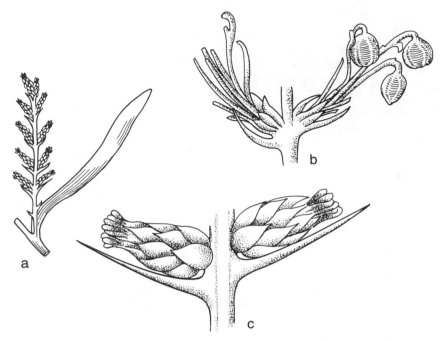

Figure 8.11. Habit and details of the inflorescence of *Cordaitanthus:* (**a**) whole inflorescence in the axil of a leaf on the inferred fundamental axis (extending to the left). The inflorescence is unisexual and has only two orders of axes. (Redrawn from Stewart and Rothwell, 1993.) (**b**) detail of a female inflorescence showing two secondary dichasia that consist of spirally arranged sterile scales, proximally, and fertile scales, distally. Mature ovules are attached to the fertile scales in the left dichasium. (Redrawn from Taylor and Millay, 1979.) (**c**) same view of a male inflorescence showing the distal fertile scales with their distal microsporangia. (Redrawn from Delvoryas, 1953.) (See also Fig. 8.18a for the inferred homologies of the cordaite inflorescence to other gymnosperms and to the angiosperms.)

Species of cordaites bore from four to six microsporangia on the distal margins of their scales (Fig. 8.11c), with the related genus *Mesoxylon* having four (Eames, 1952; Crane, 1985, Gifford and Foster, 1989). Eames homologized these structures directly with the microsporangiophores of *Ephedra* (Fig. 8.12). The megasporangiate scales are much less well known, but generally they bore a single ovule at their tip, although in the case of some species of cordaites, the fertile scale was dichotomous at the tip, appearing indeterminate, and produced two ovules (Fig. 8.11b) (Stewart and Rothwell, 1993).

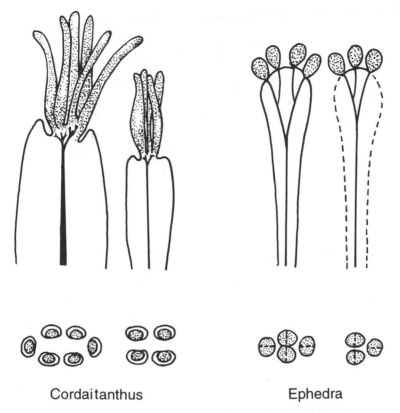

Cordaitanthus   Ephedra

Figure 8.12.   Longitudinal (top) and transverse (bottom) views of the microsporangiate structures of *Cordaitanthus* and *Ephedra* showing the similarities in their anatomy, morphology, and the shape and position of their microsporangia (stippled). (Redrawn after Eames, 1952.)

DISCUSSION

At this point, we will begin an extended heuristic argument for an evolutionary derivation of angiosperm flowers from gnetopsid inflorescences, building on the descriptions of the relevant groups given above. We will start by comparing objectively observable features in the inflorescences of the two groups, then move to other characters whose homologies are more inferential, and follow this by trying to determine the transitional stages that led from the inflorescence of the gnetopsids to the angiosperm flower. Once we have built a plausible bridge between what we hypothesize to be the most closely related members of the ancestral and derived taxa, we will attempt to reinforce our assertion of relatedness by a wider survey of relevant characters.

As we start this process, we are mindful of Sigmund Freud's (1967) caution that, "No probability, however seductive, can protect us from error; even if all parts of a problem seem to fit together like the pieces of a jigsaw puzzle, one has to remember that the probable need not necessarily be the truth, and the truth not always probable." Obviously, what we propose is only one of several plausible hypotheses, but one which, nevertheless, has considerable power to organize what has often been regarded as disparate data on angiosperm relationships and evolution.

When the inflorescence of the gnetopsids is compared to that of the eoangiosperms, especially to that of the Chloranthaceae, a number of striking similarities emerge. Some of the most important points involve the number and position of the various inflorescence organs and these are shown in Figs. 8.7 and 8.8. Especially notable is the prominence of the decussately opposite mode of attachment of the inflorescence components in the Chloranthaceae and, as in the gnetopsids, the organization of these components into dichasia that are subtended by bracts. Both taxa are inferred to be primitively bisexual and to have their sterile, male, and female organs organized in the same sequence from the base to the apex of the dichasial subunits of their inflorescences. Their inflorescences can be classified as indeterminate thyrses, consisting of an extremely limited range of organs, with no sign of a perianth to be found in their subunits. A striking difference, however, is the drastic reduction in the complexity and number of components in the chloranthoid dichasia relative to the gnetoid anthion (Figs. 8.7 and 8.8; Table 8.1). Comparing their fertile organs, the tetrasporangiate stamen of most genera of the Chloranthaceae is strikingly similar to the tetrasporangiate microporangiophore found in some of the gnetopsids (Eames, 1952), except that the anthers are paired and each pair often develops a single stomium, or line of dehiscence, in the angiosperms. When we come to the ovule, however, we find that, although it is orthotropous in both cases, it is now enclosed in an apparently new organ, the carpel, rather than occurring on a bracteate axis, as in the gnetopsid anthion.

This is about as far as most observers would be willing to admit that an objective comparison can carry them. For the rest, we must depend on a chain of phylogenetic inference that begins below. In the discussion that follows, we shall be using a method of topological analysis developed by Wilhelm Troll [as explicated by Weberling (1989)] in order to trace homologies in the inflorescences of the angiosperms and their putative sister groups. According to this method, the position of an element within the overall structure of the fertile truss is an important criterion for the determination of phylogenetic relationship (homology). This can only be done if identical structural elements are compared. Without using a topologically based method, it is easy to become confused when comparing the often highly reiterative elements of the inflorescences of various taxa. Our first task, therefore, will be to establish which

Inflorescence
axis

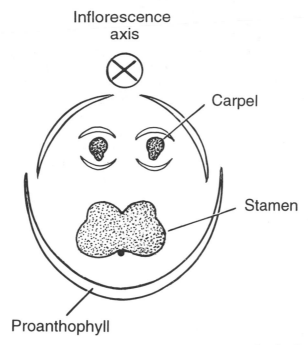

Carpel

Stamen

Proanthophyll

Figure 8.13.   Floral diagram of an unusual flower of *Ascarina lucida*. (Redrawn from Moore, 1977.)

parts of the fertile system correspond to one another within a particular taxon and then between taxa before we will be able to make any assertions about the transitions that occur in the fertile structures from the gymnosperms to the angiosperms (cf. Weberling, 1989, p. 201).

Origin of the Chloranthoid Flower

To begin with the chloranthoid flower, we hypothesize that the ovule and the bract subtending the floral unit are homologous with one of the terminal ovules and the proanthophyll subtending the anthion of the gnetopsids. This would make the floral unit thus delimited an anthion and the axis to which the subtending bract is attached the equivalent of either the fundamental, or the inflorescence axis (Figs. 8.7 and 8.8, Table 8.1). Comparison of the inferred anthions in the two groups indicates that the anthion of the Chloranthaceae has undergone considerable reduction in its number of parts, including the higher-order axes of the anthial system, as well as in its general level of its elaborateness. Even the inferred anthial axis in Chloranthaceae has undergone a drastic foreshortening and really exists only as a vascular trunk from which the other organs arise (Figs. 8.3, 8.7b). In the chloranthaceous anthion, no trace of anthial bracts remains except for the carpel.

An exception to this pattern is seen in the chloranthaceous genus *Ascarina* in which the normal unisexual flowers have a pair of bracts above the one that subtends the flower. This unusual situation is clarified by the study of Moore (1977) who found that in rare bisexual individuals of *Ascarina lucida* what appears to be a flower actually consists of a doubly compound dichasium (Fig. 8.13). Thus, the flowers of *Ascarina* are inferred to represent a condensed doubly or triply compound inflorescence in which an inflorescence bract takes the place of the proanthophyll. The chloranthoid inflorescence recently described from the Aptian of Australia also appears to have bractioles in this position (Taylor and Hickey, 1990a).

Origin of the Outer Integument in Ovules

The number of outer layers enclosing ovules of the gnetopsids varies from one to three (Fig. 8.5). A single layer is found in the nonfunctional ovules of *Welwitschia*, although they each appear to be subtended to a pair of reduced, oppositely arranged bracts. The fertile ovules of *Welwitschia*, the nonfunctional ones of *Gnetum*, and the ovules of all *Ephedra* species have two layers, whereas the fertile ovules of *Gnetum* have three. In *Ephedra*, the outer integument is supplied by two separate vascular bundles, whereas the inner one appears to originate from two primordia. In *Gnetum* and *Welwitschia*, the second envelope appears to arise from an oppositely arranged pair of primordia that fuse and become ring shaped later in development. Based on our interpretations of the anthions in the gnetophytes, the first layer is homologous to the integument found in all other seed plants, whereas the second integument originates from a pair of bracts on the ovular axis (Table 8.1). The third and outermost layer in *Gnetum* appears to originate from a pair of gynanthial bracts subtending the now pseudoterminal gynocladal axis. We hypothesize that although the outer integument of the angiosperm bitegmic ovule has a ring-like origin, it is homologous with the ovular bracts that form the second integument in the gnetopsids.

Origin of the Carpel and Its Features

The anthions of the more primitive species of *Ephedra* have a pair of ovules at their apical end with the abortive apex of the anthial axis lying in between (Martens, 1971; Gifford and Foster, 1989) (Fig. 8.14). These ovules arise in the axil of an anthial bract and are subtended by one or two additional pairs of bracts. Normally, the paired ovules lie in the lateral plane (Eames, 1952), but in rare cases, the pair will be in the medial plane or the ovules will occur in a whorl of three with one in the dorsal position (Mehra, 1950). In derived species

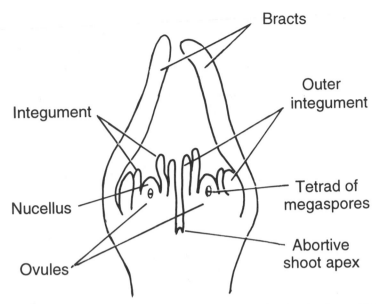

Figure 8.14.   Apex of an anthion of *Ephedra foliata* showing the abortive anthial shoot apex, paired ovules with subtending anthial bracts, and ovules with bracteoles forming their inferred outer integument. (Redrawn after Maheshwari, 1935.)

of *Ephedra*, and in *Gnetum* and *Welwitschia*, one of the paired ovules does not develop and the remaining one thus appears to be terminal (Fig. 8.9e–i) (Eames, 1952). The anthion in these taxa has thus become determinate.

In addition to their foliar portion, bracts in the anthions of the gnetopsids have a characteristic meristematic collar that clasps the axis to which they are attached and that sheathes the base of the axis that they subtend. Another element of importance to the argument we advance here was Taylor's (1991a) inference that the ascidiate, or cup-shaped, carpel is basic in the angiosperms. With this evidence in hand, we infer that the angiosperm carpel arose from the foreshortening and integration of the remains of the anthial axis and the gynocladal axis with the sheathing base of the single, dorsal anthial bract that subtended it. The bract, including the sheathing base, became the ascidiate carpel wall; the gynocladal axis made the placenta; the ovular axis remained as the funiculus; whereas the ovular bracts formed the outer integument of the ovule (Figs. 8.7 and 8.8, Table 8.1). The base-number of ovules in the carpel is thus one, inherited directly from the original medial lateral ovule of the gnetopsid anthion. By the time that the carpel originated, the ventral member of the ovule pair had disappeared and the apex of the anthial axis had become reduced to a vestige, as the remaining ovule assumed a pseudoterminal position.

In this model, the conduplicate portion of the carpel is thus the homolog of the foliar portion of the anthial bract and, like it, it is elaborated later in ontogeny and has its suture oriented ventrally, with regard to both the anthial axis and fundamental axis, as both of these axes lie in the medial plane. It is the retention of an ovule in the posterior position in the medial plane of the anthion that "locks" the suture into a ventral, rather than a lateral orientation, in the angiosperms (Fig. 8.7b,b_1, 8.8b).

This model also suggests that the problematic "chair structure" on the ventral side of the ascidiate portion of the carpel represents some vestige of the potential for apical growth inherited from the apex of the anthial axis, but this is now embedded within the ovary wall. This would also be consistent with the attachment of the ovule to the ventral (anthial) wall of the carpel. However, the gynocladal axis retained the potential for indeterminacy, but this was now enclosed within the carpel (Fig. 8.7b,b_1). Thus, only proliferation of ovules but not of carpels was possible within the confines of the angiosperm anthion. Subsequently, the gynocladal axis became integrated into the foliar portion of the carpel and was able, as the placenta, to proliferate ovules on the adaxial surface of the carpel inside the suture (Taylor, 1991a). This hypothesis of a bipartite origin for the carpel is compatible with Taylor's hypothesis that angiosperm carpels develop from two growth areas (Taylor, 1991a; Taylor and Kirchner, Chapter 6).

We feel that this model provides the most parsimonious solution to the problem of why the angiosperms, alone of all the seed plants, should have their ovules on the adaxial side of an organ that has been interpreted as a leaf. In this model, the angiosperm carpel is not only a leaf but is a composite structure consisting of the sheathing base and conduplicate, foliar portion of the gynanthial bract, together with the fused and enclosed anthial and gynocladal axes on its ventral margin (Figs. 8.7b and 8.8b).

Origin of the Stamen

We infer that in the precursor to the angiosperms, the paired microsporangiophores became reduced and fused laterally to the androcladal axis. This would have occurred in a fashion parallel to the evolution of the compound microsporangiophores in *Ephedra* (Eames, 1952). As in some species of *Ephedra*, we would expect to see the remains of the androcladal axis protruding beyond the microsporangia as well as separate vascular traces to the apex and pairs of microsporangia. Such a configuration occurs in the Magnoliidae, where a protruding apical tip of the anther connective and a three-trace vascular system is found in the stamens of many species. This apical tip can also undergo some elongation, showing the latent potential for indeterminacy that we infer is inherited from its ancestry as the apex of the androcladal axis. In stamens with three vascular strands, the lateral traces usually go to the pairs of locules while the medial goes to the apical tip. Pairs of microsporangiophores are found in some species of *Ephedra* and *Gnetum*, whereas three are found in *Welwitschia*. In all the gnetopsids, the androcladal axis is terete and can be elongated in *Ephedra* and *Gnetum*. Our analysis thus suggests that the androcladal axis is directly homologous with a part of the filament in the angiosperm stamen and that the potential of the filament for variable length is directly inherited from the gnetopsids.

The standard view of the origin of stamens is that they were derived from leaf-like structures attached to the receptacle (Cronquist, 1988; Takhtajan, 1991). To be sure, the stamens of a number of magnoliid families are dorsoventrally flattened and leaf-like but so also are some gnetopsid microsporangiophores (Martens, 1971). In addition, if the stamen had a foliar origin, the anthers should occur on its abaxial side, whereas, in reality, they occur marginally and adaxially as well as abaxially in magnolids (Cronquist, 1988; Takhtajahn, 1991).

If the carpel marks the original insertion of the gynocladal axis and subtending gynanthial bract in the posterior-medial position within the anthion, then the decussately opposite phyllotaxy of the anthial appendages should result in a pair of stamens in the lateral position at the first node below the carpel (see Fig. 8.8a). However, in the Chloranthaceae, proposed as the most basal family of the angiosperms in the current discussion, we seem to see only a single stamen and this is in the posterior-medial position. This could be explained as resulting from the loss of the uppermost, lateral pair of stamens due to the foreshortening of the anthial axis that placed all of its organs in close proximity to the massive, medial vascular trunk of the subtending bract. Nevertheless, it would be better for our argument if at least one case could be found of a chloranthaceous taxon with a pair of lateral stamens inserted above the medial one.

Evidence that the stamens of primitive angiosperms did, at one time, occur in precisely this orientation comes from several sources: First, a number of species of *Chloranthus* have an anomalous eight-locular, trilobed stamen (Swamy, 1953d). These data led Swamy (1953d) and Endress (1987a) to favor the idea, first advanced by Cordemoy (1863), that the androecium of *Chloranthus* represents a fusion of three stamens rather than the lobation of what was originally a single stamen. In addition, a set of critical SEM photographs of the developing floral apex of *Chloranthus spicatus*, published by Endress (1987a, Figs. 76–86) shows a pair of stamen-primordia arising in the lateral plane and inside of the single posterior-medial primordium that will form the central lobe of the stamen. Such a configuration is predicted by the anthion model (Fig. 8.8) and is strong evidence that the ancestral angiosperm flower had a lateral pair of four-locular stamens above the one in the posterior-medial position. Finally, Friis et al. (1986) have described detached stamens of putatively chloranthoid affinites from beds of Early and Late Cretaceous age. These fossils show two lateral stamens in the process of fusing to a medial one, with the loss of the inner thecae of the laterals by Late Cretaceous time. However, it appears that very quickly in taxa more advanced than the Chloranthaceae, stamens become freed of their constraint toward an opposite-decussate insertion and vascular trunks began to divide to supply separate stamens in a manner that suggests that a considerable rearrangement of the original components of the anthion was already taking place (see Tucker and Douglas, Chapter 7).

The Anthial System and the Fundamental System in the Evolution of the Angiosperm Flower

The potentiality of the anthial axis is an important point for understanding the evolution of derived angiosperm flowers. In this hypothesis, we take the view that the apical meristem of the anthial axis lost function and became effectively determinate. This occurred as the apex became enclosed within the carpel and, although not domed over by carpellary tissue, lost its independence by becoming incorporated into the ventral margin of the collar (Figs. 8.7b, 8.8b). The anthial axis thus became determinate, with the formation of the carpel in a pseudoterminal position on it. Based on this model, the development of an effectively determinate anthial axis would have precluded the evolution of the multiparted, complex flowers of the higher angiosperms or of complex inflorescences by means of the elaboration or terminal growth of the anthial system that the eoangiosperms inherited from their gnetopsid ancestors. We concede the possibility that other models might be developed, but our discussion will be restricted to the determinate hypothesis and its implications for the evolution of various derived flowers.

Basically, the gnetopsid fertile truss that was inherited by the eoangiosperms consists of two different branching systems, a determinate anthial system and an indeterminate inflorescence system (Fig. 8.18d,g,h, Table 8.1). As defined above, the anthial system extends from the pseudoterminal ovule in the anthion to its subtending proanthophyll, whereas the inflorescence axes are interposed between the base of the anthion and the fundamental axis. These are equivalent to the "inflorescence axes" in Weberling's (1989) terminology. In the eoangiosperms, as in the higher gnetopsids, the anthial system appears to be simple and determinate. The hypothesis that the eoangiosperm anthion is determinate narrows the possibilities for the floral evolution of the eoangiosperms and derived groups. In contrast to the anthial system above the level of *Ephedra*, the inflorescence axes of *Gnetum* and the eoangiosperms are indeterminate and, although that of *Welwitschia* is determinate, lateral axes can be elaborated by cymose branching. Both of these branching methods provide the inflorescence system with the possibility of an indefinite replication of their axes and, with them, the determinate anthial units and their subtending bracts. The result is to proliferate the anthions and to carry them out from the vegetative axis of the plant on progressively higher orders of inflorescence branches.

The inflorescence of the Chloranthaceae is inferred to be made up of reduced, determinate anthions, composed of terminal pistils, one or two nodes of stamens, and the proanthophyll. These anthions are, in turn, attached to the fundamental axis or to one of a variously branched set of inflorescence axes. The inflorescence system of the Chloranthaceae is a double thryse (or a double spike) in *Sarcandra, Chloranthus,* and some species of *Hedyosmum,* triple in *Ascarina,* and up to three orders in *Hedyosmum* (in part from Endress, 1987a).

Eoangiosperm families above the Chloranthaceae have flowers with multiple carpels which often show some level of adnation (Cronquist, 1981; Raju, 1961; Tucker and Douglas, Chapter 7). Inflorescences of this group are mostly simplified into racemose or spicate configurations that place the flowers on, or within, one order of the main ascending axis of the plant. We infer that the flowers of this group are pseudanthia consisting of multiple anthions brought together by the reduction of higher-order branches of the inflorescence system. Among this group, *Piper* is presumably syncarpous because it has separate, often three, stigmas that arise from separate primordia. We hypothesize that the flowers of Piperaceae, as well as of *Anemopsis* and *Houttuynia* in the Saururaceae arose through the suppression of the system of inflorescence axes and bracts to bring either a single distal and one more proximal pair of anthions together above the subtending bract of a second-order inflorescence axis (Table 8.1). In the case of *Piper,* the occasional presence of a remnant bract or pair of bracteoles adaxial to the carpel (Fig. 8.4c) (Engler, 1894) argues for the origin of its apparently simple anthion through the condensation of an inflorescence. Using the developmental data from Tucker and Douglas (Chapter 7), we infer that the four-carpellate flowers of *Saururus* and *Gymnotheca* of the Saururaceae are the result of the reduction of an inflorescence axis to a penultimate and ultimate pair of anthions.

All of the eoangiosperm families above the Chloranthaceae display an increasing decoupling of the phyllotactic constraints present in the original anthion. Thus, the stamens are out of line with regard to carpel position in many genera (cf. Burger, 1977; Tucker and Douglas, Chapter 7); the higher-order inflorescence bracts disappear with the condensation of the inflorescence axes. Even the proanthophylls tend to be lost as the remnant of the anthial axis disappears and its remaining organs become fused directly to the fundamental axis, or to the inflorescence axes. This process is accompanied by the loss of the decussately opposite phyllotaxy in leaves and axes as well—a trend first seen in the distal portions of the inflorescence in Chloranthaceae (Fig. 8.15).

Our inference that the determinate nature of the angiosperm anthion precludes the apical elaboration of further carpels within its confines has some

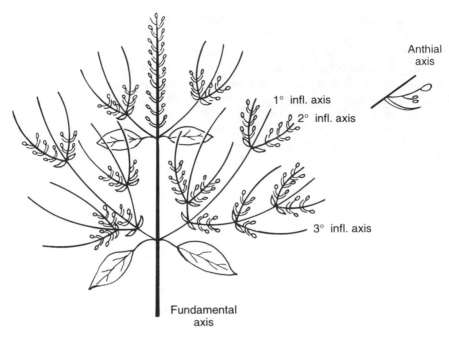

Figure 8.15.   The probable limit in the degree of elaboration in the inflorescence system of the eoangiosperms based on the inflorescence of *Hedyosum*. Note the three orders of inflorescence axes in addition to the fundamental axis, which is terminated by a first-order inflorescence. Insertion of anthions becomes spiral above the base where they are decussately opposite. Many anthions were removed to simplify this drawing.

important implications for the evolution of the angiosperm flower. There are only a limited number of possibilities for multiplication of floral units within the constraints of the inflorescence system proposed here and all of them seem to have been pursued early in angiosperm evolution. The Chloranthaceae replicated anthions by elaborating the ramifications of their inflorescence axes, ending up with as many as three orders of inflorescence axes in *Hedyosmum* (Endress, 1987a). In Saururaceae and Piperaceae, the reduction of the system of inflorescence axes finally brought their flowers on to the main axis of the plant or one of its first-order branches. Both of these trends took advantage of the capacity for indeterminate apical growth and anthion replication inherent in the inflorescence and fundamental axes and overcame the limitations of axial attenuation that constrained the number of axis orders in the cymose inflorescences of the Chloranthaceae. In that family, however, the receptacle of the flower is the anthial axis, which remained limited to producing a single carpel. The next step in floral evolution within the eoangiosperms involved the disappearance of the anthial axis altogether and the transference of the remaining complement of anthially attached organs (carpels, stamens) to either the fundamental axis or to the distalmost inflorescence axis that retained indeterminate growth. This axis now became the receptacle for all angiosperm flowers above the level of the eoangiosperms.

Now, for the first time, the angiosperm flower possessed radial symmetry. Because of the apical determinism inherent in the anthion, this transitional structure was soon destroyed and its organs transferred to the receptacle in which they were arranged in a spiral or in whorls, retaining a remnant of the old opposite-decussate arrangement only at the base, as in the flowers of the Magnoliales (Cronquist, 1981, Endress, 1987a, 1987b). The radial flower built on a receptacle that is the old fundamental axis or one of its derivatives, the inflorescence axis, became the ground plan for the flowers of all higher angiosperms. This new configuration was established very early in the record of angiosperm evolution, reaching some fairly extreme formulations with the late Albian to Cenomanian forms *Lesqueria* (Crane and Dilcher, 1984) and *Archaeanthus* (Dilcher and Crane, 1984).

By the model proposed here, the complex, radial flower is a transformation from, and not directly homologous to, the truly primitive angiosperm flower that was inherited from the gnetopsids. In this interpretation, the perianth of such derived flowers represents sterilized proanthophylls or inflorescence bracts.

THE CHLAMYDOSPERMS AS THE ANCESTORS OF THE ANGIOSPERMS

Although we have extensively discussed the homologies between angiosperms and gnetophytes, especially in relation to their reproductive parts, we are also interested in the homologies and transformations in state for other characters. In addition, much of our argument for the evolution of the angiosperm flower rests on the validity of our inference of the relationship of flowering plants to the gnetopsids. This necessitiated a phylogenetic analysis of angiosperms and their outgroups. To investigate this relationship, we collected characters from cordaites, bennettitaleans, *Ephedra*, *Gnetum*, and *Welwitschia*, eoangiosperms, and the higher angiosperms (Tables 8.2 and 8.3). Our character coding partially reflects the influence of the anthion hypothesis in the interpretation of structure.

Figures 16 and 17 show a selection of the cladograms resulting from our analysis of 57 characters using the parsimony algorithm PAUP, version 3.0 (Swofford, 1991a) and run on a Macintosh LC II in the exhaustive mode. The single most parsimonious tree (Fig. 16), of 85 steps, resulted from using all unordered characters and placed the angiosperms as the sister group of *Gnetum,* making the gnetopsids a paraphyletic group (cf. also Nixon et al., 1994). The

Table 8.2.   Data Matrix for the Phylogenetic Analysis of Certain Gymnosperms and the Angiosperms.

Data Matrix:

```
                    11111111112222222222233333333334444444444455555555
Taxon               12345678901234567890123456789012345678901234567

ANCES               0000000000000–0000000–0——000000000000–0–000000000000000
CORDA               0000000000010–000110100——000000000000–0–001000000000000
BENNE               0010?00000010–1??1101100——100000102000–1–01100100?0100??
EPHED               1011100010021011111111010000100101110010100110010011000 11
WELWI               1011101110131011111111211100120111100111001111111112111 1
GNETU               1011112111131111111111111100120112210011100111111112121 11 1
EOANG               21122121111412111112111111001111232110111012211111012011 1
HGANG               2112222111141211111211111–1011102322111111202111110120211 1
```

The characters and their codings are given in Table 8.3. Taxa used are conifers and cordaites (CORDA), Bennettitales (BENNE), *Ephedra* (EPHED), *Welwitschia* (WELWI), *Gnetum* (GNETU), eoangiosperms (EOANG), and "higher" angiosperms (HGANG).

next tree occurred at 87 steps (Fig. 8.17a). Its only difference was that it reordered *Gnetum* and *Welwitschia* together as the sister group of the angiosperms. The next tree occurred at 88 steps and gave the same configuration, except that it treated *Gnetum, Welwitschia*, and the angiosperms as an unresolved trichotomy (Fig. 8.17b). The strict consensus tree for our five shortest trees appears in Fig. 8.17d. This shows *Gnetum* and *Welwitschia* together as the sister group to the angiosperms and also treats the cordaites and the Bennettitales together as the unresolved common ancestor of the gnetopsid-angiosperm clade. Histograms of the distributions of the trees for each set of data show that the most parsimonious group occurs well isolated from the distribution of most of the possible trees.

We feel that this analysis provides considerable support for an origin of the angiosperms from well above the base of the gnetopsids, or chlamydosperms, unlike the most widely accepted of current treatments, which regard flowering plants either as the sister group of the chlamydosperms (Crane, 1985; Loconte and Stevenson, 1990; Loconte, Chapter 10), or even roots them below the level of the Bennettitales (Doyle and Donoghue, 1986a, 1986b, 1993). However, it must be emphasized that, despite the apparent robustness of this analysis, the main purpose of our character set was to explore the transitions that may have taken place and that some of our characters, such as phyllotaxy, were not as independent as is optimal for such an analysis.

Even though our analysis was not designed to exhustively evaluate relationships at the base of the tree, we infer that a doubly compound, racemose fertile truss is a plesiomorphy for the clade. Although the complete fertile truss of the

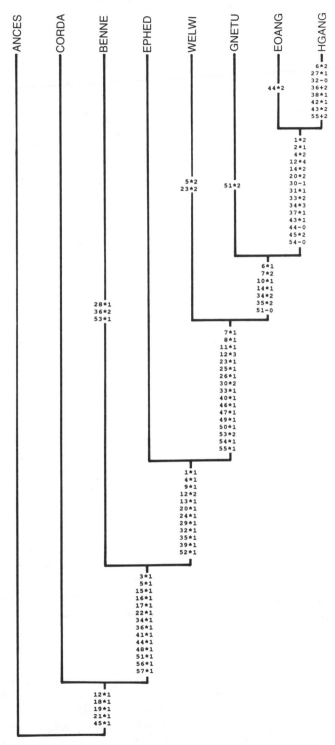

Figure 8.16. The shortest of the phylogenetic trees resulting from our analysis of selected gymnosperms and angiosperms. The tree included 7 taxa and 55 characters and had a length of 85 steps, a consistency index of 0.941, and a retention index of 0.937. Changes from the primitive state or first appearance of a character (0) are given on the figure and are keyed to Table 8.3. Synapomorphies are indicated by an asterisk (*), parallelisms with a plus (+) sign and reversals with a minus (−) sign. The character matrix for this analysis is given in Table 8.2.

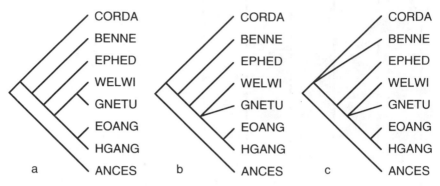

Figure 8.17. Additional cladograms resulting from our phylogenetic analysis of certain gymnosperms and the angiosperms: (**a**) our second shortest tree (87 steps; c.i. = 0.920 and r.i. = 0.911; (**b**) our third shortest tree (88 steps; c.i. = 0.909 and r.i. = 0.899); and (**c**) a strict consensus tree of our five shortest trees.

cordaites is not known, we surmise, based on its putative derivatives, that its fertile and sterile organs were arranged with the sterile ones proximally, and the secondary, microsporangiate axes above these, with the ovulate axes distally.

The next branch above the cordaites in our analysis is the Bennettitales (Figs. 8.16 and 8.17). Although the Bennettitales and the angiosperms share a number of advanced characters such as embryos with two cotyledons, heterogeneous, multicellular wood rays, a lateral fertile truss, bisexual inflorescences, a micropylar tube, monosulcate pollen, a thin megaspore membrane, and cellular early embryogenesis, the number of homoplasies between the angiosperms and the gnetopsids would be unacceptably high if the flowering plants were rooted below this node, as in the option preferred by Doyle and Donoghue (1986a, 1986b, 1993).

Twelve characters support the position of *Ephedra*, from which both the "higher" gnetopsids and, later, the angiosperms emerge. This finding favors Crane's (1985) and Loconte and Stevenson's (1990) hypotheses that the angiosperms were rooted at this level in seed–plant phylogeny. It also supports the general consensus among botanists that the gnetopsids represent a highly diverged group within the gymnosperms and lends weight to proposals to recognize it as a separate class of plants, the Chlamydospermopsida. Among the advanced characters of gnetopsids plus the angiosperms are the following: a tendency toward the rhizomatous habit, decussately opposite phyllotaxy, a discrete tunica and a corpus, rarely nontorate pits, vessels, anthions, microsporangiate connectives filamentous; two integuments; double fertilization; and short pollination periods and generation times. Although the equivalency of a

Table 8.3.   Table 8.3. List of Significant Characters

Sterile:

1 Rhizomatous habit (0) absent, (1) incipient, (2) common (Martens 1971; Cronquist, 1981; Crane, 1985; Taylor and Hickey, 1992; Stewart and Rothwell, 1993; Wheeler and Bass, 1993).

2 Perennial herbs (0) absent, (1) present (Cronquist, 1981; Taylor and Hickey, 1992).

3 Embryos with (0) > 2 cotyledons, (1) 2 cotyledons (Cronquist, 1988; Gifford and Foster, 1989; Stewart and Rothwell, 1993).

4 Phyllotaxy (0) spiral, (1) opposite, decussate or whorled, (2) loss of opposite-decussate to whorled habit (Eames, 1952; Crane, 1985; Cornet, 1986; Endress, 1987b; Gifford and Foster, 1989; Friis and Endress, 1990; Stewart and Rothwell, 1993; Taylor and Taylor, 1993).

5 Apical meristem (0) without tunica, (1) w/ tunica and a corpus, (2) lost after three nodes (Gifford, 1943; Martens, 1971; Gifford and Foster, 1989; Doyle and Donoghue, 1992).

6 Leaf organization (0) fundamentally parallelodromous, either apically or marginally; (1) costate; (2) with a unitary midrib (Martens, 1971; Hickey and Peterson, 1978; Crane, 1985; Cornet, 1986; Stewart and Rothwell, 1993; Taylor and Taylor, 1993).

7 Laminar venation (0) open, (1) reticulate at one order, (2) reticulate, with anastomoses at several orders (Rodin, 1967; Hickey and Wolfe, 1975; Doyle and Hickey, 1976; Gifford and Foster, 1989; Doyle and Donoghue, 1992).

8 Laminar vein-orders (0) one, (1) several (Doyle and Donoghue, 1992).

9 Leaf sheath (0) absent, (1) present (Rodin, 1967; Martens, 1971; Taylor and Hickey, 1992).

10 Plate meristem in leaves (0) absent, (1) present (Rodin, 1967; Kaplan, 1973; Hickey and Wolfe, 1975).

11 Astrosclerids (0) absent, (1) present (Bierhorst, 1971, p. 475).

12 Pit membranes (0) lacking a torus; (1) torate; (2) torate, rarely nontorate; (3) mostly nontorate, with torate pits occasional; (4) nontorate with only very rare torate pits (Bliss, 1921; MacDuffie, 1921; Bierhorst, 1960; Martens, 1971; Carlquist, 1975; Mauseth, 1988; Rothwell, 1988).

13 Vessels (0) absent, (1) present (Bierhorst, 1971; Cronquist, 1981).

14 Perforations (0) only foraminate, (1) foraminate dominant, (2) scalariform dominant (Bliss, 1921; MacDuffie, 1921; Muhammad and Sattler, 1982; Gifford and Foster, 1989).

15 Wood rays (0) homogeneous, uniseriate; (1) heterogeneous, multiseriate (Metcalfe and Chalk, 1983; Martens, 1971; Carlquist, 1975; Muhammad and Sattler, 1982; Taylor and Hickey, 1992; Stewart and Rothwell, 1993).

16 Lignin with (0) no Maule reaction, (1) Maule reaction (Martens, 1971; Gifford and Foster, 1987; Doyle and Donoghue, 1992).

17 Chromosomes (0) large, symmetrical; (1) small, asymmetrical (Ehrendorfer, 1976).

Fertile:

18 Fertile truss (0) monosporate, (1) bisporate (Crane, 1988a; Stewart and Rothwell, 1993; Taylor and Taylor, 1993).

Table 8.3. (*cont.*)

___

19 Vegetative appendages of the fertile truss (0) simple, all telomic; (1) complex, with secondary axes in the axils of scales or bracts (Crane, 1985; Gifford and Foster, 1989; Stewart and Rothwell, 1993; Taylor and Taylor, 1993).

20 Phyllotaxy of the appendages of the fertile truss (0) all spiral; (1) all whorled or decussately opposite; (2) mixed, decussately opposite, to whorled, to spiral (Chamberlain, 1935; Crane, 1985, 1988; Gifford and Foster, 1989; Endress, 1987a, 1987b; Rothwell, 1986, 1988; Stewart and Rothwell, 1993; Taylor and Taylor, 1993).

21 Fertile truss with (0) irregular distribution of organs; (1) male proximal, female distal (Chamberlain, 1935; Eames, 1952; Crane, 1985, Stewart and Rothwell, 1993; Taylor and Taylor, 1993).

22 Organization of the complex fertile truss (0) medial, with the fundamental axis forming the truss axis; (1) lateral, with the truss axis arising from the fundamental axis (Chamberlain, 1935; Crane, 1985; Endress, 1987; Stewart and Rothwell, 1993).

23 Inflorescence axes (0) absent, (1) indeterminate, (2) determinate (Pearson, 1929; Chamberlain, 1935; Endress, 1987a; Gifford and Foster, 1989; Stewart and Rothwell, 1993; Taylor and Taylor, 1993).

24 Lateral fertile truss (0) a simple raceme with spirally arranged appendages, (1) a doubly compound raceme with whorled to opposite-decussate appendages (an anthion) or its derivative (Wieland, 1906; Chamberlain, 1935; Martens, 1971; Endress, 1987a; Rothwell, 1988; Stewart and Rothwell, 1993; Taylor and Taylor, 1993).

25 Anthions (0) indeterminate, (1) determinate (Bierhorst, 1971; Martens, 1971; Endress, 1987a; Gifford and Foster, 1989).

26 Anthial reduction (0) doubly compound, (1) single (Eames, 1952; Martens, 1971; Crane, 1985, 1988; Gifford and Foster, 1989).

27 Anthial axis (0) present, (1) lost and the anthial components fused to a lower-order axis (Eames, 1952; Martens, 1971; Crane, 1985, 1988; Endress, 1987a; Gifford and Foster, 1989).

28 Microsporangia borne on (0) simple structures, (1) pinnate structures (Wieland, 1906; Crane, 1985; Stewart and Rothwell, 1993).

29 "Anther" connective (0) scale-like, (1) filamentous (Eames, 1952; Martens, 1971; Crane, 1985; Cronquist, 1988; Stewart and Rothwell, 1993; Taylor and Taylor, 1993).

30 Microsporangia number (0) > 4, (1) 4, (2) < 4 (Eames, 1952; Martens, 1971; Crane, 1985; Endress, 1987a; Gifford and Foster, 1987; Stewart and Rothwell, 1993; Taylor and Taylor, 1993).

31 Microsporangiate structure (0) simple, (1) thecate (Hooker, 1886; Swamy, 1953; Martens, 1971; Takhtajan, 1980; Friis et al., 1986; Endress, 1987a; Cronquist, 1988; Foster and Gifford, 1989; Stewart and Rothwell, 1993).

32 Microspore bearing appendages (0) spirally arranged, (1) decussately opposite or whorled (Eames, 1952; Martens, 1971; Crane, 1985; Friis, Crane, and Pedersen, 1986; Endress, 1987a; Gifford and Foster, 1989; Stewart and Rothwell, 1993; Taylor and Taylor, 1993).

___

Table 8.3.   (*cont.*)

33 Number of cells or nuclei in pollen when shed (0) 4 to 5 cells, (1) 3 (2?) nuclei, (2) 2 cells (Wieland, 1906; Chamberlain, 1935; Martens, 1971; Bold, 1973; Gifford and Foster, 1989).

34 Pollen exine structure (0) alvaeolar, (1) granular, (2) incipiently tectate-columellate, (3) tectate-columellate (Doyle, 1969; Walker and Walker, 1984; Doyle and Donoghue, 1992).

35 Exine striations (0) absent, (1) present, (2) secondarily lost (Erdtman, 1957; Traverse, 1988; Doyle and Donoghue, 1992).

36 Pollen apertures (0) with a tetrad scar, (1) acolpate or forate, (2) monosulcate or monosulcate derived (Crane, 1985; Doyle and Donoghue, 1992; Osborne et al., 1993; Brenner, Chapter 5.)

37 Pollenkitt (0) absent, (1) present (Hesse, 1984).

38 Ovules (0) orthotropous, (1) anatropous (Taylor, 1991; Taylor and Hickey, 1992).

39 Ovules with (0) one integument, (1) two integuments, both derived from gynocladal bracts (Bierhorst, 1971; Crane, 1985; Stewart and Rothwell, 1993).

40 Vascular strand of outer integument (0) double, (1) single (Bierhorst, 1971; Martens, 1971; Takaso and Bouman, 1986; Doyle and Donoghue, 1992).

41 Integument(s) differentiated (0) after megasporogenesis, (1) by the time of megasporogenesis (Pearson 1906, 1929; Coulter, 1908; Martens, 1971; Crepet and Delevoryas, 1972; Crepet, 1974).

42 Integumentary initiation (0) centripetal, (1) centrifugal (Johnson, 1902; Quibell, 1941; Murty, 1960; Raju, 1961; Roth, 1977; Bouman, 1984; Takaso, 1985; Takaso and Bouman, 1986; Gifford and Foster, 1989, p. 576; Friis and Endress, 1990).

43 Micropyle formed from (0) the inner integument; (1) the inner, or both integuments; (2) mixed (both and other) (Endress, 1987a; Crane, 1988a; Taylor and Taylor, 1993).

44 Micropylar lips of inner integument (0) unmodified, (1) forming a tube, (2) swollen (Edwards, 1920; Quibell, 1941; Murty, 1960; Raju, 1961; Bhandari, 1963, 1971; Martens, 1971; Crane, 1985; Takaso and Bouman, 1986; Gifford and Foster, 1989; Stewart and Rothwell, 1993).

45 Pollination drop (0) absent, (1) present, (2) vestigial (Tilton, 1980; Lloyd and Wells, 1992; Willense and Franssen-Verheijen, 1992).

46 Andranthial bracts (0) present, (1) lost (Chamberlain, 1934; Bierhorst, 1971; Martens, 1971; Gifford and Foster, 1989).

47 Megagametophyte (0) monosporic, (1) tetrasporic, at least occasionally (Martens, 1971; Crepet and Delevoryas, 1972; Crepet, 1974; Crane, 1985; Doyle and Donoghue, 1986b).

48 Megaspore membrane (0) thick, (1) thin (Crane, 1985; Loconte and Stevenson, 1990).

49 Megagametophyte (0) with archaegonia, (1) without archaegonia (Chamberlain, 1935; Martens, 1971; Gifford and Foster, 1989).

50 Megagametophyte cellularity at the time of fertilization (0) complete, (1) incomplete (Bierhorst, 1971; Martens, 1971; Crepet and Delevoryas, 1972).

Table 8.3.   (*cont.*)

---

51 Egg (0) cellular, (1) free nuclear (2) both in the same genus (Chamberlain, 1935; Johansen, 1950; Martens, 1971; Gifford and Foster, 1989; Doyle and Donoghue, 1992).

52 Fertilization (0) single, (1) double (Chamberlain, 1935; Johansen, 1950; Martens, 1971; Friedman, 1990a, 1990b, 1992; Friis and Endress, 1990).

53 Early embryogenesis (0) free nuclear; (1) cellular, generalized; (2) cellular, onagrad type (Wieland, 1906; Johansen, 1950; Maheshwari, 1950; Davis, 1966; Stebbins, 1974, p. 203; Haig, 1990; Loconte and Stevenson, 1990, 1991; Friedman, 1994).

54 Embryo (0) without feeder, (1) with feeder apparatus (Crane, 1985).

55 Sclerotic layer in the (0) inner integument, (1) outer integument, (2) ovary wall (Endress, 1987a).

56 Time from pollination to fertilization (0) long [months], (1) short [10 hours] (Chamberlain, 1935; Martens, 1971; Mulchahy, 1979).

57 Generation time (seed-set to seed-set) (0) > 1 year, (1) < 1 year (Chamberlain, 1935; Martens, 1971; Mulchahy, 1979).

---

number of these features with those in the angiosperms has been the source of controversy at some time in the past, their convergence at the base of the chlamydosperm–angiosperm clade adds a strong contextual argument for considering that they represent valid homologies. One note of caution here is that among the characters used in our analysis, the type of fertilization cannot be determined for the Bennettitaleans, which potentially narrows the gap between them and the angiosperms.

The question of homology of the gnetopsid and angiosperm vessel has also generated a large amount of controversy over the years, with the general consensus being that they represent parallelisms, because of the difficulty of explaining the transition to scalariform perforation plates starting from the polyforate types found in the gnetopsid vessel (Thompson, 1918; Bierhorst, 1960; Martens, 1971; Carlquist, 1975, Chpater 4; Cronquist, 1988; Gifford and Foster, 1989; Takhtajan, 1991). Nevertheless, several authors (Bliss, 1921; MacDuffie, 1921; Muhammad and Sattler, 1982; Mauseth, 1988) have maintained that scalariform and simple perforation plates are derived from the foraminate type seen in the gnetopsids, with a transition involving the aggregation of lateral foraminate pores resulting in occasional cases of scalariform end walls in *Gnetum*. These authors also report occasional foraminate pores in angiosperms but not, it must be noted, in any of the eoangiosperms. Muhammad and Sattler (1982) explain the transition as the result of early cessation of secondary-wall deposition in the vessel elements of *Gnetum* and illustrate the presence of a ghostly pattern reminiscent of scalariform pitting under a lightly deposited secondary wall coating that is foraminate.

With regard to pitting, the divergence in pattern between the vessels of the chlamydosperms and the angiosperms may result from considerations of strength. Essentially, the chlamydosperms would have had to maintain the strength of their vessels by retaining circular pitting, whereas, in the primitively herbaceous eoangiosperms, strength would not have been an especially important consideration. Later, when angiosperms started to radiate toward shrubby and arborescent habits (Taylor and Hickey, Chapter 9), they solved the strength problem by using their late wood-tracheids with circular-bordered pitting (cf. Carlquist, Chapter 4). Another consideration working against the occurrence of scalariform pitting in the chlamydosperms was their broad vessel elements where scalariform pitting would have made the stems a good deal weaker than its occurrence in narrow vessels packed in with masses of fibers and tracheids. We therefore suggest, based on this phylogenetic analysis, that the apparent differences in the vessel elements of the chlamydosperms and angiosperms may have been a consequence of the drastic reduction in the stature and vegetative body of the line leading from the gnetopsids to the angiosperms.

Seventeen characters support the derivation of the clade leading to *Welwitschia, Gnetum*, and the angiosperms above the node for *Ephedra* in our shortest tree (Fig. 8.16), thus pushing the rooting of the angiosperms into the gnetopsids themselves at, or above, the level of divergence for *Welwitschia* and *Gnetum*. Among the vegetative characters that favor this position are leaves possessing several orders of reticulate veins and astrosclerids and the presence of mostly nontorate pit membranes in the xylem. These are paralleled by an array of fertile characters, including reduced, determinate anthions, loss of the andranthial bracts, microsporangial number reduced to four or less, reduction of the male gametophyte to three or fewer cells or cell nuclei, a single vascular strand in the outer integument, occasionally tetrasporic origin of the megagametophytes, loss of the archaegonia, megagametophytes incompletely cellular at the time of fertilization, early embryogenesis onagrad, and a sclerotic layer in the outer integument. This rooting of the angiosperms from within the gnetopsids is well supported by our cladistic analysis. On the other hand, the shortest tree in our array that kept the chlamydosperms strictly monophyletic by treating them as the sister group of the angiosperms was 15 or 20 steps longer.

Obviously, these results depend on the quality of the characters upon which they were built and some of those used here have been the subject of considerable controversy in the past. We point to our homologizing of the ring-like second envelope of the ovule of the higher chlamydosperms (*Gnetum* has three envelopes around its ovule) to that of the outer integument in angiosperms. The outer integument of the angiosperm ovule is a single structure and shows no sign of division (Bierhorst, 1971), whereas in *Gnetum* (Takaso and Bouman, 1986) and *Welwitschia* (Martens, 1971), the second envelope appears to arise from a pair of opposite primordia even though the mature envelope is ring-like.

Nevertheless, because we regarded the outer ovular envelopes as forming a continuum from the bistranded ring in *Ephedra* through the primordially bracteate ring in the higher chlamydosperms to fully ring-like in the angiosperms, we coded for the mature state of this structure.

Another such character is the occurrence of megagametophytes of tetrasporic origin in this clade. The exact equivalence of this feature between the higher chlamydosperms and the angiosperms is a matter of some controversy (Maheshwari, 1950; Bierhorst, 1971; Martens, 1971; Crane, 1985; Doyle and Donoghue, 1986a, 1986b) and it is only of sporadic occurrence in the eoangiosperms, being found in the piperaleans but not in the Chloranthaceae (Johansen, 1950; Maheshwari, 1950; Davis, 1966). However, its presence in the higher gnetopsids, within an already strong consensus of angiospermous features, is at least suggestive evidence for the existence of a general tendency toward tetraspory among the plants originating at this node. Another feature coded as a synapomorphy is the development of a cellular embryo without the intervention of the free-nuclear condition, as is the case in all lower seed plants. Again, however, some equivocation exists on this point because one species of *Gnetum*, *G. uli*, retains a single free-nuclear division at the initiation of its embryo (Martens, 1971; Gifford and Foster, 1989).

Probably the most disparate element in the results of this analysis for deriving the angiosperms from the higher clamydosperms is the presence of a well-developed feeder on the embryos of both *Gnetum* and *Welwitschia*. However, the presence of cellular-onagrad early embryogenesis (Loconte and Stevenson, 1991) is a striking synapomorphy supporting this clade. Another highly characteristic advancement shared with the basal angiosperms involves the development of a sclerotic layer in the outer integument. In most of the angiosperms, when a sclerotic layer is present, it develops within the ovary wall, but in the Chloranthaceae (Endress, 1987a), a number of other eoangiosperms, and some magnoliids, it arises in the outer integument of the ovule, despite their enclosure in the ovary, just as it does in the seeds of the higher chlamydosperms (Martens, 1971; Murty, 1960; Quibell, 1941; Takaso and Bouman, 1986; Raju, 1961; Edwards, 1920; Bhandari, 1963, 1971). We believe that this character represents a direct inheritance from the higher chlamydosperms. Less compelling, but interesting nonetheless, are the occurrence of a micropyle formed by the inner integument, swollen micropylar lips of the inner integument (references as in the previous citation), and a vestigial pollination drop (Tilton, 1980; Lloyd and Wells, 1992; Willense and Fransen-Verheijen, 1992) that all appear to be remnants of the pollination system of the gymnosperms.

As can be seen from the cladograms (Figs. 8.16 and 8.17), our analysis only slightly favors rooting the angiosperms as the sister group of *Gnetum* over the alternative that *Gnetum* and *Welwitschia* together represent the sister group, with the two trees differing by only two steps. A simple phenetic analysis somewhat more strongly favors *Gnetum*, which shares 34 advanced characters with the angiosperms, over *Welwitschia*, with 30, as the sister group. Nevertheless, several integrated sets of characters that define consistent trends in the evolution of the angiospermous plant body or of its organs lend some support to the hypothesis of a common origin for the eoangiosperms with *Gnetum*. Such integrated groups of characters are not emphasized in the more binary context of modern phylogenetic analysis but are sometimes more useful in supporting assertions of relatedness than the sum of the individual characters.

One such set involves the leaves of the gnetaleans and those of angiosperms. These share broad laminas with a costate organization, vein anastomoses at several orders, plate meristematic growth, and intercostal venation that exhibits a mainly monopodial rather than a dichotomous pattern of branching (Rodin, 1967; Martens, 1971; Hickey and Wolfe, 1975; Gifford and Foster, 1989). Broad leaves with pinnate venation that arises from a costate mid-vein are unique to *Gnetum* and the angiosperms, with the possible exception of some of the gigantopterids (Li et al., 1994; Mamay, 1989). We use *costa* in the sense of Hickey and Peterson (1978) who defined it as a mid-vein that is made up of a number of closely spaced vascular strands. Rodin (1967) claimed that the manner in which the secondary veins of *Gnetum* leaves arose, by a series of low-angle dichotomies of the costal strands, was not seen in the angiosperms. On the contrary, this is precisely the pattern of secondary vein departure in the eoangiosperms, including the Chloranthaceae and other basal angiosperms, persisting well into the base of the monocotyledons where it was illustrated by Hickey and Peterson (1978, Figs. 8.6–9). Similarly, although the somewhat irregular pattern of areole organization in the intercostal venation of *Gnetum* has been contrasted to the generally more regular pattern in higher-rank angiosperm leaves that belong to derived groups; this pattern replicates the irregular, or low-rank intercostal venation seen in the more basal angiosperms, including the eoangiosperms (Hickey, 1977).

A second reason for favoring *Gnetum* as the sister group of angiosperms rests on its spherical pollen grains with nonplicate sculpturing, in contrast to the ellipsoidal, plicate grains that characterize the other two genera of the chlamydosperms (Brenner and Bickoff, 1992; Brenner, Chapter 5). In addition, *Welwitschia* grains have a distinct sulcus, whereas the grains of *Ephedra* and *Gnetum* are inaperaturate or have thin areas in their walls (Crane, 1985; Osborn et al., 1993), as do the generally spherical grains of *Sarcandra* in the Chloranthaceae (Endress, 1987a; Brenner, Chapter 5). Figures in Erdtman (1957) also suggest the presence of incipient columellae in at least two species of *Gnetum*, although this observation is not repeated in the summary by Osborne et al. (1993). Our suggestion that angiosperm pollen was primitively spherical and acolpate would also corroborate Brenner's (Chapter 5) inference that the spherical, acolpate, nonstriate grains that he observes in very early Cretaceous (Valanginian to lower Hauterivian) strata in Israel may represent the earliest angiosperm pollen grains. A third point in favor of this relationship is that the inflorescence axes to which the anthions are attached are indeterminate in both *Gnetum* and the eoangiosperms, whereas these axes are strictly determinate in *Welwitschia.*

Both *Gnetum* and *Welwitschia* developed a number of autapomorphies like the structure of their "strobili," whereas the branch that led to the line of perennial herbaceous eoangiosperms developed the syndrome of advanced characters that were to define the group (Taylor and Hickey, 1992; Chapter 9). In addition to stigmatic germination, these included the acquisition of the rhizomatous herbaceous habit, more discrete demarcation of a corpus layer in the apical meristem, thecate anthers, reduction of the male gametophyte to two cells upon shedding, tectate–columellate pollen, the pollenkitt (Hesse, 1984), a micropyle from both the inner and the outer integument with reduction of the pollen tube formed by the inner integument, and, we surmise, the loss of the feeder apparatus. Later advances among the more derived angiosperms involved development of a unitary midrib in the leaves, the origin of the radially symmetrical flower by the absorption of the anthial axis into the former fundamental axis or its derivatives, the origin of monosulcate pollen for a second time among the seed plants, the development of anatropous ovules, a shift in integumentary development from centrifugal to centripetal (Bouman, 1984; Johnson, 1902; Murty, 1960; Quibell, 1941; Raju, 1961, Takaso, 1985; Takaso and Bouman, 1986; contra Friis and Endress, 1990), and further reduction of the inner integument, with the outer one playing an increasingly important role in the formation of the micropyle and a shift of the sclerotic layer of the seed coat to the ovary wall or endocarp.

STAGES IN THE EVOLUTION OF THE ANGIOSPERM FLOWER

In this section, we attempt to reconstruct the stages in the evolution of the angiosperm flower, based on the homologies that we propose and on our resulting phylogeny. This sequence is recapitulated in Fig. 8.18 where we have labeled the homologous units of the fertile truss through the sequence.

The first stage in the evolution of the flower (Fig. 8.18a) began when the fertile truss became bisporangiate, a development that we speculate had occurred with the cordaites or, possibly, their immediate ancestors. As we reconstruct it, the fertile truss of the cordaites would have consisted of a *primary fertile axis* (labeled as Roman numeral I in Fig. 8.18a), which, in turn, bore monosporangiate fertile units or "strobili" along its length. Each of these was a homothetic, doubly compound raceme consisting of *secondary* and *tertiary*

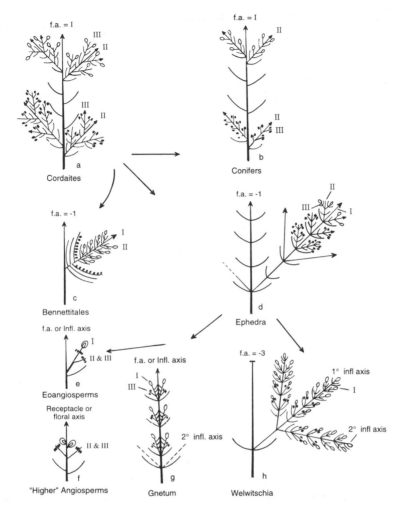

Figure 8.18. Stages in the evolution of the fertile truss from the cordaites to the angiosperms. Indeterminate axes are indicated by the arrow symbol. Roman numbers identify homologous axes; f.a. indicates the fundamental axis; shifts in the position of the fundamental axis are shown as a negative number indicating the number of axial orders that the fundamental axis has been shifted downward from I. Microsporangia indicated by the filled symbols; megasporangia and other female parts by the open symbols. Additional symbols: fusion of the anthial axes to the inflorescence axis in *Gnetum* is shown by curved diagonal lines connecting the two. In most cases, all organs and appendages have been rotated into the medial plain except for the upper pair of stamens in **e**. Details of anthions in *Gnetum* and *Welwitschia* are considerably simplified for this scale. The inflorescence of *Welwitschia* is shown as bisexual, based on comparative evidence, even though the modern species is unisexual or pseudomonoecious.

fertile axes, or paraclades (II and III in Fig. 8.18a) subtended by and bearing bracts or scales. The axes and scales were arranged in an alternate decussate pattern. Proximal scales on the tertiary axes were sterile, but the distal ones bore sporangia on their distal margins. By inference from their descendants, the chlamydosperms and the conifers, we further speculate that the monosporangiate fertile units were arrayed on the primary fertile axis with the microsporangiate ones proximally and the megasporangiate ones distally, a configuration that may reflect the beginnings of the genetic system that controls the position and type of organs found in angiosperm flowers today (cf. Chasan, 1991). The primary fertile axis was also the *fundamental axis* (f.a. of Fig. 8.18a) of the fertile truss, in that it appears to have been capable of producing further vegetative growth of the plant by apical growth, as in conifers today. This coincidence of the fertile and the vegetative roles in the same axis means that the fertile truss would have had an axial orientation on the fundamental axis, as in the modern conifers.

In the conifers (Fig. 8.18b), the fertile truss remained axial, with the same axis playing both primary fertile and fundamental roles. The unisexual, homothetic, doubly compound racemose organization of the cordaitalean fertile units was retained in the strobili of the conifers but underwent varying degrees of condensation, with the most important of these trends being the foreshortening and fusion of the tertiary fertile axis to the adaxial side of its subtending bract (Florin, 1950; Eames, 1952; Clement-Westerhof, 1988).

Above the phylogenetic level of the cordaites and conifers in the lineage leading to the flowering plants, the fertile truss is inferred to have been shifted from an axial to a lateral position. This change probably took place by suppressing the apical vegetative growth of the combined primary fertile/fundamental axis. This had the effect of holding the position of the primary fertile axis constant but moved the vegetative activity of the fundamental axis down to the next lower order of axes. This shift is indicated in Fig. 8.18c,d by the designation f.a. for fundamental axis, followed by the designation of minus 1. In the Bennettitales, this shift was accompanied by the atrophy of the higher-order, fertile, paracladal axes resulting in the reduction of the inflorescence or strobilus to a simple raceme or, in the cycadeoids, to a capitulum. We thus infer that the "strobilus" of the Bennettitales is the homolog of the entire fertile truss of the conifer–cordaite line, now displaced to a lateral position on the fundamental axis.

In *Ephedra,* the full complement of higher-order fertile, paracladal axes has been retained and the primary and secondary fertile axes are now terminated by a simple raceme (Fig. 8.18d). The resulting indeterminate, heterothetic, doubly compound racemes are termed *anthions.* In addition, all units of the gnetopsid inflorescence show a pervasive tendency to be arranged in decussately opposite pairs or in whorls. In the higher gnetopsids and the angiosperms, the anthion becomes determinate by suppression of all but a single ovule, which then assumes a pseudoterminal position on its subtending axis. The primitive anthion also became simplified through loss of the higher-order fertile axes. The result of these trends reduced the anthions of *Gnetum* and *Welwitschia* to determinate simple panicles. In both of these genera, we also see the interpolation of a set of inflorescence axes between the base of the anthion and the fundamental axis. These are inferred to have developed from fundamental axes that had lost the ability to produce normal vegetative leaves. In *Gnetum*, the inflorescence axes are indeterminate, occur in up to two orders, and bear whorls of unifacial anthions [*drepaniums* by Weberling's (1989) terminology] whose adaxial faces are fused to the inflorescence axis (Figs. 8.6c,d and 8.18g). In *Welwitschia*, the inflorescence axes are determinate, occur in up to two orders to form triply compound thryses, and bear simply paniculate anthions that are amalgamated into compound strobili (Figs. 8.6 and 8.18h).

The simple anthions of the ancestor of the angiosperms became determinate because apical growth of their anthial axis was terminated by the carpel. The principal trends in the floral evolution of the basal angiosperms appear to have been for the reduction of the anthion and its axis and the replacement of the opposite-decussate arrangement of the gnetopsid inflorescence with a spiral arrangement. Eventually, the process of reduction did away with the anthion as an integrated structure. Instead, its organs were transferred to the distal-most inflorescence axis that had retained the capacity for indeterminate growth, or in some cases to the fundamental axis itself. In any event, this indeterminate axis now became the receptacle of the flower above the level of the eoangiosperms. This changed the floral symmetry from bilateral to radial and allowed for the nearly indefinite proliferation of carpellary units within the newly reconfigured angiosperm flower.

CONCLUSIONS

This comparison of the gnetopsids and the Chloranthaceae uses a topological method of analysis, in part derived from Wilhelm Troll (Weberling, 1989), to examine the possible homologies between the structure and position of the organs making up the inflorescences of the two groups. This, in turn, produces numerous insights regarding the transition from the gymnosperms to the flowering plants. Several longstanding problems relating to the orientation and arrangement of floral organs are explained in this model; among them were ovules on the adaxial side of the carpel, the so-called "chair" structure in certain apocarpous carpels, the ventral orientation of the carpellary suture, the relative positions of the male, female, and sterile parts of the flower, the apiculate tip of the anther filament, and, finally, the apparent simplicity and lack of perianth, the frequent occurrence of decussately opposite phyllotaxy, and the presence of bilateral symmetry in flowers of taxa frequently regarded as lying at or near the base the angiosperms.

From a systematic standpoint, our data suggest that the ancestors of the angiosperms may nest well within the gnetopsids, arguably closest to *Gnetum*. There is, thus, no need to invoke a long, latent, separate lineage leading far back into geological history for the flowering plants. The cordaites appear in the Carboniferous, the gnetopsids possibly as early as the Permian or the Triassic at the latest, and the angiosperms in the earliest Cretaceous (Traverse, 1988; Doyle and Donoghue, 1993; Stewart and Rothwell, 1993; Taylor and Taylor, 1993; Brenner, Chapter 5). In many respects, this interpretation is closer to the timing and types of early angiosperm fossils seen in the fossil record than are models invoking long latency of the lineage. Our character analysis confirms the marked separation of the gnetopsids from the remainder of gymnosperms and justifies the use of the designation Chlamydospermae or Chlamydospermopsida for this group.

The Chloranthaceae may well be the most basal living family of angiosperms, even though a number of other eoangiosperm families, notably among the piperaleans, retain a significant number of inferred basal character-states, some of which are not found in the Chloranthaceae. Our model of the eoangiosperm inflorescence closes the apparent gap between the floral plan of the monocotyledons and the basal angiosperms and is reminiscent in this regard of the hypothesis of Burger (1977; 1981). The eoangiosperms are defined as that group of the flowering plants that retains the anthion and the anthial axis as the floral axis or receptacle.

In addition, the surprising insight emerges that the angiosperms belong to a lineage of vascular plants in which both the male and female sporangia are attached laterally to essentially indeterminate axes or modified telomic scales rather than to leaves, thus retaining an arrangement first seen in the trimerophytes of the Devonian. This should help further to sort out distant ancestors; for instance, we can eliminate the cycads and many of the so-called seed ferns from any search pattern. Finally, this analysis reinforces the insight that the fundamental adaptive style of the immediate ancestors of the angiosperms was the condensation and elimination of elaborate, energetically costly structures, both fertile and vegetative, and their replacement by simpler, more rapidly growing, and energetically cheaper structures. This coordinated suite of changes produced, in the earliest angiosperms, the first and only group of seed plants to achieve the herbaceous habit. The adaptive advantages conferred by this process may well have been the prime reason for the later success and vast diversification of the flowering plants.

ACKNOWLEDGEMENTS

The authors wish to thank Linda Raubeson, Una Smith and Dorian Q. Fuller for assistance with our cladistic analysis; and Barbara Ertter for a stimulating exchange of ideas. Finally, we wish to thank Virginia Simon who made most of the drawings for this paper.

Evidence for and Implications of an Herbaceous Origin for Angiosperms

David Winship Taylor and Leo J. Hickey

For over a century, botanists have speculated on the ancestral morphology of the angiosperms and on the adaptive characteristics that led to their success. Most discussions have centered on the ancestral form of the flower and associated organs. At present, there are two major alternative views regarding the morphology of the first angiosperm. The first of these, which we term the Magnolialean hypothesis, is concordant with the Anthostrobilus or Euanthial floral hypothesis (Fig. 8.1) of Arber and Parkin (1907) and pictures early angiosperms as having been woody shrubs or small trees with large, complex, multiparted flowers. Living species regarded as similar to the archetype are members of the winteroids and Magnoliales (Fig. 11.1 and 11.2). The second view parallels the Pseudanthial hypothesis (Fig. 8.1) of Wettstein (1907a). In this formulation, the ancestral angiosperm is perceived as a rhizomatous herb with small, simple flowers as in living herbaceous magnoliids such as Chloranthaceae and Piperaceae.

The current consensus regarding the features of the ancestral angiosperm lies somewhere within the variants of the Magnolialean hypothesis (e.g., Cronquist,

1968, 1988; Takhtajan, 1969, 1980, 1991; Thorne 1968, 1992a, Chapter 11; Donoghue and Doyle, 1989a; Doyle and Donoghue, 1993). An intermediate hypothesis has also been proposed, suggesting the ancestral form was a small, weedy shrub (Stebbins, 1974; Doyle and Hickey, 1976; Hickey and Doyle, 1977; Retallack and Dilcher, 1981a; Tiffney, 1984; Crane, 1987; Wing and Tiffney, 1987a, 1987b; Loconte and Stevenson, 1991; Loconte, Chapter 10). From these ancestors evolved the tree and herb habits. For the most part, the flower of this plant is thought to be similar to that suggested by the Magnolialean hypothesis.

From these proposed ancestral suites, numerous workers have suggested adaptive scenarios for angiosperm success (e.g., Stebbins, 1974; Doyle and Hickey, 1976; Cronquist, 1988; Takhtajan, 1991; Thorne, Chapter 11; and reference therein). These models have stressed the value of the reproductive innovations of the angiosperm and the presumed benefits of traits such as insect pollination (e.g., Regal, 1977), compatibility systems (e.g., Whitehouse, 1950), fruit dispersal (e.g., Tiffney, 1984), and endosperm (e.g., Westoby and Rice, 1982) with only a few recent studies emphasizing the importance of vegetative characters (e.g., Bond, 1989; Taylor and Hickey, 1990a, 1992; Midgley and Bond, 1991).

Recently, we used the term Paleoherb hypothesis for the alternative view (Taylor and Hickey, 1992) because many of the basally placed taxa belong to the so-called paleoherbs of Donoghue and Doyle (1989a). The hypothesized ancestral states have generally come from suggestions that small, simple flowers were ancestral (e.g., Burger, 1981; Meeuse, 1987) or from the placement of herbaceous magnoliids at the root of the angiosperm phylogenetic tree. Several living taxa have been proposed to be basally placed in the angiosperm phylogeny, including Chloranthaceae, Lactoridaceae, Aristolochiaceae, Nymphaeales (sensu stricto), Ceratophyllaceae, Piperales, and monocotyledonous plants (e.g., Burger, 1981; Meeuse, 1987; Donoghue and Doyle, 1989a; Zimmer et al., 1989; Taylor and Hickey, 1990a, 1992; Taylor, 1991a; Hamby and Zimmer, 1992; Chase et al., 1993 Doyle et al., 1994), although several of the these lack some of suggested ancestral characteristics (e.g., Lactoridaceae are not herbaceous).

We now call our model the Herbaceous Origin hypothesis because of the difference in our circumscription of herbaceous magnoliids and that of Donoghue and Doyle (1989a) for paleoherbs (Hickey and Taylor, Chapter 8). Unlike previous analyses suggesting herbaceous origin, our hypothesis was based on an analysis of flowering plants and their sister groups. It included ancestral states for both vegetative and floral characters and indicated the importance of the herbaceous habit for explaining angiosperm evolution (Taylor and Hickey, 1990a, 1992). We suggested that the ancestral angiosperms were small, herbaceous plants with a rhizomatous to scrambling, perennial habit.

They had simple leaves that were reticulately veined and had a primary venation pattern that would have been indifferently pinnate to palmate, whereas the secondary veins branched dichotomously. We postulated that the leaf base was laterally extended to form a sheath and the nodal anatomy was multi-trace–multilacunar with the medial lacuna having two traces. The vegetative anatomy included sieve-tube elements and elongate tracheary elements with both circular-bordered and scalariform pitting and oblique end walls. These would have graded imperceptibly into vessel elements by the loss of primary wall between the bars of the scalariformly pitted end walls. Secondary growth, although present, would have been limited.

Our analysis also indicated that the flowers occurred in cymose to racemose inflorescences. The stamens would have been basifixed with an apical extension of the connective, four microsporangia and two loculi. The small, monosulcate pollen would have had perforate to reticulate sculpturing and a tectate–columellate ektexine. Finally, we suggested that the free, ascidiate carpel had one or two proximally attached orthotropous, bitegmic, crassinucellar ovules and dicotyledonous embryos.

In this chapter, we expand on our Herbaceous Origin hypothesis and begin by summarizing the phylogenetic evidence that supports it. We then list ancestral character states, both previously proposed (Taylor and Hickey, 1992) as well as some that are new for our hypothesis. The distribution of these states is examined to see if they are shared among anthophytes or restricted to angiophytes or even crown angiophytes [= angiosperms; i.e., Doyle and Donoghue (1993) and see below] in order to identify those ancestral characteristics that are unique to angiosperms. This is followed by a review of additional evidence that is compatible with the Herbaceous Origin hypothesis. We then speculate on the adaptive value of some of the character states that are unique to angiosperms in order to show how they better fit the herbaceous ancestral form. Finally, we produce a scenario for angiosperm origin that describes their ancestral habitat, ancestral morphology, and early ecological radiation.

Phylogenetic Support for the Herbaceous Origin Hypothesis

A number of recent studies lend support to our Herbaceous Origin hypothesis. Among these are an analysis of angiosperms by Taylor and Hickey (1992) using structural data (Fig. 6.2) and results from a structural analysis of seed plants by Nixon et al. (1994). Even though the shortest trees in other structural analyses support the Magnolialean or other hypotheses, trees only a few steps longer are consistent with a rooting among the herbaceous magnoliids (e.g., Dahlgren and Bremer, 1985; Donoghue and Doyle, 1989a; Doyle, 1994; Loconte, Chapter 11).

Support is also found from DNA sequence data. Analyses of 18S rRNA (Zimmer et al., 1989; Hamby and Zimmer, 1992; Doyle et al., 1994) suggest that the Nymphaeales (sensu stricto) and Piperales are most basal. When this dataset is combined with a morphological dataset, Nymphaeales remain basally placed (Doyle et al., 1994). Analyses of *rbcL* show variable results, although they frequently support the Herbaceous Origin hypothesis with Ceratophyllaceae and sometimes other herbaceous magnoliids as most basal (Chase et al., 1993; Qiu et al., 1993; Sytsma and Baum, Chapter 12). Finally, molecular phylogenies of other sequences using limited suites of taxa also indicate that herbaceous taxa, as represented by monocotyledons, are placed basally (Troitsky et al., 1991; Martin et al., 1993).

Both the structural and sequence datasets show that the herbaceous magnoliids are probably ancestral. Yet it is still unclear from current analyses which living taxon is most basal. In addition, we suggest that the most basally placed taxon in the available trees does not by itself necessarily determine the ancestral states. Structural characters must be optimized on topologies that include angiosperm sister groups and basal angiosperms, and outgroup comparison and parsimony must be used to determine the ancestral states. This is because any living basal taxon may have undergone considerable anagenesis since it diverged from the other angiosperms up to 130 million years ago. We feel that this is particularly true in addressing the question of whether the ancestral angiosperm was aquatic. It is not sufficient to find basal placement for or large numbers of presumed ancestral character states in currently aquatic clades such as Nymphaeales and Ceratophyllaceae to state that angiosperms are ancestrally aquatic.

EVOLUTIONARY PLACEMENT OF ANCESTRAL ANGIOSPERM CHARACTER STATES AND THE IDENTIFICATION OF ANGIOSPERM SYNAPOMORPHIES

When first formulating the Herbaceous Origin hypothesis we provided a description of the ancestral character states (Taylor and Hickey, 1992) but did not explicitly state which characteristics were angiosperm synapomorphies. The growing volume of data on anthophyte relationships now allows assessment of character distributions in order to determine whether a given state is plesiomorphic or synapomorphic. In this analysis, we decided to determine the character state distribution based on the phylogenetic ideas of Doyle and Donoghue (1993). They suggested that there are three nested groups of importance in the lineage leading to and including flowering plants: anthophytes, angiophytes, and angiosperms by themselves (Fig. 9.1).

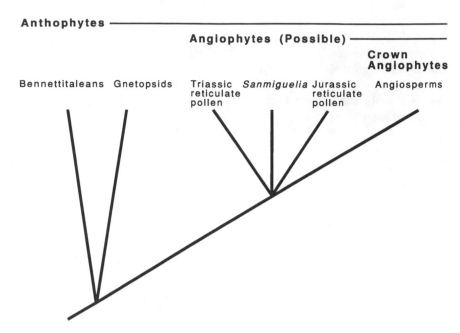

Figure 9.1.   A consensus phylogeny showing the Anthophytes including the angiosperm sister groups, Angiophytes including the possible stem angiophytes, and the Angiosperms, the crown angiophytes. The conceptual basis for this hierarchy is based on the work of Doyle and Donoghue (1993).

Anthophytes include the angiosperm lineage and its sister groups. For our purposes these sister groups will include the bennettitaleans and gnetopsids. We exclude the less well-known *Pentoxylon*, as the homologies to bennettitaleans appear weak (Crepet et al., 1991; Rothwell and Serbet, 1994; Nixon et al., 1994). *Pentoxylon* lacks key anthophyte characteristics such as distal, medial, and proximal positioning of the female, male, and sterile organs on the reproductive axis, as well as the enclosure of the ovules by bract derived organs (Hickey and Taylor, Chapter 8). Since research shows variability with the relationships within anthophytes (Crane, 1985; Doyle and Donoghue, 1986b, 1992; Rothwell and Serbet, 1994; Nixon et al., 1994; Hickey and Taylor, Chapter 8, Fig. 8.16) we chose to use a consensus phylogeny for our analytical scheme (Fig. 9.1).

The angiophytes make up the angiosperm lineage and include the stem angiophytes and the crown angiophytes [= angiosperms, sensu Doyle and Donoghue (1993)]. Stem angiophytes include members of the angiosperm lineage that do not have all the synapomorphies that are found in living angiosperms today. Note that in Chapter 8 we propose a heuristic model of the possible homologies between gnetopsids and Chloranthaceae. If all these homologies are supported then early angiosperms (stem angiophytes) would be quite similar and intermediate to members of these two groups. If, on the other hand, the ancestral morphology is somewhat different (e.g., alternate phyllotaxy), then many of our evolutionary transformations remain supported, but potential transitional forms could have a broader range of morphologies (cf. Crane, 1993).

The angiosperms (crown angiophytes) are thought to be monophyletic, and it is this direct ancestor that the Magnolialean and Herbaceous Origin hypotheses propose to reconstruct. Because of the definition of the angiophytes, the character states found as synapomorphies for the crown group may not be synapomorphies for the entire angiosperm lineage because they are restricted to the distal portion of the tree. However, these states may have been important innovations for the successful radiation of the angiosperms.

Potential Angiophytes

Sanmiguelia (Cornet, 1986; 1989b) and Triassic (Cornet, 1989a) and Jurassic (Cornet and Habib, 1992) reticulate and tectate–columellate pollen species have been attributed to the angiosperm lineage (e.g., Cornet, Chapter 3) but remain problematic (Doyle and Donoghue, 1993). Although angiosperms are thought to be unique in having tectate-columellate pollen, the Triassic pollen species are not accepted as angiospermous by Doyle and Donoghue (1993), in part, because of their thick endexine. The Jurassic pollen is also problematic, as there have been no other reports of this pollen type from this age. Nevertheless, for the purpose of this analysis, we designate these fossils potential stem angiophytes.

We do not accept as an angiophyte the recently described Triassic reticulate-veined leaf *Pannaulika triassica* (Cornet, 1993). This species is lacking in many angiosperm characteristics and appears to similar to reticulate-veined leaves with numerous vein orders that exist in several groups of ferns (e.g., Tryon and Tryon, 1982; Trivet and Pigg, Chapter 2). In addition, several characteristics of *P. triassica* are not found in basal angiosperms or gnetopsids. These include excurrent origination of secondaries, a single strand in the mid-vein, regular higher-vein orders and less regular lower orders, and higher order veins running to and ending at the margin. Yet ferns have these same characteristics, and, in addition, fern leaflets are often sterile, have asymmetrical constricted bases, and vary considerably in size.

Methods for Determining Ancestral States

In our previous analysis (Taylor and Hickey, 1992), we used two methods to determine the ancestral states for angiosperms. The phylogeny was determined using only those characters polarizable by the outgroups, and a hypothetical ancestor was constructed based on the polarized characters. The advantage of developing an explicit hypothetical ancestor was that it could be constructed from the consensus of several outgroup topologies. If a character state was polarized as ancestral using different designated outgroups, then it was likely to be ancestral for the ingroup. Once the most parsimonious trees were obtained, the ancestral states of the uniquely angiospermous characters could then be determined. These characters were unpolarizable by outgroup comparison because the homologies with sister-group characteristics were unclear. These ancestral states were determined by optimizing them on the tree using the distribution of states in the six most basally placed taxa (Chloranthaceae, Piperaceae, Saururaceae, Barclayaceae, Lardizabalaceae, and Amborellaceae).

In this analysis, the ancestral states and whether they are synapomorphies or symplesiomorphies are determined by their distribution within the anthophytes. A character state is considered to have anthophyte distribution if the structurally homologous state is found in any sister group and in most of the basal angiosperms of Taylor and Hickey (1992). Such states are symplesiomorphies for angiosperms. A state is considered to have an angiophyte distribution if it is found in any stem angiophyte and most basal, herbaceous magnoliids. Such states are considered angiosperm synapomorphies. Finally, the ancestral states of unpolarizable characters are determined by the method we previously used (Taylor and Hickey, 1992) and are also angiosperm synapomorphies. Many of these characters are discussed in detail elsewhere (Taylor and Hickey, 1992) and the ancestral nature of many of these states has been suggested previously from numerous perspectives (e.g., see discussion in Cronquist, 1968, 1988; Takhtajan, 1969, 1991; Stebbins, 1974).

Vegetative Growth and Habit

Several apomorphic features for angiosperms are related to vegetative characters (Table 9.1A). A divergence from the tree habit is clearly found both in the Bennettitales and gnetopsids. The former range from small trees to short-stature shrubs, whereas the gnetopsids are perennial and tuberous shrubs, small trees, and lianas, and occasionally even have a rhizomatous habit (Martens, 1971). *Sanmiguelia* is also a shrub-sized, perennial plant with axes that appear to grow from an underground rhizome.

Angiosperms are the first seed plants to develop the herbaceous growth habit. These herbaceous plants would have been rhizomatous or scrambling perennials with limited secondary growth. Determination of the herbaceous state as ancestral is based on finding the habit in the basally placed taxa of Taylor and Hickey (1992). Other unique characters include the potential for small size at reproductive maturity and rapid vegetative growth. Angiosperms include the smallest

Table 9.1. Angiosperm Symplesiomorphies and Synapomorphies for Vegetative Growth and Habit (A) and Vegetative Morphology (B) Characteristics. The possible adaptations for some of the characteristics are given. The synapomorphies are related to a new habit for seed plants: that of the rhizomatous herb.

Character-States	Adaptations
A. Vegetative Growth and Habit	
Symplesiomorphies: perennial, rhizomatous.	Allow successful colonization and dominance in disturbed habitats, flexible resource allotment as necessary depending on the degree of competition or loss, and rapid seedling establishment and growth.
Synapomorphies: herbaceous, small size, rapid vegetative growth.	
B. Vegetative Morphology	
Symplesiomorphies: ethereal oil cells, tannins, silica bodies, asterosclerids, small asymmetrical chromosomes with base number = 12-19.	In part, generalized defense strategies.

of all seed plants and this small size is the final expression of a series of size reductions that began with the bennettitaleans and gnetopsids. All members of the basally placed herbaceous angiosperms grow rapidly in comparison to seedlings of living conifers (Bond, 1989; Midgley and Bond, 1991) and young fern sporophytes.

Vegetative Morphology

General Characters. Several of these general vegetative characters are symplesiomorphies in angiosperms (Table 9.1B) as they are also found in the angiosperm sister groups. These include the presence of ethereal oil cells, tannins, silica bodies, asterosclerids, and small asymmetrical chromosomes (Ehrendorfer, 1976) with numbers between 12 and 19 (cf., Masterson, 1994). The presence of tannins, asterosclerids and small asymmetrical chromosomes are new character states for our hypothesis. Tannins are found in the outgroups as well as at least five of the six basal families of Taylor and Hickey (1992), whereas the latter two states are found in basal angiosperms and in the gnetopsids.

Leaf Characters. Previously, we have noted the great similarity between the leaves of angiosperms and other anthophytes (Taylor and Hickey, 1992; Table 9.2). The anthophytes share simple leaves that have a laterally extended base that sheaths the shoot and appears to be the structure from which stipules

Table 9.2.   Angiosperm Symplesiomorphies and Synapomorphies for Leaf Characteristics. The possible adaptations for some of the characteristics are given. Although only some characters are synapomorphies, the suite of characters would have been adaptive for rapidly growing plants with high proportional investment in leaves.

Character-States	Adaptations
Leaves	
Symplesiomorphies: broad, simple, evergreen, sheathing base, two medial strands at node, multistranded midrib, two pairs of diverging veins in lamina base, secondaries dichotomous, freely ending veinlets, hypodermis, chloranthoid-type teeth, preformed leaves.	Such leaves are important to plants of small size. The multistrand allows for redundancy of water transport system; the sheath protects young axillary organs. Preformed leaves and teeth with water hydathodes permit rapid leaf expansion; teeth also allow for guttation. Reticulate venation permits efficient water transport even when damaged.
Synapomorphies: pinnate–palmate primary venation, incomplete areoles.	
Uncertain: reticulate venation at multiple vein orders, festooned brochidodromous loops, plate meristematic type growth, lateral chloranthoid teeth.	

form in some angiosperm groups. The node has at least two medial traces and forms a multistranded midvein. These veins form at least two pairs of diverging veins in the base of the lamina, whereas the secondaries dichotomize. The leaves have a hypodermis, and the leaf tips are glandular with a vein pattern similar to that of a chloranthoid tooth.

We now include several addition states that are shared among anthophytes. Anthophyte leaves are also broad, evergreen, and preformed in the bud [at least in some bennettitaleans (Wieland, 1906) and gnetopsids]. The leaves of several gnetopsids (Rodin, 1967; Taylor and Hickey, 1992) have freely ending veinlets. The leaves of the ancestral angiosperm were likely to have been evergreen, based on the distribution of this state in conifers and gnetopsids (Loconte and Stevenson, 1991).

There are also several leaf characters found in both *Gnetum* (Fig. 2.2) and angiosperms (Hickey and Taylor, Chapter 8), including reticulate venation between several vein orders, festooned brochidodromous primary venation, and plate meristematic growth. These homologies have generally been presumed to be due to convergent evolution (Rodin, 1966; 1967) but may have been inherited from a common ancestor. The leaves of the other two living gnetopsid taxa clearly have adaptations to growth in dry areas. Thus, their leaf morphol-

Table 9.3.   Angiosperm Symplesiomorphies and Synapomorphies for Phloem (A) and Xylem (B) Characteristics. The possible adaptations for some of the characteristics are given. The synapomorphies are adaptations for rapid translocation of photosynthates in the phloem, and water and nutrients in the xylem.

Character-States	Adaptations
A. Phloem	
Symplesiomorphies: S-type plastids, large-diameter plastids, over 20 variably sized starch grains.	Sieve tubes allow for more efficient and rapid translocation of photosynthates, and possibly better protection from injury.
Synapomorphies: sieve tube members and companion cells, increased callose.	
B. Xylem	
Symplesiomorphies: sympodial system of two rings, limited secondary growth, heteroseriate rays, tracheary elements with circular bordered or scalariform pits, Maule reaction.	Vessels are advantageous when water needs are periodically high but are less strong and more likely to develop cavitation than tracheids.
Uncertain: vessels, vessels present initially in roots.	

ogy could be due to the loss of the reticulate-related characteristics or the needle-like leaves of *Ephedra* could be the ancestral type proposed by Doyle and Hickey (1976) as the groundplan from which angiosperm reticulate leaves evolved.

Angiosperms also have several unique leaf states that we consider to be ancestral. The first is pinnate–palmate primary venation. This appears throughout the basal angiosperms and is the logical ancestral form from which leaves with either pinnate or acrodromous venation evolved. All these primary-venation patterns are found in the early Cretaceous (Hickey and Doyle, 1977; Hickey, 1978, 1986). Another synapomorphy is the presence of incomplete areoles, a state widely found in the Magnoliidae.

A problematic characteristic in early angiosperms is the presence of teeth along the margin of the leaf. Leaves with laterally placed chloranthoid teeth are found among the earliest suites of angiosperm leaves (Hickey, 1978, 1986). Although this character is not shared among taxa hypothesized as basal by Taylor and Hickey (1992), this is because none of the living magnoliid taxa with pinnate–palmate venation have leaves with lateral chloranthoid teeth. Thus, leaves with this character are from a group of angiosperms that are extinct. The fossil axis with leaves and flowers described by Taylor and Hickey (1990a) shows that at least some of these leaves with lateral teeth are from plants that would be considered herbaceous magnoliids. Lateral teeth have not yet been found in other angiophytes or their sister groups.

Phloem Characters.   Although angiosperms are variable in some phloem characteristics, outgroup comparison shows which states are ancestral (Table 9.3A). Within phloem, the ancestral plastids are the S-type, large in diameter (approximately 2 µm) and have 20 or more variably sized starch grains.

A clear angiosperm synapomorphy is the presence of sieve-tube members and companion cells. These differ from sieve cells and albuminous cells found in the outgroups by having a shorter length, a sieve plate, and a common ontogenetic origin of the two cells (Esau, 1979; Behnke, 1989). The amount of callose in the pores of the plate appears to increase with the trend from scalariform to simple (Esau, 1979).

Xylem Characters.   We suggest that several xylem characters are shared among the anthophytes (Table 9.3B). One is a sympodial system composed of two rings in the stem that may be overlapped to make a single ring composed of stellar vascular bundles and leaf traces (Metcalfe, 1979). Two rings may occur when leaf traces extend beyond a single node. In comparison to conifers, anthophytes have limited secondary growth, heteroseriate rays, and their tracheary elements have circular-bordered or scalariform pitting and oblique end walls. A final state is the presence of the Maule reaction, a characteristic of both gnetopsid and angiosperm wood.

An important but equivocal character state is the presence of vessels in ancestral angiosperms (Fig. 4.1-4.3, 4.5, 4.6). One hypothesis is that they are structurally homologous with those found in the gnetopsids (e.g., Muhammad and Sattler, 1982; Fig. 4.6-4.10). This would be supported from simple parsimony if gnetopsids are the sister group to angiosperms (e.g., Hickey and Taylor, Chapter 8). Similarities are found between the tracheary elements of the two groups. Angiosperms have a two-phase pattern of secondary wall deposition (Bierhorst and Zamora, 1965) and this appears to be also found in gnetopsids (Bierhorst, 1960; Muhammad and Sattler, 1982). Both have early tracheary elements with secondary wall arranged in annular rings and helixes. The transition from helices to more continuous secondary wall is due to the deposition of additional secondary wall between the gyres. The resulting wall patterns can be considered reticulate (Bierhorst, 1960; Bierhorst and Zamora, 1965) or even scalariform (Muhammad and Sattler, 1982). In the areas between the primary helical framework, pits can develop. In angiosperms, these can be simple or bordered, have a variety of shapes (Bierhorst and Zamora, 1965), and appear to be similar to those found in gnetopsids (see Bierhorst, 1960; Muhammad and Sattler, 1982). In addition, some gnetopsid vessels have laterally elongate perforation plates.

However, angiosperm and gnetopsid tracheary elements are not identical

Table 9.4.   Angiosperm Symplesiomorphies and Synapomorphies for Timing of Reproductive Phases (A) and General Reproductive Organization (B) Characteristics. The possible adaptations for some of the characteristics are given. This suite of characters appears to be related to the herbaceous habit and permits reproductive success in unstable environments.

Character-States	Adaptations
A. Timing of Reproductive Phases	
Symplesiomorphies: short interval between pollination and fertilization, short interval between development of reproductive organ and mature seed, insect pollination.	Quicker and more efficient reproduction, thus able to survive in unstable areas with high plant loss.
Synapomorphies: short interval between germination and reproduction.	
B. General Reproductive	Organization
Symplesiomorphies: pseudanthial origin of reproductive unit, bisexual, anthion derived, branched inflorescence.	High number of small seeds and functionally bisexual reproductive axes.

(e.g., Carlquist, Chapter 4; Fig. 4.1-4.10). In gnetopsids, bordered pits frequently have a torus, whereas the scalariform pits rarely even have borders. Most of the perforation plates appear to be developmentally derived from circular bordered pits as opposed to scalariform bordered pits. The scalariform perforation plates frequently have many more bars in angiosperms than in gnetopsids. The distribution of vessels also differs, with gnetopsid vessels found throughout the plant body, whereas in some basal angiosperms (including monocotyledons), they are either restricted to shoots and roots, to roots only, or are lacking altogether. In fact, vessels appear to be lacking in the leaves of both herbaceous and woody magnoliids in contrast to *Gnetum*. Finally, the evolution of accessory characteristics to vessels are different between the two groups (Carlquist, Chapter 4). These differences have led to the common view of the separate origin(s) of angiosperm vessels and that they are not homologous (e.g., Bierhorst, 1960; Bierhorst and Zamora, 1965; Carlquist, 1975, 1983, 1988a, Chapter 4; Young, 1981).

Timing of Reproductive Phases

We have identified several additional anthophyte characters that we theorize are important for understanding the trends in angiosperm evolution. These characters relate to the relative timing of the reproductive phases and emphasize the differences between conifers and angiosperms (Table 9.4). The first of these is the interval between pollination and fertilization. Although in conifers this interval is long (up to a year), in *Ephedra* it may be as short as 10 hours (Bold, 1973; Friedman, 1990b; 1992a). In angiosperms this interval ranges from hours in herbs to months in trees (Bond, 1989). Another symplesiomorphy is the interval between the formation of the reproductive organ to production of mature seeds. In *Ephedra*, this may be as short as 5 months, whereas in conifers, it can be up to 2 years (Bold, 1973). Lastly, both the bennettitaleans (Crepet, 1972) and gnetopsids (e.g., Lloyd and Wells, 1992) indicate that insect pollination is a plesiomorphic state for angiosperms. The unique feature for angiosperms is the potentially short period of time between germination and reproduction.

Reproductive Morphology

There is no doubt that reproductive morphology has been considered the most important factor in angiosperm success (e.g., Stebbins, 1974; Cronquist, 1988; Takhtajan, 1991), yet we find that many of the supposed angiosperm states are found in the sister groups. Most of these states are discussed in detail in Hickey and Taylor (Chapter 8), whereas some suggested reproductive homologies among bennettitaleans, gnetopsids, and angiosperms are discussed in Taylor and Kirchner (Chapter 6).

General Characters.   The hypothesis that angiosperms and gnetopsids are closely related allows for a more detailed determination of the ancestral states in angiosperms (e.g., Hickey and Taylor, Chapter 8). These data suggest that the ancestral angiosperm flower was bisexual (e.g., Bernhardt and Thien, 1987; Lloyd and Wells, 1992) and had a pseudanthial origin. Hickey and Taylor (Chapter 8) propose that angiosperms shared a derived reproductive unit called the anthion with gnetopsids. In it basic form, an anthion is a compound, bisexual axis composed of a primary axis with anthial bracts. The most proximal bracts are sterile, the medial subtend secondary male axes, and the distal ones subtend secondary female axes. Gnetopsid anthions can be arranged alone or in threes in a cymose arrangement and subtended by a leaf, or can occur in opposite-decussately arranged or whorled groups on branched axes to form inflorescences. Thus, the fundamental organization of the simple flower and of the inflorescence (cf., Weberling, 1988) existed in the anthophytes and allowed for the compact reiteration of large numbers of ovules and functionally bisexual reproductive structures.

Stamens.   We find character states shared between gnetopsid and angiosperm pollen organs (Hickey and Taylor, Chapter 8; Table 9.5). Gnetopsids have male organs with filaments bearing sporangia in pairs. In some *Ephedra*, the

Table 9.5. Angiosperm Symplesiomorphies and Synapomorphies for Stamens (A) and Pollen (B) Characteristics. The possible adaptations for some of the characteristics are given. As in female reproductive organs, pollen has a reduction in the number of cells.

Character-States	Adaptations
A. Stamens	
Symplesiomorphies: four microsporangia in two pairs, basally fixed anthers, connective that extends past microsporangia, filament with multiple vascular strands.	Possibly advantageous for pollen presentation to or attraction of pollinators.
Synapomorphies: lateral dehiscence, two loculi.	
B. Pollen	
Symplesiomorphies: small size, simultaneous microsporogenesis, final division of microgametophyte occurs after pollination.	Lower energetic cost and modifications related to stigmatic germination (see Table 9.6).
Synapomorphies: perforate to reticulate pollen sculpturing[a], tectate-columellate ultrastructure[a], male gametophyte reduced to two nuclei, pollinkitt, reduced endexine.	
Uncertain: monosulcate, granular sculpturing of sulcus membrane.	

[a] Characteristic found in potential stem angiophytes.

male axes have been reduced to two pairs of sporangia with the remains of the axis making a small projection past the apices of the sporangia and separate vascular strands (Eames, 1952). Hickey and Taylor (Chapter 8) propose that the microsporangiate structures in *Ephedra* are directly homologous with the basifixed, four locular angiosperm anther with its apical tip, connective, and three vascular strands. Uniquely angiospermous states include the lateral dehiscence of the microsporangia and the opening of each of the two pairs of microsporangia into shared loculi.

Pollen. There are several angiospermous pollen character states that are also found in the outgroups (Table 9.5). The pollen of the anthophytes is generally of small size. The existence of an aperture is unclear (Brenner, Chapter 5) as Bennettitales are monosulcate, yet the gnetopsids are apparently

Table 9.6.   Angiosperm Symplesiomorphies and Synapomorphies for Carpels (A) and Ovules/Seeds (B) Characteristics. The possible adaptations for some of the characteristics are given. The adaptations of these states are related to the advent of stigmatic germination and reduction of the embryo size.

Character-States	Adaptations
A. Carpels	
Symplesiomorphies: ovules one or two.	Allows for recognition and compatibility systems, pollinator rewards and pollen competition. Also later elaboration allows evolution of advanced pollination systems, fruit dispersal, and high seed number.
Synapomorphies: small fruit size, ascidiate morphology, apocarpous, stigmatic germination on dry stigmas, progyny.	
B. Ovules and Seeds	
Symplesiomorphies: nucellus and funiculus attached opposite micropyle, outer integument free from funiculus, micropyle formed from inner integument, thick nucellar cuticle, bitegmic, onagrad embryo development, epigeal germination, embryo dicotyledonous.	Reduction in period of growth and size of embryo may be related to small size of seed and late development of storage tissues.
Synapomorphies: embryo minute, suspensor does not elongate to push embryo into gametophyte storage tissues.	

not (Erdtman, 1952). The stem angiophytes have monosulcate pollen, and if the stem angiophytes' affinities are confirmed, then this state would be ancestral along with granular sculpturing of the sulcus. We agree with Loconte and Stevenson (1991) that simultaneous microsporogenesis and postpollination timing for the final division of the microgametophyte are ancestral, based on the distribution of these conditions in conifers and gnetopsids.

Several pollen states are synapomorphies for angiosperms including two new states for our hypothesis: reduction of the male gametophyte to two nuclei (Heslop-Harrison and Shivanna, 1977) and the existence of pollenkitt (Hesse, 1984). These are found in all basal angiosperms that have been examined. Two other characteristics, perforate to reticulate pollen sculpturing and tectate–columellate ultrastructure, are found in the paleoherbs (Doyle and Hotton, 1991). If the pollen from the Triassic (Cornet, 1989a) is from an angiophyte, then the major evolutionary changes in the pollen had occurred. The only remaining character needed to achieve the suite found in the basal angiosperms would have been the reduction of the endexine.

Carpel Characters.   Speculation on the evolution and time of origin of the angiosperm carpel has stimulated considerable debate. Our polarization is compatible with Taylor's (1991a; Taylor and Kirchner, Chapter 6) analysis of carpel homologies and evolution. In Taylor and Kirchner's (Chapter 6) analysis of outgroup structure, they considered a low ovule number (one to two) per reproductive unit as ancestral (Table 9.6A). This would be homologous with one or two ovules in each carpel. The remaining characters are related to the carpel, a structure not found in the outgroups. Some of these states may have an angiophyte rather than an angiosperm distribution if *Sanmiguelia* has a carpel homologous to that of angiosperms. These states would be the presences of ascidiate morphology and basal to lateral placentation. The alternative method of polarization based on the distribution of states in the base of the phylogeny yields the same polarization (Taylor and Kirchner, Chapter 6).

Several other carpel and fruit states are distributed at the base of the phylogeny of Taylor and Hickey (1992). These ancestral states include carpels with dry stigmas (Heslop-Harrison and Shivana, 1977) and small fruit size. Although incompletely sampled, 13 families of the Magnoliidae have dry stigmas compared to four with wet (and one with both; Heslop-Harrison and Shivanna, 1977). Also, angiosperms ancestrally have ascidiate carpels that are free from one another. A final state is the presence of progyny based on its widespread occurrence in the basal Magnoliidae, although further work is needed to determine if this state is found in most basal herbaceous taxa (Bernhardt and Thien, 1987).

Ovule/Seed Characters.   Most ovule characters are shared among all of the outgroups and the basal angiosperms (Table 9.6B). The outgroups have ovules with the nucellus attached opposite the micropyle which is opposite the funiculus (0° to 5° curvature), an outer integument not developmentally integrated with the funiculus, and the micropyle formed by the inner integument (terminology from Taylor, 1991a). By conventional terminology this would be an orthotropous ovule and we agree with Taylor's (1991a) polarization that this condition is ancestral. Some of the outgroups appear to have two integuments and our evidence suggests the integuments of gnetopsids and angiosperms are homologous (Hickey and Taylor, Chapter 8). The embryo is dicotyledonous. We also accept Loconte and Stevenson's (1991) polarizations of the thick nucellus, epigeal germination, and onagrad embryo developmental states as ancestral because of their presence in gnetopsids. There are two angiosperm synapomorphies, based on the distribution in our basally placed angiosperms. One is the minute size of the embryo. The other is that the suspensor in

Table 9.7. Angiosperm Symplesiomorphies and Synapomorphies for Embryo Sac Characteristics. The possible adaptations for some of the characteristics are given. The synapomorphies are adaptations allowing energetic efficient reproduction.

Character-States	Adaptations
A. Embryo Sac	
Symplesiomorphies: archesporal division, free-nuclear division of megagametophyte, proliferation of nonreproductive gametophytic tissue, double fertilization, development of second fertilization product, no archegonia.	For small size and lower energetic cost until after fertilization. Triploid endosperm as an intermediate transfer between diploid mother and embryo, and to prevent proliferation of genetically identical sibling (polyzygoty).
Synapomorphies: reduction in megagametophyte size, formation of triploid endosperm, postfertilization production of stored resources.	

angiosperms does not enlarge to extend the embryo further into the gametophyte.

Embryo Sac Characters. Several embryo sac characters appear to be shared by all anthophytes (Table 9.7). These include an archesporal cell that divides before it goes through meiosis and the proliferation of cells that are homologous with the antipodals in angiosperms. In the gnetopsids, these form most of the gametophyte. Also shared by gnetopsids and angiosperms is free-nuclear division of the megaspore as it begins to forms the embryo sac. Friedman's (1990a, 1992b, 1994) recent work clearly shows that gnetopsids also have double fertilization resulting in cellular proliferation and development of a second fertilization product that can produce viable offspring.

Other character states are unique to angiosperms. First is the reduction in the number of cells of the female gametophyte to seven cells. It would appear that the embryo-sac type with the greatest number of free-nucellar divisions is the *Polygonum* type, suggesting that it is ancestral (Battaglia, 1989; Haig, 1990). The second state is the formation of a triploid tissue called endosperm that is frequently used as food for the developing embryo. The final state is the formation of the embryo's food reserves after fertilization, a condition unlike that found in conifers and gnetopsids. All these characters are common in the basally placed angiosperms.

EVIDENCE COMPATIBLE WITH THE HERBACEOUS ORIGIN HYPOTHESIS

In addition to phylogenetic data, other evidence is compatible with the Herbaceous Origin hypothesis. These data either support the basal placement of herbaceous Magnoliidae or show the early occurrence in the fossil record of the characteristics we propose are ancestral. We realize that fossils provide only minimum times of origin for clades and character states, but congruence with our hypothesis at least shows that the requisite morphology existed early on (Taylor and Hickey, 1990a; cf. Doyle and Donoghue, 1993). The sedimentology of deposits in which early angiosperms remains are found also provide secondary data on the paleobiology, paleoecology, and even habit of the early angiosperms, and these are also reviewed below.

Fossil Record

The growing number of unequivocal early angiosperm fossils lends support to our hypothesis of herbaceous origin. The earliest fossil axis with attached leaves and inflorescences most likely represents the remains of a diminutive plant (Taylor and Hickey, 1990a). The extreme rarity of Early Cretaceous angiosperm wood is in contrast to the abundance of well-preserved wood from other groups such as conifers (Doyle and Hickey, 1976; Hickey and Doyle, 1977). Examples of Early Cretaceous wood arc rare and small in size, a situation that has led many researchers to infer that angiosperms were of small stature and posessed limited secondary growth until Campanian times (Wheeler et al., 1987; Wing and Tiffney, 1987a, 1987b; Herendeen, 1991; Wheeler and Herendeen, 1993; Wheeler and Baas, 1993). These results are compatible with our hypothesis that the ancestral angiosperm was of small size with limited secondary growth.

Early Cretaceous angiosperm leaves are simple, variable in form, and have both pinnate and pinnate–palmate lower-order venation, and reticulate higher-order veins (Doyle and Hickey, 1976; Hickey and Doyle, 1977; Hickey, 1978, 1986; Upchurch and Wolfe, 1987; Taylor and Hickey, 1990a). The pinnate–palmate leaves can have a smooth margin or lateral chloranthoid teeth. Many early angiosperm leaves have a multistranded midrib, basal secondary veins that have a different course and behavior from those above, ethereal oil cells and incomplete areoles. One specimen has a laterally extended leaf-base (Taylor and Hickey, 1990a). Although several contemporaneous forms are found, a number of these are compatible with our hypothesis that the ancestral angiosperm leaf had pinnate–palmate lower order principal veins. Many also support the polarizations that the ancestral angiosperm leaves were simple and had incomplete areoles, chloranthoid teeth, a sheathing base, ethereal oil cells and a multistranded midrib. In addition, the physiognomy of the suite of early leaves is typical of plants with semiaquatic, herbaceous, or scrambling habit that occur today among early successional forms of stream-side and understory plants (e.g., Doyle and Hickey, 1976; Hickey and Doyle, 1977; Upchurch and Wolfe, 1987).

Recent discoveries of Early Cretaceous reproductive organs (Friis et al., 1986; Taylor and Hickey, 1990a; Friis, Pedersen and Crane, 1994) also allow discussion of the early states of anthers and flowers. The stamens have four microsporangia and a connective that extends past the locules. The carpels are usually free with few ovules. The flowers are simple and can be subtended by bracts and bracteoles, clustered into racemes with the inflorescence subtended by a leaf. These data support our hypothesis for the structure of the early angiosperm flower (Taylor and Hickey, 1992; Hickey and Taylor, Chapter 8) and for our contention that it had stamens with basifixed anthers, an apically extended connective, four microsporangia, and two loculi; had a carpel with ascidiate morphology and few ovules; and as the flowers were simple and arranged in cymose to racemose inflorescences.

The Cretaceous to Tertiary record of seeds and fruits (Tiffney, 1984; Wing and Tiffney, 1987a, 1987b) is also informative by showing that throughout the Cretaceous, these organs were small. Based on seed and fruit size in living angiosperms, these fossils plants were likely to have been herbs (e.g., Harper, 1977; Tiffney, 1984) even with the possible changes related to the evolution of new dispersal agents between the Cretaceous and Tertiary. This record suggests that the tree habit did not become well established among angiosperms until the Tertiary (Tiffney, 1984; Wing and Tiffney, 1987a, 1987b).

Early angiosperm pollen has been reported from the Valanginian in Israel (Brenner and Bickoff, 1992; Brenner, Chapter 5) and Hauterivian of England (Hughes and McDougall, 1987, 1990; Hughes et al., 1991). These grains are small, with perforate to reticulate sculpturing. Some are monosulcate, whereas others, including the earliest in the sequence, are inaperturate. Again, these states include those we suggest are ancestral and are found in living herbaceous magnoliids.

Sedimentology and Geochemistry

Paleoecological studies also support the hypothesis that the early angiosperms were herbaceous. Doyle and Hickey (1976; Hickey and Doyle, 1977) noted that the earliest angiosperm fossils from U.S. Barremian/Aptian localities appear in near-stream facies, although they proposed that these plants were descended from a weedy, shrub-like, ancestor. Retallack and Dilcher's (1981a) analysis of western U.S. Cenomanian deposits also supports a fluvial or coastal paleoecology. Hickey and Taylor (1992) carried out a detailed depositional analysis of the facies and fossil plants from the Barremian–early Aptian Dutch Gap locality in the Potomac Group of Virginia. We found that the angiosperms there were restricted to levee environments where they were associated with several species of ferns. Bennettitaleans were found mostly in backswamp deposits, and although some conifers occurred there as well, they were more diverse and abundant in channel deposits, presumably washed in from eroded river terraces. When evaluated within the context of ecological theory (Harper, 1977; Tilman, 1988), the available sedimentological data indicate that early angiosperms were small and rapidly growing to survive on such unstable, nutrient-rich sites.

Recent paleoecological studies of an early Maestrichtian flora from Wyoming also provide indirect evidence on angiosperm stature (Wing et al., 1993). These data show that although dicotyledonous angiosperms are diverse (61% of the species), they made up only 12% of the cover of the site, with nearly half of the cover composed of free-sporing plants. In addition, most of the angiosperm species whose affinities or habit could be determined represent taxa that are herbs, scramblers, or shrubs. Thus, even at the midpoint of their fossil record, angiosperms iñ at least some localities appear to be predominately small, opportunistic plants.

Additional data suggesting that large size and increased biomass were late acquisitions for flowering plants are provided by the record of the biomarker oleanane. This triterpanoid compound is found in oil, deltaic sediments, fluvial sediments and fossil plants (see Moldowan et al., 1994 and references therein). Oleanane is putatively an angiosperm biomarker based on its affinity to the beta-amyrin group of natural products and its abundance in Tertiary oil and sediments. Oleanane and related compounds are currently found in many different clades of angiosperms, with the only other extant occurrences from a lichen and several ferns such as *Polypodium* and *Marsilea*. Recent studies show that it is found both in fossil angiosperms (Stout, 1992) and bennettitaleans (Taylor et al., 1992).

Low levels of oleanane (in relation to the bacterial biomarker hopane) are found from the Carboniferous to the Early Cretaceous (Moldowan et al., 1994). Based on extensive sampling of marine sediments from the Jurassic to Recent, oleanane concentrations increase slightly during the Early Cretaceous and then increase dramatically during the early Tertiary (Moldowan et al., 1994). The frequency of oleanane occurrence is low in Jurassic and Early Cretaceous sediments with a major increase occurring in sediments from the end of the Late Cretaceous.

A surprising aspect of this trend of oleanane increase is that it does not directly follow the Late Cretaceous angiosperm diversification curves suggested by Muller (1981) for appearance of modern families or the mid-Cretaceous increase of fossil taxa documented by Lidgard and Crane (1988; Crane and

Lidgard, 1989, 1990), but appears to be intermediate. There are several possibilities (Moldowan et al., 1994), one of which is that the delay in the increase of the amount of oleanane compared to the appearance of taxa occurs because oleanane is tracking angiosperm dominance and biomass. Another possibility is that oleanane producing groups of angiosperms did not diversify until late in the Cretaceous, but this inference is contradicted by the pollen diversification curves (Muller, 1981). Bond (1989) has also noted that taxonomic diversity does not necessarily correlate with the dominant plant type or with the plant group producing the greatest biomass. If the biomass of angiosperms was low, then little oleanane was likely to reach the marine record. The oleanane record, thus, is compatible with the paleoecological studies described above in suggesting small stature for the early angiosperms.

Summary of Compatible Evidence

Considerable data from a variety of sources are consistent with the Herbaceous Origin hypothesis. Many of the characteristics found early in the fossil record are the same as those proposed by our herbaceous hypothesis. Fossils, especially flowers that are similar to those proposed by the Magnolialean hypothesis are not found until later (Friis and Crepet, 1987; Friis, Pedersen and Crane, 1994). Paleoecological studies and the geochemical record support our hypothesis that early flowering plants lived in the unstable fluvial systems, were of small size, and were not dominant either as groundcover or biomass. It is notable that the aspects of the fossil record from which habit can be assessed suggest that the angiosperms were small during much of the Cretaceous and that the tree habit did not become dominant until the Late Cretaceous to Early Tertiary. At a minimum, the record points to the herbs as a diverse group among the early angiosperms (Taylor and Hickey, 1990a; Doyle and Donoghue, 1993; Doyle, 1994).

ADAPTATIONS OF ANGIOSPERM ANCESTRAL CHARACTERISTICS

Just as the discussion of ancestral angiosperm morphology has preoccupied evolutionary botanists, so has the discussion of the adaptive importance of those states unique to the group. In Table 9.1–9.7 we summarize the ancestral states and the possible adaptations of some of these characteristics. We also propose a suite of synapomorphies for angiophytes and angiosperms. Some of these shared derived states are final stages in trends inherited from the outgroups; for example, reduction in size of the mature plant. Others are potentially nonadaptive states associated with innovations, such as the apocarpous carpel as compared to the evolution of the carpel with stigmatic germination. Finally, some, such as sieve-tube members, appear to be major innovations that have allowed angiosperms to function differently from any other plant group. In this section, we provide scenarios under which these states can be understood as adaptations best suited to herbaceous plants. We propose adaptive scenarios for synapomorphies from five character suites: vegetative growth and habit, leaf characters, phloem and xylem characters, timing of reproductive phases, and reproductive morphology.

Vegetative Growth and Habit

Herbaceous habit, small size, and rapid growth are synapomorphies for angiosperms and appear to be advantageous in certain environments (Table 9.1). Generally, perennial herbs are found in pioneer habitats (Stebbins, 1974) or disturbed localities that experience a periodic loss of plant cover (Tilman, 1988). This habit allows for a clonal strategy in which a successful colonization event is followed by a rapid attainment of local dominance by clonal growth (Corner, 1964; Harper, 1977). The clonal strategy itself is linked to disturbed habitats (Eriksson, 1993), and, depending on environmental conditions, clonal plants can invest in vegetative propagation, sexual reproduction, or vegetative growth. Limited secondary growth permits relatively greater energetic investment into photosynthetic area and earlier reproduction (Tilman, 1988).

Although such features as herbaceous habit and small size are clearly related to the Herbaceous Origin hypothesis, so is rapid vegetative growth. Angiosperm herbs and seedlings, but not trees, grow more rapidly than conifers (Bond, 1989). Bond (1989; Midgley and Bond, 1991) suggests that angiosperms are successful against conifers because angiosperm seedlings grow more rapidly and are thus more competitive at this stage, at least on nutrient-rich soils. Rapid growth would have allowed them to outcompete other seed plants, whereas the flexibility accorded by seeds would have allowed them to become established more rapidly than their vegetatively equivalent competitors, the ferns. Thus, we would suggest that, although rhizomatous growth may have existed earlier, acquisition of the herbaceous habit by a seed plant, accompanied by small size and rapid growth, constituted a key innovation for angiosperms.

Leaf Characters

The leaf may have been a crucial element in the development of the angiosperm herb. Although a number of the angiosperm foliar characteristics are already found in the outgroups, these may have been of greater adaptive value to small, short-lived plants or rapidly growing plants with short-lived leaves (Table 9.2). Carlquist (1975) notes that the multiple bundles in the leaf provide a redundant system of water transport. This would be significant in rapidly expanding leaves or in plants with limited secondary growth. Preformed leaves would allow for rapid leaf growth but need an adequate water supply to allow for this expansion (Bond, 1989). Evidence of the importance of continual or periodic water stress is the existence of a hypodermis (Stebbins, 1974) and water hydathodes. The latter are related to foliar water balance (Wilkinson, 1979) and are frequently found at the glandular tips of teeth. Chloranthoid-like teeth with an apical papilla are found throughout the anthophytes at the leaf apex. The gland is of relatively large size in chloranthoid teeth and this tooth type is widespread in basal angiosperms (Hickey and Wolfe, 1975, Taylor and Hickey, 1992). These glands may have allowed sufficient flow of water and nutrients for rapid leaf expansion.

Other characters that may be synapomorphies for angiosperms include lateral teeth, reticulate venation at multiple vein orders, and festooned brochidod-romous loops. Lateral teeth may represent one of a syndrome of characters acquired by angiosperms at about the time they achieved the herbaceous habit. They would have increased the efficiency of water and photosynthate flow over that found in other anthophytes. In addition, the glandular portion of the tooth may have helped maintain water balance in the plant by allowing guttation to overcome cavitation in the vessels (Carlquist, 1975, 1980). Reticulate venation may also have been adaptive because it allows for increased rates of flow (see Bond, 1989) and efficient water transport in the leaves that were damaged (Plymale and Wylie, 1944). These benefits are particularly great in leaves with multiple primary veins at the base, as we suggest are ancestral in angiosperms. Finally, festooned brochidodromous loops may have provided marginal strengthening for broad, thin leaves.

Phloem and Xylem

Several characteristics of the vascular tissue may have been related innovations for the herbaceous habit of the early angiosperms (Table 9.3). Although the functional adaptations of some of the phloem states are unclear (Esau, 1979), we concur with the suggestion that sieve-tube members are more efficient in rapid translocation (Carlquist, 1975). Carlquist (1975) also notes that such efficiency would be important in plants with a finite amount of conducting tissue that must function for long periods without replacement, as in plants with limited secondary growth. Interestingly, several rhizomatous, herbaceous, spore-bearing taxa (*Isoetes*, *Psilotum*, and *Equisetum*) have highly evolved photosynthate transport cells that closely resemble sieve tube elements (Behnke, 1989). In addition, we speculate that the increase of callose along with the production of slime may be related to increased ability to seal off damage (Esau, 1965). More efficient translocation and damage control would be important adaptations for rapidly growing, small herbs. Because such analogous characteristics have evolved in parallel in herbs but not in trees, these states would again point to an herbaceous origin.

Several aspects of angiosperm vessel evolution remain unclear, including whether they are homologous with those found in gnetopsids or had separate origin(s). It is still debated whether the absence of vessels in vesselless angiosperms is ancestral or due to their evolutionary loss. Finally, the significance of vessel distribution in angiosperm organs is unclear. Nevertheless, vessels do appear to be adaptive as they are found in a number of unrelated groups. Vessels have evolved in parallel in several spore-bearing groups including *Selaginella* (lycopsid), *Equisetum* (sphenopsid), and ferns (Bierhorst, 1960; Carlquist, 1975). In these groups, the evolution of vessels is clearly associated with the rhizomatous, herbaceous habit (Carlquist, 1975). In gnetopsids, vessels may have evolved in response to water needs related to the specialized desert-shrub and tropical liana habits (Carlquist, 1975). From these associations, Carlquist (1975) concludes that the existence of vessels is due to periodic water stress. The stress may be due to variable periods of rapid growth, of higher transpiration rates, or of lower water availability.

Several potential problems occur with vessel members in comparison to tracheids. Vessel members have lower mechanical strength and are more susceptible to air embolisms that can result in cavitation of the xylem (Carlquist, 1975). Thus, many researchers have noted that the evolution of angiosperm vessel members is paralleled by the evolution of the fiber-tracheids and libriform fibers for support. Many angiosperms also have phloem fibers (Esau, 1965) although these are lacking in the gnetopsids and many basal angiosperms. Problems of cavitation can be overcome by several means, including root pressure in smaller plants (Carlquist, 1975, 1980). These disadvantages of vessel members are less severe for perennial herbs. These plants do not need to have the mechanical strength necessary for the shrub or tree habit, and root pressure can be used to overcome air embolisms due to their small size. Evidence for this strategy is found in the high guttation rates in herbs (Carlquist, 1975), rates that may have been augmented by the evolution lateral teeth with water hydathodes in angiosperm leaves.

Based on the factors associated with vessels in other plant groups, it appears that angiosperm vessels evolved in response to variability in growth, transpiration, and water availability. Yet we need to further develop this scenario in order to explain why vessels first appear in the roots of herbaceous angiosperms and why some angiosperms are vesselless. Carlquist's (1975) summary of xylem evolution in the monocotyledons is informative, as it provides an understanding of the trends in vessel evolution in a strictly herbaceous angiosperm group. These same trends are also found in ferns. Evidence suggests that the basal monocotyledons include the Dioscoreales (Dahlgren, et al., 1985) and Araceae (Duvall, Learn et al., 1993). As with herbaceous magnoliids, these species are herbs, vines, and epiphytes from aquatic habitats, marshes, and wet forests (Carlquist, 1975). Vessel elements are restricted to roots in these mono-cotyledonous species as they are in perennial herb magnoliids such as *Sarcandra* (Carlquist, 1987) and *Nelumbo* (Carlquist, 1975). A possibility is that vessels appeared first in the roots in the early angiosperms. The increase water flow may have been particularly adaptive for these plant that lived in conditions where water availability was variable.

The evolution of derived types of vessel elements and elements in other organs is likely due to the interplay between problems of strength and cavitation. Vessel members with simple end walls probably arose quickly in early perennial herbs, and they are the type currently found in most herbs (Carlquist, 1975). In addition, derived groups evolved the placement of vessels throughout all their organs. These changes further increased the conductive efficiency that began with the appearance of vessels. Yet many trees live in conditions where the trade-off between the advantages conferred by derived vessel types and the loss of strength are insufficient for the further evolution. This may explain why conifers compete well against angiosperms in some habitats and why plants with scalariform vessel members still exist (Carlquist, 1975; Bond, 1989; Midgley and Bond, 1991).

Based on our hypothesis that vessels were initially restricted to the roots of perennial herbs, we propose a new explanation for vesselless angiosperms. We would hypothesize that the loss of vessels is due to the blocking of vessel development in the roots. Carlquist (1975) has suggested such a mechanism for some aquatic monocotyledons and we would extended it to all vesselless aquatic angiosperms (e.g., Nymphaeaceae). The evolution of vessels in woody plants would have depended on whether the benefits of vessels (efficient water flow) outweighed the problems (weakness and cavitation) in comparison to tracheids. Indeed, vesselless angiospermous trees live in habitats comparable to those where tracheid-bearing conifers are found, suggesting that tracheids are equally effective in these habitats (Carlquist, 1975). Thus we suggest that the vesselless condition of Winteraceae and others may not be the ancestral condition for angiosperms but may be due to the loss of vessels in the roots. The fact that the xylem in vesselless angiosperm groups has other independent specializations (Carlquist, 1975, 1980) suggests multiple origins of the woody habit from perennial herbs that lacked vessels in the stem. Once dicotyledonous plants evolved vessels in the stems, these would have been retained because the problems of strength and cavitation were overcome by different means (Carlquist, 1975).

Timing of Reproductive Phases

Another suite of important adaptations for herbaceous angiosperms is related to the interval between reproductive phases (Table 9.4). The phylogenetic trend toward decreased reproductive intervals reaches it culmination in angiosperms. In comparison to other seed plants, anthophytes benefit from the short interval between pollination and fertilization, and between production of reproductive organs and seed formation. They have also developed more efficient means of pollination by using biotic pollinators. Herbaceous angiosperms have an even shorter interval between seed germination and reproduction in comparison to other seed plants. In comparison to spore-bearing plants such as ferns, rapid fertilization may be similar, yet, angiosperms have more rapid completion of the gametophyte stage and have the other benefits of seeds.

Although ancestral angiosperms are inferred to have shared a similar habitat and growth form with ferns and other spore-bearing plants, their advantage would have lain in reproduction by seeds, higher efficiency of fertilization through pollination, and shortened reproductive intervals. Although insect pollinators may have been an important factor in the later radiation of the angiosperms (e.g., Regal, 1977; Crepet, 1983; Bernhardt and Thien, 1987; Crepet and Friis, 1987; Crepet et al., 1991; Kiester, Lande and Schemske, 1984; Eriksson and Bremer, 1992), they do not appear to be crucial for the initial evolution and diversification of the group (Bond, 1989; Midgley and Bond, 1991; Lloyd and Wells, 1992). This is supported by the existence of insect pollination in the sister groups and analysis of the insect fossil record that shows that important insect pollinators do not radiate during the initial radiation of the angiosperms (Labandeira and Sepkoski, 1993).

It appears to us that a drastically shortened reproductive cycle was of great importance to rhizomatous herbs living in unstable environments. Unlike other seed plants, angiosperms hold most of the energetic investment in the next generation until after fertilization. This would be advantageous in sites where resources for reproduction are limited and when it is uncertain if fertilization will occur or if reproduction will be completed. In our view, the rapid development of reproductive organs coupled with the herbaceous growth habit would have been of optimal value for exploiting ephemeral environments that were the initial angiosperm niche and for allowing them to invade surrounding sites.

Reproductive Morphology

Although a number of floral characteristics appear to be shared among anthophytes, several angiosperm synapomorphies could be related to their herbaceous habit (Tables 9.5 to 9.7). We note that the existence of carpels implies stigmatic germination. The evolution of the carpel allowed for adaptations relating to pollen, or more specifically, sperm competition (Mulcahy, 1979; 1983; Mulcahy, Mulcahy and Searcy, 1992), fruit dispersal systems (e.g., Regal, 1977; Tiffney, 1984), compatibility and recognition systems (e.g., Whitehouse, 1950; Zavada and Taylor, 1986; Bernhardt and Thien, 1987), and pollinator rewards (e.g., Lloyd and Wells, 1992). Pollen structure, such as reticulate to perforate ornamentation, may also be related to the evolution of recognition and incompatibility proteins (e.g., Whitehouse, 1950; Beach and Kress, 1980; Zavada and Taylor, 1986; Lloyd and Wells, 1992). Other evidence suggests that the evolution of elaborate or large fruits is not directly related to the diversification of the angiosperms (Herrera, 1989; Eriksson and Bremer, 1992). The evolution of the carpel may also have been necessary to delay the accumulation of stored food in the seed until after fertilization.

Thus, the evolution of the carpel may have been important in pollen competition, compatibility and recognition systems, and protection of the developing seed. The first two states, along with dichogamy, may have resulted in increased speciation rates (Doyle and Hickey, 1976; Doyle and Donoghue, 1986b). These increased rates would have allowed angiosperms to radiate in this new adaptive niche for seed plants. Compatibility and recognition systems would have been particularly important in these clonal herbaceous plants especially for allowing outcrossing.

Other reproductive adaptations may be related to the small size of the angiosperm fruit, seed and embryo. Small diaspores are correlated with high dispersal ability and a long dormancy period (Harper, 1977; Rees, 1993). Long term dormancy appears to be related to a short life span (Rees, 1993) as is found in the herbaceous magnoliids. The small embryo and suspensor size in angiosperms may be related to their growth within the simultaneously developing food storage tissue, in contrast to the case in other seed plants where the embryo has to be forced into the earlier formed food reserves.

The major evolutionary trends and adaptations for the angiosperm gametophyte are a delay in producing stored energy reserves until fertilization is completed and the production of triploid endosperm. The first results in reduction in the energy invested by the plant until fertilization has occurred. As for triploid endosperm, there have been many suggestions about why it is adaptive. These include multiple dosage of genes, heterozygous and polyploid vigor, more favorable environment for maternal investment, and sibling–maternal relationships (e.g., Westoby and Rice, 1982; Donoghue and Scheiner, 1992; and references therein).

We note several additional implications of double fertilization that develop out of ecological and evolutionary models on genetic similarity and resource investment. Because gymnosperms produce stored food for the embryo before fertilization, the amount they can accumulate is restricted and the energetic committment is already made in the haploid gametophytic tissue before fertilization. One means to increase fitness of the final, and usually single zygote in the seed, when stored resources are limited, is to produce multiple, genetically different embryos, a condition called polyzygosis. The fastest growing embryo will sequester the greatest amount of food. One strategy by which a single zygote can increase its success is polyembryony, in which a single zygote makes additional embryos. This group of embryos act as resource sink against competing, genetically different embryos, but because they are genetically indistinguishable, it does not matter which individual survives. The evolution of double fertilization in *Ephedra* carries polyembryony to the extreme. Although there are multiple archegonia and, thus, the possibility of polyzygosis, the first pollination results in two fertilization events and eight genetically identical embryos from the initial divisions of the two fertilized cells (Friedman, 1992a). So, in the gymnosperm system the diploid mother plant invests in the haploid gametophyte because the gametophyte only has maternal genes. The gametophyte can invest in any embryo because all of its genes are found in all of the new sporophytes. Thus, the gametophyte is an important intermediate between the mother plant and the zygote.

The situation clearly changes in angiosperms because the mother's investment in resources for the embryo now occurs after fertilization (Westoby and Rice, 1982) and the ovules have only a single egg. With a single egg, polyembryony should be repressed as is the case in angiosperms. We suggest that endosperm is important for both these new developments. First, with little or no food storage in the gametophyte [although this is variable and occasionally gametophyte tissue does proliferate (e.g., Battaglia, 1989; Haig, 1990)] and there is no intermediate genetic stage to transfer resource from the diploid mother to the diploid zygote. Triploid endosperm accomplishes this by having two-thirds of the genetic information identical to that found in diploid mother whereas all of the genetic information is found in the zygote. Thus, triploid endosperm can act as an intermediate sink between the mother sporophyte and the zygote. Of additional importance is the fact that the second fertilization that produces triploid endosperm cannot be viable because of its chromosome complement. Thus, by being triploid, the product of double fertilization can sequester energetic resources (Haig and Westoby, 1989) but it cannot produce a viable offspring to compete with the zygote.

In any event, the angiosperm embryo sac which has evolved in response to the reduction in size and the shifting of embryo food production until after fertilization, results in lower energetic cost for angiosperms seeds as compared to other seed plants (Haig and Westoby, 1991). It is not obvious that the evolution of triploid endosperm is adaptive over similar food reserves in other seed plants (Westoby and Rice, 1982). Yet the increase in energetic efficiency of the seed and embryo sac due to postfertilization production of energetic resources and endosperm formation would be adaptive in an herbaceous plant. The plant could efficiently invest in offspring if fertilization occurred whereas otherwise reserving energetic resources for vegetative growth. This would allow highly flexible resource allocation in an herbaceous plant.

A COMPREHENSIVE HYPOTHESIS OF ANGIOSPERM ORIGIN

Solution of the problem of angiosperm origin concerns more than just predicting the ancestral form. Models of origin should examine and hypothesize the adaptive milieu of the early angiosperms. Most recently, Doyle and Donoghue (1993) have suggested that herbaceous angiosperms are a successful but derived group. In fact, they agree that some of the innovations that we suggest are crucial were important for herbaceous magnoliid success and diversity. Yet, in contrast to our hypothesis, their scenario does not regard these traits as ancestral for all angiosperms.

The argument that angiosperm innovations predate angiosperm diversification is put forth by maximum likelihood arguments (Sanderson and Donoghue, 1994; Nee and Harvey, 1994). These analyses suggest the rate of branching remained the same until after the herbaceous magnoliids diverged. We disagree with several points of their analysis. They assume that the relationships of the basally placed angiosperms are restricted to one of four groups with the basal herbaceous members represented by the Nymphaeales, and suggest that the woody magnoliids (sensu Donoghue and Doyle, 1989a) are monophyletic. We disagree (see also Sytsma and Baum, Chapter 12) and suggest the most basally placed herbaceous group is in doubt. The most compatible result between Hamby and Zimmer (1992) and Taylor and Hickey (1992) would have the herbaceous magnolialeans represented by the Piperaceae. Piperaceae has similar species diversity (2000 species) to the wood magnoliids.

In general, herbaceous plants do show higher speciation rates than woody plants (Eriksson and Bremer, 1992). In addition, higher diversity is found in disturbed sites, an environment in which perennial herbs are commonly found (Reice, 1994). The use of Nymphaeaceae as a representative clade may skew the results because aquatic angiosperms may simply be less diverse than other angiosperms. In addition, we agree with Sanderson and Donoghue (1994) that this method ignores differences in branching rates due to differing rates of extinction. We have previously argued (Taylor and Hickey, 1990a) that herbaceous angiosperms are missed in the fossil record because of their small size, thus, Chloranthaceae may have been considerably more diverse. In fact, there is a large group of Cretaceous fossil leaf species with lateral chloranthoid teeth that may represent a diverse, herbaceous clade. Finally, the change in branching rate may have been more complex, with stepwise acquisition of additional derived character states. These states would include advanced insect pollination (e.g., Crepet and Friis, 1987), more elaborate defense compounds (e.g., Kubitzki and Gottlieb, 1984), and coevolution with seed dispersers (e.g., Tiffany, 1984; Herrera, 1989). Even if angiosperm radiation is not associated with such key innovations, the Herbaceous Origin hypothesis is not excluded. If, indeed, the angiosperms were initially of low diversity like their sister groups, extrinsic factors may have been important.

In the sections above, we have listed our proposed ancestral characteristics and their possible adaptations. A great deal of current phylogenetic and paleobotanical data support or are compatible with our Herbaceous Origin hypothesis. In the next section, we propose our general scenario of angiosperm origin. In it we include their ancestral habitat and habit as well as the subsequent morphological and ecological radiations from the ancestral states.

Ancestral Habitat

For some years now, sedimentological data have suggested that the early angiosperms lived in a fluvial regime on sites of relatively high disturbance with moderate amounts of alluviation (Doyle & Hickey, 1976; Hickey and Doyle, 1977; Crane, 1987; Hickey and Taylor, 1992). We now propose that rather than just being an early site, that this was where the ancestral angiosperms first evolved. These sites would have been characterized by high nutrient levels and frequent loss of plant cover due to periodic disturbances. Ecological models of living plants and ecosystems can be used to predict the types of adaptation and habit that would be present in such environments. Tilman (1988) has developed and tested a comprehensive ecological model of plant interactions and life-history traits. Based on his analyses, he finds that plants that grow on nutrient-rich, high-plant-loss sites have the following characteristics: low total biomass, optimized physiology for high light penetration, small size at maturity, high vegetative growth rate, young age at reproduction, low allocation to stems, and high allocation to leaves. Stebbins (1974) also predicted which characteristics might be found in plants radiating into a deteriorating, unstable environment. We suggest that a climatically stable but periodically disturbed environment found on the floodplain would be closely comparable to the unstable environment envisioned in the Stebbins's (1974) model. He suggests that the plants in such environments should have physiological specializations, dormancy mechanisms, rapid growth during favorable conditions, specialized resting organs, and defense systems and should be reduced in size. The characteristics from both models are the same traits suggested by our Herbaceous Origin hypothesis.

The ecological literature (e.g., Harper, 1977) also has shown empirically that perennial herbs are well adapted to unstable environments. Such plants have a suite of characteristics that allow them to exploit a variety of niches due to the highly flexible nature of their vegetative and reproductive systems and to their patterns of growth. These include multiple dispersal methods and resting stages that use both seeds and rhizomes. Small seeds allow long distance dispersal and colonization of new or favorable sites, whereas rhizomes can act as large seeds that allow the plant to survive shading, herbivory, and inundation until favorable conditions return (Harper, 1977). Flexible allocation of resources to roots, stems, leaves, and sexual reproduction is possible and varies depending on competition and environmental conditions experienced by the plant. Thus, a single species can vary its life history strategy from that of a colonizing herb controlled by density-independent factors to that of a tall shrubby herb competing with shrubs and under density-dependent control. Finally, the perennial, herbaceous habit allows survival until sexual reproduction is successful (Stebbins, 1974). Thus, early angiosperms would have been able to tolerate disturbance both from physical factors and herbivory, including dinosaurs (Bakker, 1978, 1986).

This ecological and adaptive scenario is compatible with evidence from the fossil record. First, the oleanane increase attributed to angiosperms is delayed in comparison to the taxonomic record (Moldowan et al., 1994), as would be predicted if angiosperm biomass was initially low. Second, under our scenario, although angiosperms had become dominant in some niches by the mid-Cretaceous, they did not reach overall dominance as trees until the end of the period. This is supported by paleoecological studies and the fossil wood record. Finally, the fossil record of angiosperms is dominated by a large number of their leaves, especially in comparison to their wood. This suggests an ecological strategy that placed a premium on energetic investment in leaves.

Another prediction is that angiosperms evolved in high-nutrient habitats in which efficient nutrient retention was unimportant. Habitats dominated by early successional and deciduous angiosperms lose nutrients through runoff at a faster rate than do conifer-dominated habitats (Knoll and James, 1987). This change in nutrient efficiency is noted in the geochemical record with an increased weathering rate beginning during the Late Cretaceous (Knoll and James, 1987). Again, the achievement of angiosperm biomass dominance appears to have occurred later than taxonomic diversification and yields a similar pattern to that found from the oleanane record. Thus, part of the success of the angiosperms was their exploitation of nutrient-rich environments from the equator to the poles.

Ancestral Morphology

As we have previously suggested, many of the angiosperm synapomorphies may be adaptive for herbaceous perennials. This is not to say that some of these are not adaptive as well to angiosperms with shrubby or tree-like habits. Rather, they are most fully developed in perennial herbs (see also Doyle and Donoghue, 1993). To summarize, these plants would have been small at maturity and have had rapid growth, allowing them to live in unstable environments. In such a setting, secondary growth is often limited because energetic investment is to other organs. The presence of vessels in the roots would have permitted rapid

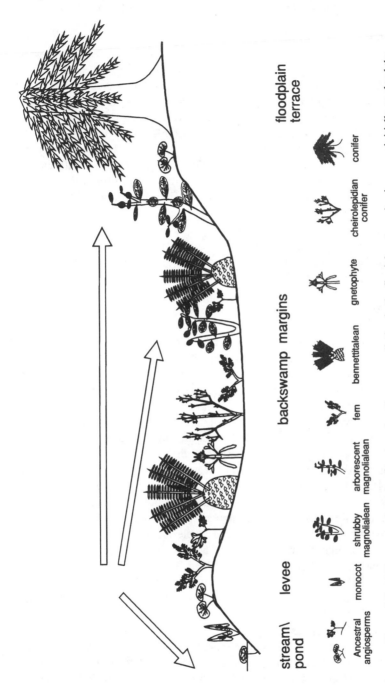

Figure 9.2. Diagrammatic representation of angiosperm radiation in the fluvial system. Angiosperms initially evolved the herbaceous habit and reduced, efficient reproductive systems. They lived on levees in competition with ferns and other spore-bearing herbaceous perennials. From there, they radiated to the aquatic setting, to the margins of the backswamp and to floodplain terraces. During this time, angiosperms gained new adaptations, , including the wood habit, to survive in competition with other seed plants. Other than conifers, most of the other seed-plant groups became extinct.

263

water and mineral transport during periods of rapid growth. In addition, sieve tubes would have been advantageous because during periods of rapid growth, these plants would have needed the capacity for rapid and efficient translocation of photosynthates. The reticulate venation of angiosperm leaves would have been a consequence of maintaining a relatively low energetic cost for a large foliar unit and to allow water movement in the event of damage.

We agree with Bond (1989; Midgley and Bond, 1991) that the overall superiority of the angiosperm reproductive system is related to the ability to grow rapidly, and only secondarily to pollination biology and advanced dispersal systems (see also Herrera, 1989). Changes in the duration and onset of reproductive phases, reduced size, and postfertilization production of stored food in the seed allow these plants to reproduce early and with energetic efficiency. The perennial herb habit would have been a generalist strategy including good dispersal ability. This may explain the rapid spread of angiosperms over the world during the Early Cretaceous (Lidgard and Crane, 1988; Crane and Lidgard, 1989, 1990).

Early Ecological Radiation

Early angiosperms followed several ecological/adaptive pathways during their initial radiation (Fig. 9.2). Angiosperms' initial competitive interactions were with ferns and sphenopsids in disturbed areas, such as the levee environment. Our proposal is in contrast to the ideas of Doyle and Hickey (1976; Hickey and Doyle,1977) who proposed a similar pattern but who, with Stebbins (1974), thought that the earliest stages of angiosperm evolution may have occurred in the xeric environment. We infer that these plants all shared the perennial, rhizomatous herb habit; the only real advantage that angiosperms had was mode of reproduction (cf. Haig and Westoby, 1991). Seeds were more effective dispersal agents than spores and would have allowed more rapid establishment of angiosperms than ferns on unstable sites. Paleoecological data show that initially ferns and angiosperms occur on similar sites but that subsequently the ferns become restricted to more stable areas.

We suggest that adaptations to unstable habitats not only led to angiosperm success in areas underexploited by seed plants but was instrumental to the success of angiosperms in invading other niches. The flexibility of the rhizomatous, perennial herb habit facilitated the evolution of other growth forms that were adapted to adjacent habitats.

A niche invaded by angiosperms during the Early Cretaceous diversification was the aquatic setting (Fig. 9.2; Doyle and Hickey, 1976; Hickey and Doyle, 1977). Angiosperms are the most successful tracheophyte group to invaded aquatic sites, although a few fern and lycopsid lineages developed aquatic forms. Only rhizomatous perennials can develop a submerged aquatic habit because they are the only plants that are able to concentrate a large portion of their metabolic investment in leaves with a relatively minor investment in stems and roots. The evolution of aquatics occurred several times in both monocotyledonous and dicotyledonous herbaceous plants.

Another habitat that angiosperms invaded early was the more stable regions of the floodplain (Fig. 9.2; Doyle and Hickey, 1976; Hickey and Doyle, 1977). In this setting there was competition for light and the plants evolved extensive secondary growth. This is where angiosperms finally began to compete against the major clades of conifers. Bond (1989) has made a compelling argument that angiosperm trees out-compete conifers through seedling competition. Thus, in any forests that are periodically disturbed through fire, herbivory (Bakker, 1978, 1986; Wing and Tiffney, 1987a, 1987b) or other causes, angiosperms can compete in the gaps. Yet angiosperms do less well on nutrient poor sites and these are places that conifers still dominate (Bond, 1989). The Magnoliales, Laurales and Illiciales are examples of early evolution of the woody, probably shrubby habit in angiosperms.

Angiosperms also penetrated the distal margins of the stream levees early in their evolution (Fig. 9.2; Doyle and Hickey, 1976; Hickey and Doyle, 1977). During the late Triassic to early Cretaceous, this environment appears to have been occupied by shrubs and small trees such as bennettitaleans, *Caytonia*, corystosperms, cheirolepidian conifers like *Frenolopsis* and *Pseudofrenolopsis* and some gnetopsids. When angiosperms evolved the woody habit they could penetrate this habitat, especially with their advantage of reduced and efficient reproductive system, and the ability to grow more rapidly as seedlings. It may be significant that many of the probable competitors of flowering plants became extinct during the middle of the Cretaceous (e.g., Retallack and Dilcher, 1981a; Crane, 1987), although conifers continued to dominate the backswamps until the Eocene (Hickey, 1977; Johnson and Hickey, 1990).

The subsequent history of flowering plants during the Cretaceous was for radiation away from the fluvial setting. Their ancestral habit as perennial, seed bearing herbs opened the door to deciduous strategies in niches with fluctuating rainfall, temperature, or sunlight. Derived growth forms such as annual and biennial herbs, and various geophytes are abundant in these environments (Harper, 1977).

The success of angiosperms in tropical forests is in some ways the hardest aspect of this radiation to understand. Tropical soils are typically poor in nutrients in comparison to temperate soils. On the other hand, gaps produced by tree falls are typically rich in nutrients, and we suggest that the success of early angiosperm herbs and seedlings allowed angiosperms to gain a foothold here. Small angiospermous plants and seedlings would have had considerably faster growth rates than conifers as they do today (Buckley, 1984). Equally important may have been the evolution of endomycorrhizal–angiosperm relationships that are found in many herbs, shrubs and trees. These are much more common, especially in the tropics, than the ectomycorrhizae found in conifers and some dicotyledonous trees (Longman and Jeník, 1987). Mycorrhizae are fundamental in allowing nutrient recycling and may have allowed angiosperms to penetrate environments with poor soils. Interestingly, in the poorest tropical soils, conifers such as podocarps are found competing successfully with angiosperms. Thus, the inferred original adaptations of angiosperms to nutrient-rich soils have effectively prevented angiosperms from extirpating conifers from their last refuges on nutrient-poor soils.

Another setting in which angiosperms have been less successful is the arid habitat. There, shrubs are the usual life-form, although angiosperms also exploited the habitat by means of the quick-growing annual herbs. In arid areas gnetopsids and conifers such as pinion pine, juniper, and even cycads are still successful. On such sites resource acquisition is important and these gymnosperms are as successful as angiosperms.

CONCLUSIONS

Angiosperms appear to have evolved from a line of seed plants in which a trend toward condensation in vegetative growth and reduction in generation time was carried to its extreme. We infer that the unique innovation of the early angiosperms was that they were the only seed plants to have reduced their vegetative body to the habit of perennial herbs, an advance that placed a premium on the rapidity and efficiency of their reproductive cycle. Vegetative features such as rhizomes, reticulately-veined leaves, sieve tubes and vessels made possible rapid growth and the herbaceous habit. These developments were coordinated with the reduction and condensation of the fertile organs that placed emphasis on rapid seed production and conservation of energetically expensive reproductive material. The first angiosperms seem to have been perennial herbs of low stature, growing in ephemeral habitats on stream levees in the fluvial system. Essentially, these traits would have placed the earliest angiosperms in competition with ferns in open sites. Only later, with the acquisition of the shrub and tree habit did they compete with the shrubby seed plants and woody conifers where rapid establishment and growth of their seedlings gave the angiosperm a marked competitive advantage. The primary impetus for the success of the angiosperms may thus have been the result of vegetative and reproductive innovations that allowed them to compete as perennial seed-plant herbs on unstable, stream-margin, sites during the early Cretaceous.

ACKNOWLEDGMENTS

We thank the reviewers, including Robyn J. Burnham and Richard Olmstead, for their critical comments that have much improved our thinking on these ideas.

Comparison of Alternative Hypotheses for the Origin of the Angiosperms

Henry Loconte

The development of a phylogenetic system of classification for the flowering plants is dependent on the objective resolution between alternative hypotheses for their origination. Here six hypotheses have been selected for comparison. Contemporary systems are unanimous in their support for the "ranalean theory," which was developed by Delpino (1890), Hallier (1912), and Bessey (1915). According to this theory, the order Magnoliales (Takhtajan, 1987; Dahlgren, 1989) or the family Winteraceae (Cronquist, 1981; Thorne, 1992a) are considered to be the most primitive angiosperms. The new Calycanthales hypothesis (Loconte and Stevenson, 1991) also conforms to the "ranalean theory." These three hypotheses are compared to three other hypotheses that contradict the "ranalean theory." Recently, Taylor and Hickey (1990a, 1992) have elaborated the Chloranthaceae hypothesis. Even more recently, Chase et al. (1993) have supported the Ceratophyllaceae hypothesis (Les, 1988). Finally, the historical Casuarinaceae hypothesis of Engler (Melchior, 1964) will be considered.

The Magnoliales *sensu* Loconte and Stevenson (1991) is a large group of woody plants including eight families: *Austrobaileya*ceae (1 sp.), *Hinastar-*

*drac*eae (2 spp.), *Degeneria*ceae (2 spp.), Magnoliaceae (7 genera, 184 spp.), *Eupomatia*ceae (2 spp.), Annonaceae (128 genera, 2300 spp.), Canellaceae (6 genera, 16 spp.), and Myristicaceae (17 genera, 370 spp.). These magnolialean taxa are considered to exhibit significant angiosperm plesiomorphies, such as primitive wood and strobilar flowers (Takhtajan, 1991). Corner (1967) hypothesized the seeds of Myristicaceae as the most plesiomorphic among angiosperms because of their complex seed coat anatomy and enveloping arils; however, this is based on ingroup comparison and is not supported by outgroup comparison. Another reputed magnolialean plesiomorphy is their monosulcate pollen with a columellaless exine (Walker and Skvarla, 1975).

Winteraceae comprise 4 genera and 65 species of woody plants (Vink, 1993). Although most systems consider the family to be an early branch of Magnoliales, the Winteraceae have been hypothesized to share a common ancestry with Illiciaceae (Walker and Walker, 1984; Doyle and Donoghue, 1993) and Amborellaceae (Loconte and Stevenson, 1991). The outstanding structural characteristic of Winteraceae is their vesselless wood, which has been considered to represent an unambiguous angiosperm plesiomorphy (Carlquist, 1987; Takhtajan, 1991). Another putative plesiomorphy of the Winteraceae is their unsealed carpels (Bailey and Nast, 1943b). However, Winteraceae have also been interpreted as exhibiting neoteny (Young, 1981) and, therefore, a series of apomorphic reversals.

The Calycanthales is a small group of woody plants including *Idiospermum* (1 sp.) and Calycanthaceae (3 genera, 6 spp.). Typically, these taxa have been classified as part of Laurales (Cronquist, 1981; Takhtajan, 1987, Dahlgren, 1989; Thorne, 1992a), and this was followed by Doyle and Donoghue (1993). However, the character states utilized for Laurales *sensu* Doyle and Donoghue, such as unilacunar two-trace nodes and opposite leaves are symplesiomorphies shared with the gymnospermous outgroups. In contrast, Endress (1990), who has studied Laurales extensively, excludes the calycanths from Laurales in favor of Magnoliales. Loconte and Stevenson (1991) have concluded that Calycanthales lacks the autapomorphies of Laurales or Magnoliales; they also hypothesized that Calycanthales exhibits a series of vegetative and reproductive angiosperm plesiomorphies such as the shrub habit, unilacunar two-trace nodes, opposite leaves, adplicate ptyxis, strobilar flowers with leaf-like bracteotepals, and few-ovulate carpels.

The Chloranthaceae comprise 4 genera and 75 species of woody plants or herbs (Todzia, 1993). Their phylogenetic classification has been controversial with some favoring a position in Laurales (Endress, 1987a; Loconte and Stevenson, 1991; Thorne, 1992a; Doyle and Donoghue, 1993), and others considering Piperales (Cronquist, 1981; Taylor and Hickey, 1992). Chloranthaceae exhibit similarities with Ephedrales, such as opposite leaves and small reproductive structures. LeRoy (1983) has argued for homology of the reproductive structures of Chloranthaceae and *Pinus*. The outstanding structural aspect of the Chloranthaceae is their simple flowers, which lack a perianth and are composed of a single stamen or simple pistil or both (Endress, 1987a). Accordingly, Crane (1989) has positioned Chloranthaceae as the sister group of the Hamamelidae. In a phylogenetic study, Taylor and Hickey (1992) hypothesized Chloranthaceae as a first branch of the angiosperms, stressing leaf characteristics and the herbaceous habit.

The Ceratophyllaceae include only *Ceratophyllum* with six species (Les, 1993). *Ceratophyllum* species are submerged aquatic herbs with numerous specializations including the lack of roots, reduced vasculature, dissected leaves, the lack of stomata, hydrophilous pollination, and exineless pollen. Typically, Ceratophyllaceae are classified in Nymphaeales as the sister group of Cabombaceae (Cronquist, 1981; Ito, 1987). Based on a phenetic analysis of 28 characters for 10 water-lily genera, Les (1988) concluded that *Ceratophyllum* is phenetically isolated and phylogenetically unrelated to Nymphaeales and other angiosperms, but phylogenetic interpretation of phenetics is unfounded (Ax, 1987). Les (1988) then utilized the putative isolation of *Ceratophyllum* to hypothesize that certain characteristics are angiosperm plesiomorphies. Additionally, fossil *Ceratophyllum* fruits from the early Aptian have been discovered (Herendeen et al., 1990). The Ceratophyllaceae hypothesis has been supported by *rbcL* sequence data that positions *Ceratophyllum* as a first branch of the angiosperms and unrelated to Nymphaeales (Chase et al., 1993).

The Casuarinaceae, composed 4 genera and 96 spp. (Johnson and Wilson, 1989), are vegetatively distinctive xerophytes with scale leaves and equisetoid branches that resemble gymnosperms such as *Ephedra*. Casuarinaceae flowers are unisexual and the plants are often dioecious, as in gymnosperms. Their infructescence is analogous to a gymnosperm cone, but is actually a multiple fruit composed of a head of samaras enclosed by corky bracteoles. The ovule, which was originally considered to be homologous with gymnosperm ovules with multiple archegonia, has multiple embryo sacs; however, in gymnosperms, multiple archegonia are derived from one megaspore (Singh, 1978), whereas in angiosperms, multiple embryo sacs are derived from multiple megaspores (Johri et al., 1992). The primary evidence supporting the Casuarinaceae hypothesis was the simple nature of the flowers, which lack a perianth and are composed of either a single stamen or a pistil; however, the pistil is pseudomonomerous and composed of two fused carpels with the posterior locule being sterile (Flores and Moseley, 1982). Modern systems are unanimous in their interpretation of Casuarinaceae as derived through simplification and related to higher hamamelids (Cronquist, 1981).

The objective of this study was to develop a cladistic analysis that accounts for the taxa proposed as first branches of the angiosperms. In order to compare Casuarinaceae, this analysis attempts a simultaneous resolution of Hamamelidae and Magnoliidae. Character evolution is considered in the context of the relative parsimony of the alternative hypotheses for the origin of the angiosperms. As several of these taxa have documented fossil histories, a few *a posteriori* comparisons are developed.

METHODS

The data matrix includes 69 taxa scored for 84 characters and 151 apomorphic character-states (matrix available upon request). The scope and scale of taxonomic representation in Loconte and Stevenson (1991) was adjusted to accommodate greater diversity of taxa with tricolpate pollen (eudicots). This required a reduction in taxonomic scale for magnoliids. For example, the Magnoliales was represented by *Austrobaileya* and Eumagnoliales [= Clade A in the work of Loconte and Stevenson (1991)]. As a consequence, the Myristicaceae were excluded as an advanced member of Magnoliales. The Laurales is represented by Chloranthaceae and Eulaurales [= Clade D in the work of Loconte and Stevenson (1991)]. Therefore, this analysis does not reconsider putative similarities between calycanths and Monimiaceae. The order Stemonales was coded as a primitive exemplar for the monocots (Stevenson and Loconte, in press). All 24 families of Hamamelidae sensu Cronquist (1981) were included, as well as *Nothofagus* and *Ticodendron*. The four families of Cunoniales that Dickison (1989) described as archetypic were included as exemplars for the Rosidae. For the Dillenidae, the families of Dilleniales and Theales *pro parte* were coded.

Phylogenetic Analysis Using Parsimony (PAUP, Swofford, 1993) was utilized for the numerical analysis, and character evolution was studied with MacClade (Maddison and Maddison, 1993). In PAUP, heuristic searches of MacClade topologies were conducted, as well as searches based on addition sequences. The alternative hypotheses for the origin of the angiosperms were considered by rerooting the ingroup topology. This was accomplished in Mac-Clade by rerooting the entire topology at the alternative archetype (equally parsimonious) and then repositioning the outgroup taxa at the base of the cladogram.

RESULTS AND DISCUSSION

Parsimony analysis resulted in 10 trees at 590 steps, CI = 0.26, RI = 0.59, and RC = 0.15 (counting polymorphisms, 976 steps, CI 0.55, RI = 0.59, and RC = 0.32). The same 10 trees were found with the different addition sequences (Fig. 10.1).

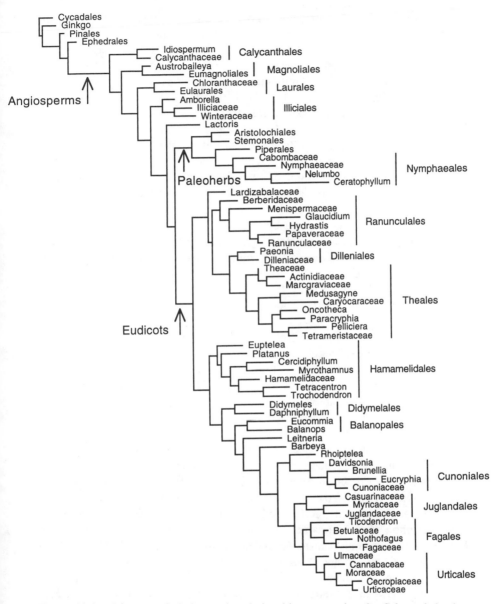

Figure 10.1. Diagram of phylogenetic relationships supporting the Calycanthales hypothesis for the origin of the angiosperms. This is one of 10 trees, of which the strict consensus is unresolved only within Illiciales (2 trees) and Hamamelidales (5 trees).

As magnoliids are paraphyletic (Loconte and Stevenson, 1991), so are ranunculids and hamamelids (Fig. 10.1). The hamamelids are paraphyletic in reference to the Rosidae, and Thorne (1992a) has already classified the hamamelids as basal rosids; this relationship has also been supported by Dickison (1989) and Endress and Stumpf (1991). The ranunculids are paraphyletic in reference to the Dillenidae and, therefore, ranunculids can be considered as basal dillenids (Crepet et al., 1991).

The hamamelids are a paraphyletic assemblage of nine ordinal branches: Hamamelidales, Didymelales, Balanopales, Leitneriales, Barbeyales, Rhoipteleales, Juglandales, Fagales, and Urticales. These correspond well to the three hamamelid pollen groups recognized by Zavada and Dilcher (1986): Group I is represented by Hamamelidales; Group II is represented by Didymelales, Balanopales, Leitneriales, and Barbeyales; Group III is represented by Rhoipteleales, Juglandales, Fagales, and Urticales.

The Hamamelidales includes *Euptelea, Platanus, Cercidiphyllum, Myrothamnus*, Trochodendraceae, and Hamamelidaceae. Most authors consider Hamamelidaceae and Platanaceae to be sister groups (Cronquist, 1981; Takhtajan, 1987; Hufford and Crane, 1989; Loconte and Stevenson, 1991; Schwarzwalder and Dilcher, 1991; Thorne, 1992a). However, in this analysis, Platanaceae and Eupteleaceae are associated, and they are positioned as early branches of Hamamelidales and Rosidae *sensu lato*, as Wolfe (1989) illustrated using leaf architecture. The genera *Cercidiphyllum* and *Myrothamnus* are sister groups (Hufford and Crane, 1989). Trochodendraceae and Hamamelidaceae share the synapomorphy of laterocytic stomata; Hamamelidaceae are polymorphic with laterocytic, paracytic, and cyclocytic, but not the plesiomorphic anomocytic stomata. Based on Trochodendraceae, Hamamelidaceae plesiomorphies would include laterocytic stomata, supervolute leaf vernation, plicate carpel ontogeny, inferior gynoecium, and parietal placentation.

The "middle hamamelids" include *Daphniphyllum, Didymeles, Balanops, Eucommia, Leitneria*, and *Barbeya*, which are each treated as isolated by most authors. Some or all of these taxa have been considered to be unrelated to hamamelids. For example, Kubitzki (1993) includes only Barbeyaceae in his circumscription of the hamamelids. Thorne (1992a) related Leitneriaceae to Rutales, and Eucommiaceae to Cornales. In Chase et al. (1993), *Eucommia* groups with Cornales, and *Leitneria* groups with Sapindales (= Rutales), supporting Thorne's treatment. In this analysis (Fig. 10.1), the sister-group relationship between *Daphniphyllum* and *Didymeles* is based on phytochemical similarities. *Balanops* and *Eucommia* share a unique type of intercalary inflorescence. In considering the putative position of "middle hamamelids", they form a connecting grade between Hamamelidales and higher hamamelids. Further analyses are necessary to reconsider the relationships of these taxa to more derived taxa.

The Rhoipteleaceae are isolated as the Rhoipteleales, based on parallel evolution with Juglandaceae and putative homology with Cunoniales. The southeast Asian endemic, *Rhoiptelea chiliantha*, has been classified within Urticales (Melchior, 1964) and Juglandales (Cronquist, 1981). There are at least nine significant morphological distinctions between Rhoipteleaceae and Juglandaceae (Stone, 1989), and these differences are typically interpreted as juglandaceous plesiomorphies. In this analysis, the similarities between Rhoipteleaceae and Juglandaceae are interpreted as nonhomologous parallelisms. For example, the samaroid fruit of *Rhoiptelea* is derived from outgrowths of the ovary wall, whereas juglandaceous samaras are composed of accrescent bracteoles (Withner, 1941). Putative homologies between Rhoiptelales and Cunoniales (particularly Davidsoniaceae) include the stipulate, pinnately compound leaves and bisexual flowers. The alternative position of Rhoipteleaceae as the sister group to Juglandaceae is five steps less parsimonious. Some of the oldest fossil "Normapolles" pollen genera are similar to *Rhoiptelea* pollen (Batten, 1989), and this is congruent with the more primitive position in reference to Juglandaceae. The remainder of the "Normapolles" complex is assignable to various branches of the clade (Juglandales-(Fagales-Urticales)). Some "Normapolles" pollen may also be assignable to Cunoniales and even *Barbeya*.

The Juglandaceae are related to Myricaceae and secondarily to Casuarinaceae. Therefore, the Myricales and Casuarinales are subsumed under Juglandales, which has priority (Dumortier, 1829). A common ancestry between Myricaceae and Juglandaceae has been previously considered (Cronquist, 1981; Takhtajan, 1987; Thorne, 1992a; Hufford, 1992). Hufford (1992) also concluded that Casuarinaceae is the sister group of (Myricaceae-Juglandaceae).

Ticodendron incognitum is a newly discovered hamamelid from Central America (Gomez-Laurito and Gomez P., 1989), which has been classified as a monotypic family (Gomez-Laurito and Gomez P., 1991). The taxon has been intensely studied for wood anatomy (Carlquist, 1991c), sieve-element characters (Behnke, 1991), leaf architecture (Hickey and Taylor, 1991), floral anatomy (Tobe, 1991), palynology (Feuer, 1991), and chromosome number (Snow and Goldblatt, 1992). The consensus opinion from these studies is that Ticodendraceae are related to Fagales, as substantiated by this analysis, which places *Ticodendron* as a first branch of Fagales. However, a phylogenetic position of Ticodendraceae as a first branch of Juglandales is only one step less parsimonious and should be reconsidered. Embryological data, missing for Ticodendraceae, could be decisive because Juglandales are embryologically distinctive.

The phylogenetic relationships within Urticales are congruent with the results of Humphries and Blackmore (1989), with Ulmaceae as a first branch, (Cecropiaceae-Urticaceae) as derived, and Moraceae as intermediate. The putative paraphyletic groups, Ulmaceae (Wiegrefe et al., 1993) and Moraceae (Humphries and Blackmore, 1989) were not addressed by this analysis. The inclusion of Urticales among higher hamamelids gains plausibility by this analysis, especially with the position of Ulmaceae relative to Fagales. Yet, the possible relationship of Urticales to malvalean dillenids still requires consideration. Furthermore, Chase et al. (1993) relate Urticales to Rhamnales of the Rosidae.

The ranunculids are recognized as a paraphyletic assemblage representing the basal branches of the Dillenidae. The Lardizabalales [1] is isolated as a first branch (Fig. 10.1), and this is congruent with the recognition of several lardizabalaceous plesiomorphies (Taylor, 1967; Takhtajan, 1987; Qin, 1989). The remainder of ranunculids, including Papaveraceae, are referable to Ranunculales. Berberidaceae are reaffirmed as primitively woody (Loconte and Estes, 1989). A herbaceous clade is composed of Hydrastidaceae, Papaveraceae, and Ranuunculaceae (Loconte et al., in press). The Paeoniaceae are unambiguously positioned in Dilleniales as concluded by Eames (1961), Cronquist (1981), and Nowicke et al. (1986).

Within the Paleoherbs, a common ancestry between *Ceratophyllum* and *Nelumbo* (Fig. 10.1) is supported by putative homologies of the ovule and seed. During development, the anatropous ovule of *Nelumbo* is initially orthotropous (Johri et al., 1992), suggesting that the morphological peculiarities of *Ceratophyllum* are neotenous. In his phenetic ordination of nymphaealean genera, Les (1988) proposed that *Ceratophyllum* is isolated from other Nymphaeales. However, if a minimum spanning tree was superimposed on the ordination, *Ceratophyllum* would connect with *Nelumbo*. Moreover, if other angiosperms were included in the analysis, *Ceratophyllum* would cluster with the Nymphaeales. The phylogenetic position of *Ceratophyllum* within Nymphaeales is ascertained by the nymphaealean autapomorphies exhibited by *Nelumbo*. Yet, it has become popular to dissociate *Nelumbo* from Nymphaeales based primarily on the tricolpate pollen of *Nelumbo* (Doyle and Donoghue, 1993). Chase et al. (1993) have positioned *Nelumbo* as the sister group of *Platanus*. In this analysis, alternative positions of *Nelumbo* are less parsimonious; that is, 13 steps longer as the sister group of *Platanus*. *Ceratophyllum* is the longest branch in this analysis (13 steps), and *Nelumbo* is also morphologically autapomorphic, which could explain the difficulties in their phylogenetic systematization (Sytsma and Baum, Chapter 12).

The relationships among the most primitive angiosperms are congruent with the results of Loconte and Stevenson (1991), with Calycanthales as a first branch. The most primitive flowering plant is the monotypic *Idiospermum*, which Endress (1983b) described as "In all respects, *Idiospermum* gives the impression of a strange living fossil." The Calycanthaceae includes the mono-

---

[1] Lardizabalales Loconte, ordo novus. Typus: Lardizabalaceae Decaisne, *Archives du museum d'historie naturelle.* 1: 185 (1839).

typic *Sinocalycanthus*, which exhibits leaf-like bracteotepals (Campbell et al., 1993). Based on the Calycanthales hypothesis, the following angiosperm plesiomorphies are recognized (the superscript asterisk denotes unique to angiosperms): shrub habit; vessels with scalariform perforations*; sieve-tube elements with starch inclusions; parenchymatous oil cells*; unilacunar two-trace nodes; leaves opposite, simple, evergreen, glabrous, exstipulate, pinnately veined,* with entire margins, adplicate ptyxis, and paracytic stomata*; flowers solitary,* monoclinous,* hypogynous, with helically arranged perianth,* stamens,* and carpels*; stamens with a connective protrusion, basifixed anthers, and longitudinal dehiscence; pollen monoaperturate, sulcate, prolate, punctate, and columellate*; inner staminodia*; carpels ascidiate, numerous, free, with marginal placentation; ovules few, bitegmic,* anatropous*; pollination entomophilous; seeds large, with endotestal seed coat anatomy, lacking an aril; embryo large, lacking endosperm; germination epigeal. Calycanthales autapomorphies include sieve-tube elements with protein inclusions, pubescence, perigynous flowers, and bisulcate pollen. Thorne (Chapter 11) lists alternate leaf arrangement as an angiosperm plesiomorphy, yet Ephedrales exhibit opposite leaf arrangement that is probably homologous with Calycanthales, *Austrobaileya* of Magnoliales, and Laurales. An Early Cretaceous fossil calycanthaceous flower has been described as *Virginianthus calycanthoides* (Friis, Eklund et al. 1994). *Virginianthus* exhibits the Calycanthales homology of a perigynous cup-flower; however, the fossil pollen is of the plesiomorphic monosulcate type, rather than bisulcate.

Friis, Eklund et al. (1994) conclude that "The character combination seen in *Virginianthus* indicates that the separation of *Idiospermum* from Calycanthaceae is poorly supported, and we suggest that *Idiospermum* be included within the family." However, the recognition of *Idiospermum* as Idiospermaceae or as subfamily Idiospermoideae of Calycanthaceae is not dependent on the degree of similarity or difference, but on the phylogenetic hierarchy. If the calycanthaceous clade is part of Laurales (Chase et al., 1993) or Magnoliales *sensu lato* (see below), familial rank would be supported and *Idiospermum* would be included. On the other hand, if the calycanthaceous clade is isolated from other orders, ordinal rank would be supported and Idiospermaceae would be segregated.

Magnoliales Hypothesis

This hypothesis is two steps longer than the Calycanthales hypothesis. The topology is the same, except for the reversed sequence of Magnoliales and Calycanthales (Fig. 10.2). The primary support for the Magnoliales hypothesis is their gymnosperm-like pollen, which is prolate, monosulcate, and has a wall structure that lacks columella and an endexine. However, the pollen of *Aus-*

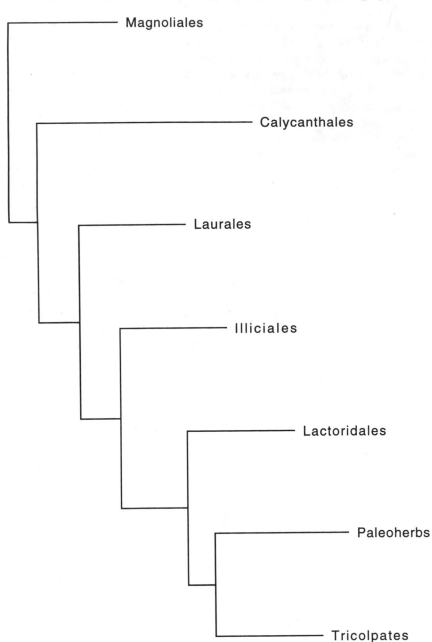

Figure 10.2.   Diagram of phylogenetic relationships according to the Magnoliales hypothesis for the origin of the angiosperms.

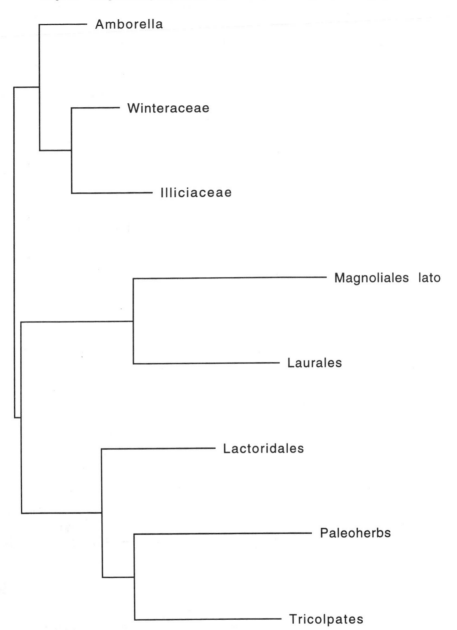

Figure 10.3. Diagram of phylogenetic relationships according to the Winteraceae hypothesis for the origin of the angiosperms, rooted at Illiciales. Magnoliales sensu lato includes the Calycanthaceae.

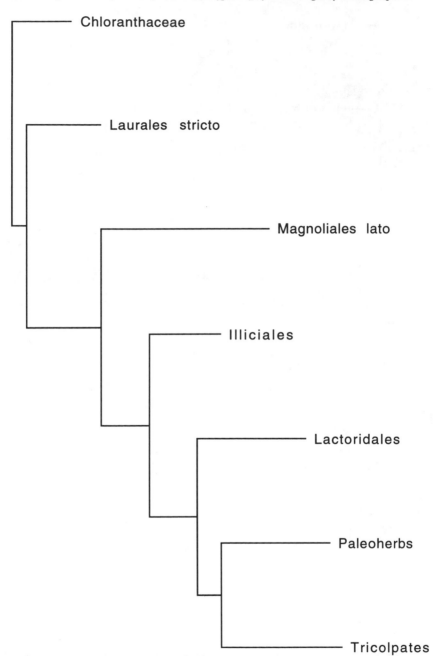

Figure 10.4. Diagram of phylogenetic relationships according to the Chloranthaceae hypothesis for the origin of the angiosperms.

trobaileya is columellate, and this taxon is clearly related to Magnoliales (Endress, 1980, 1993a) and in a primary phylogenetic position (Loconte and Stevenson, 1991). Therefore, it is equally parsimonious to consider the lack of pollen columella in Magnoliales as an apomorphic reversal or as an angiosperm plesiomorphy. Other reputed plesiomorphies of Magnoliales are based on ingroup comparisons, which are unsupported by gymnospermous outgroup comparisons. The primary example of such a putative plesiomorphy is the multiovulate condition of Magnoliales. Even with the Magnoliales hypothesis (Fig. 10.2), the character-state optimization for ovule number on the angiosperm branch is few-ovulate and the multiovulate condition is apomorphic. Character states such as carpels with numerous ovules and leaves with alternate phyllotaxy are not necessarily part of the "ranalean theory," but later became associated with the theory by commonality among the considered taxa.

Winteraceae Hypothesis

This hypothesis is two steps longer. Rather than rooted directly at Winteraceae, it is more parsimonious to consider a rooting at the base of Illiciales, of which Winteraceae is a subordinate. This topology differs in that Magnoliales *sensu lato* (including Calycanthaceae) and Laurales are sister groups (Fig. 10.3). Putative angiosperm plesiomorphies supported by the Winteraceae hypothesis include vessels absent; leaves alternate (rather than opposite); stomata anomocytic (rather than paracytic, but as in outgroups); inner staminodia absent; carpels unsealed, cyclic (rather than helical); seeds small, exotestal (rather than endotestal). Based on carpel and ovule structure, Taylor (1991a) concluded that *Amborella* is one of the most primitive angiosperms, and it is conceivable that a detailed comparison with the ovules of Ephedrales could support this hypothesis.

Chloranthaceae Hypothesis

This hypothesis is two steps longer than the Calycanthales hypothesis. The topology differs in that Laurales is paraphyletic (Fig. 10.4). The Chloranthaceae hypothesis supports the following unusual angiosperm plesiomorphies: presence of an inflorescence (rather than a solitary flower); plants dioecious (rather than flowers monoclinous); inner staminodia absent; carpels solitary, cyclic; placentation apical (rather than marginal); ovule orthotropous (rather than anatropous, but as in outgroups); fruit drupaceous; seeds small. The herbaceous habit has also been interpreted as part of the Chloranthaceae hypothesis (Taylor and Hickey, 1992) but in association with Piperaceae. However, the putative relationship between Chloranthaceae and Piperaceae is quite unparsimonious (Endress, 1987a; Loconte and Stevenson, 1991). In this analysis, the Chloranthaceae hypothesis includes the shrub habit as primitive. The fossil pollen

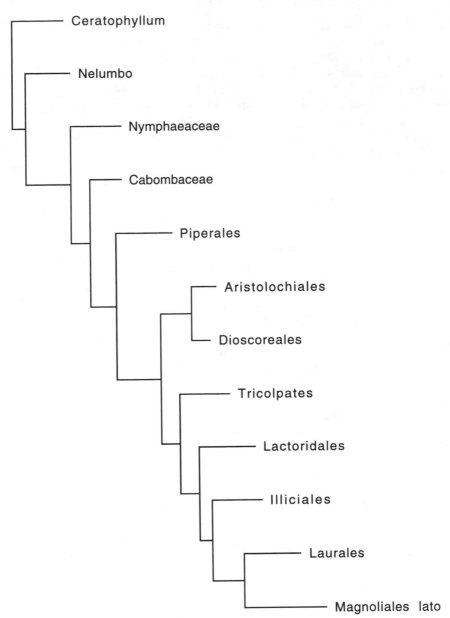

Figure 10.5.   Diagram of phylogenetic relationships according to the Ceratophyllaceae hypothesis for the origin of the angiosperms.

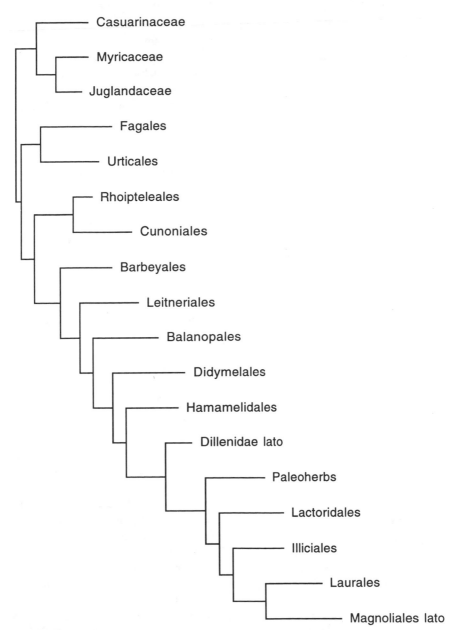

Figure 10.6. Diagram of phylogenetic relationships according to the Casuarinaceae hypothesis for the origin of the angiosperms, rooted at Juglandales.

"Clavatipollenites" has been assigned to the Chloranthaceae and represents some of the oldest known fossils assignable to angiosperms (Walker and Walker, 1984). Endress (1987a) has observed that certain "Clavatipollenites" are closer to *Austrobaileya* of the Magnoliales. Chloranthaceae and *Austrobaileya* are more closely related in the Calycanthales hypothesis (Fig. 10.1), whereas these taxa are more distant in both the Magnoliales hypothesis (Fig. 10.2) and the Chloranthaceae hypothesis (Fig. 10.4).

Ceratophyllaceae Hypothesis

This hypothesis is six steps longer than the Calycanthales hypothesis. The topology is radically different from the previous hypotheses (Fig. 10.5) in that Nymphaeales and Paleoherbs are paraphyletic at the base of the angiosperms. The "primitive angiosperms" are reconsidered as an advanced clade with the relationships: (Lactoridales-(Illiciales-(Laurales-Magnoliales sensu lato))). Unusual plesiomorphies supported by the Ceratophyllaceae hypothesis include: habit herbaceous (rather than woody); aquatic (rather than terrestrial); vessels absent; oil cells absent; leaves palmately veined (rather than pinnately veined), with involute vernation (rather than adplicate or conduplicate); carpels cyclic; placentation apical; ovules unitegmic (rather than bitegmic), orthotropous; fruit an achene. Of these, the absence of vessels in *Ceratophyllum* is certainly an apomorphic reduction and not an angiosperm plesiomorphy. Most systematists implicitly reject the Ceratophyllaceae hypothesis based on an appreciation for angiosperm diversity (Corner, 1967).

Casuarinaceae Hypothesis

This hypothesis is six steps longer than the Calycanthaceae hypothesis. Rather than rooted directly at Casuarinaceae, it is more parsimonious to consider a rooting at the base of Juglandales, of which Casuarinaceae is a subordinate. The topology is radically different from the previous hypotheses (Fig. 10.6) in that taxa with triaperturate pollen form a paraphyletic group at the base of the angiosperms. Unusual angiosperm plesiomorphies supported by the Casuarinaceae hypothesis include: arborescent habit (rather than shrubs); oil cells absent; trilacunar nodes (rather than unilacunar two-trace); inflorescence present; plants dioecious; stamens cyclic (rather than helical); filaments differentiated (rather than undifferentiated); pollen triaperturate (rather than monoaperturate), porate (rather than sulcate), triangular (rather than prolate), spinose, granular columellate (rather than columellate); carpels cyclic, syncarpous (rather than apocarpous), placentation axile; ovules unitegmic, orthotropous; pollination anemophilous; fruit a nut; seed coat crushed; endosperm development nuclear (rather than cellular).

Meeuse (1987) has hypothesized angiosperm plesiomorphies in the flowers

Table 10.1.   Revised Ordinal Classification for Gymnosperms, Magnoliids, Ranunculids, and Hamamelids.

Spermatophyta	Cycadales Dumortier (1829)
	Cladophytes
	Ginkgoales Warburg (1913)
	Mesophtes
	Pinales Dumortier (1829)
	Anthophytes
	Ephedrales Dumortier (1829)
Angiosperms	Calycanthales Martius (1835)
	Unnamed Inclusive Clade UIC 1
	Magnoliales Bromhead (1838)
	UIC 2
	Laurales Perleb (1826)
	UIC 3
	Illiciales H-H. Hu ex Cronquist (1981)
	UIC 4
	Lactoridales Takhtajan ex Reveal (1993)
	Euangiosperms
Paleoherbs	UIC 5
	Nymphaeales Dumortier (1829)
	Piperales Dumortier (1829)
	UIC 6
	Aristolochiales Dumortier (1829)
	Stemonales Loconte (1995)
Eudicots	remainder of monocots
Dillenidae	
	Lardizabalales Loconte (1995)
	UIC 7
	Ranunculales Dumortier (1829)
	UIC 8
	Dilleniales Hutchinson (1924)
	Theales Lindley (1833)
	remainder of dillenids
	Rosidae
	Hamamelidales Grisebach (1854)
	UIC 9
	Didymelales Takhtajan (1966)
	UIC 10
	Balanopales Engler (1897)
	UIC 11
	Leitneriales Engler (1897)

Table 10.1. (*cont.*)

UIC 12
Barbeyales Takhtajan ex Reveal (1993)
UIC 13
UIC 14
Juglandales Dumortier (1829)
UIC 15
Fagales Engler (1892)
Urticales Dumortier (1829)
UIC 16
Rhoipteleales Novak ex Reveal (1992)
Cunoniales Hutchinson (1924)
remainder of rosids

of Juglandaceae, but his diagrams and terminology are difficult to understand. The Casuarinaceae hypothesis is not supported by the pollen fossil record as "Normapolles" pollen is 40 million years younger than "Clavatipollenites" pollen.

A phylogenetic system for magnoliids and hamamelids includes 19 orders (Table 10.1), and coincidently Cronquist (1981) recognized 19 orders for these taxa. Not recognized are the Cronquist orders Papaverales, Trochodendrales, Daphniphyllales, Eucommiales, Myricales, and Casuarinales; segregated are Calycanthales, Lactoridales, Lardizabalales, Balanopales, Barbeyales, and Rhioptelales. Therefore, the prediction by Cronquist (1987) that a cladistic classification would be inflationary is not substantiated, although there are several unnamed inclusive clades.

The relative parsimony between these alternative hypotheses for the origin of the angiosperms could be accentuated by new characters or by exercising Hennigian reciprocal illumination on the characters considered by this analysis. As many of the floral characters are unpolarized due to a lack of comparison in the gymnosperm outgroups, it is possible to attempt a few *a posteriori* polarizations. Therefore, the character inflorescence type could be polarized as the character-state solitary flower being plesiomorphic (biasing against the Chloranthaceae and Casuarinaceae hypotheses). The character stamen arrangement could be polarized as the character-state helical being plesiomorphic (biasing against the Casuarinaceae hypothesis). The character carpel fusion could be polarized as the character-state apocarpous being plesiomorphic (biasing against the Casuarinaceae hypothesis).

Overall comparison of these six hypotheses for the origin of the angiosperms suggests that the Casuarinaceae and Ceratophyllaceae alternatives are the least probable scenarios. Of the hypotheses that contradict the "ranalean theory," only the Chloranthaceae alternative seems plausible. Of the three hypotheses that conform to the "ranalean theory," the Magnoliales alternative appears less plausible than either the Winteraceae or Calycanthales hypotheses. The Winteraceae hypothesis should be reconsidered, particularly in reference to the rare and poorly understood genus *Amborella*. Empirical investigations into the anatomy, morphology, and other systematic characters of the Calycanthales should be the most useful in our continued elucidation of the origin of the angiosperms.

CONCLUSIONS

The origin of the angiosperms is considered in the context of acladistic analysis of the families of Hamamelidae *sensu* Cronquist, with outgroup taxa including gymnosperms, magnoliids, ranunculids, monocots (Stemonales), dillenids (Dilleniales and Theales), and rosids (Cunoniales). The hamamelids are recognized as paraphyletic in reference to rosids based on a common ancestry between Rhoipteleales and Cunoniales. The phylogenetic sequence of hamamelid lineages is Hamamelidales (including Trochodendraceae), Didymelales (including Daphniphyllaceae), Balanopales (including Eucommiaceae), Leitneriales, Barbeyales, Juglandales (including Myricaceae and Casuarinaceae), Fagales (including Ticodendraceae), Urticales, and Rhoipteleales. The ranunculids are recognized as paraphyletic in reference to the dillenids: (Lardizabalales-(Ranunculales-(Dilleniales-Theales))). Therefore, the Eudicot radiation is composed of two main evolutionary lineages, which are delimited as rosids and dillenids. Within Paleoherbs, *Ceratophyllum* is the sister group of *Nelumbo* as subordinates of Nymphaeales. The most basal lineages of angiosperms are reaffirmed as Calycanthales, Magnoliales, Laurales, Illiciales, and Lactoridales. Alternative hypotheses for the origin of the angiosperms are compared among primitive angiosperms, as well as one paleoherb (*Ceratophyllum*) and one eudicot (Casuarinaceae). Rootings of the angiosperms among Paleoherbs or Eudicots necessitate radical topologies, which overturn nearly all commonly accepted character-state transformations. The latter two alternatives are also substantially unparsimonious and could be considered even more complex if certain a posteriori polarizations are adopted.

ACKNOWLEDGMENTS

This research was supported by NSF (BSR-8800188). The cladistic analysis was conducted with Lisa M. Campbell. Arthur Cronquist, Kevin Nixon, Dennis W. Stevenson, Armen Takhtajan, and David W. Taylor provided valuable discussion.

The Least Specialized Angiosperms

Robert F. Thorne

To decide which angiospermous groups are the least specialized, one has to form a conception of what features characterized the ancestors of the living angiosperms. For many reasons, this is not an easy task. The relative paucity of the paleobotanical record has resulted in our reconstruction of the evolutionary history of the angiosperms being largely inferential and hypothetical. Fortunately, the flowering plants do supply us with many "living fossils" or nonmissing links, that is, those plants with a plethora of primitive or unspecilized characteristics as recognized by experts such as those listed immediately below. The comprehensive studies by many botanical specialists of these phylogenetic relicts and conservative, that is, archaic, features in their more specialized relatives [Bailey, 1944a, 1956, 1957; Bailey and Nast, 1943a, 1943b, 1944a, 1944b, 1945, 1948; Bailey et al., 1943; Bailey and Smith, 1942; Bailey and Swamy, 1948, 1949, 1951; Canright, 1952, 1953, 1955, 1960, 1963, 1965; Carlquist, 1964; Money et al., 1950; Nast, 1944; Nast and Bailey, 1946; Smith, 1947; Swamy, 1949, 1952, 1953a, 1953b, 1953c, 1953d; Swamy and Bailey, 1950; Wilson, 1960, 1964, 1965; Wilson and Maculans, 1967; and many others (see references in the works of Thorne, 1974, 1976, 1981, 1983, 1992a)] has enabled us to recognize probable ancestral characteristics and the direction of evolutionary trends in many angiosperm organs and tissues, especially in stem and leaf anatomy, pollen grains, ovules, and other embryological structures, floral parts, fruits and seeds, chromosomes, sieve-element plastids, wax crystalloids, and other micromorphological features, biochemical compounds, and so on.

Much new information obtained from fossils by paleobotanists (Friis, 1989; Friis and Endress, 1990) and from comparative studies of gymnospermous groups and ferns has recently been very helpful in developing our knowledge of the ancestral angiosperms and their characteristics (Thorne, 1992a). More recently, cladistic and molecular investigations (Chase et al., 1993; Qiu et al., 1993; Sytsma and Baum, Chapter 12) have helped give us more confidence in our inferences. It should be pointed out here that the plesiomorphic, or ancestral, characteristics so disdained by the cladists are the very features most valuable in developing our conception of the ancestral angiosperms. These retained primitive features are the principal links among the least specialized angiosperms and between them and their now extinct ancestors.

ANCESTRAL FEATURES AND TRENDS OF SPECIALIZATION

We are still unsure when, where, and how the angiosperms originated (Thorne, 1992a). The first certifiably angiospermous fossils have been found in strata of early Cretaceous time, 140 million years ago (Appendix; Brenner, 1987, 1990; Brenner and Bickoff, 1992; Friis et al., 1986; Walker and Walker, 1984; Ward et al., 1988). They probably evolved during late Jurassic time from some group of seed ferns or other related gymnosperms (Crane, 1985; Doyle and Donoghue, 1986b) Because the angiosperms are basically a tropical class and those groups with the largest constellation of primitive, that is, least specialized or ancestral, characteristics are now found in tropical montane or summer-wet, warm-temperate forests, presumably fossil remains of the earliest angiosperms should be sought assiduously in areas that were tropical or subtropical in earliest Cretaceous or late Jurassic time.

Archaic living angiosperms, those relicts with primitively vesselless wood or unspecialized tracheid-like vessels and other primitive features, are mostly restricted to highly mesic sites with minimal seasonal water stress (Carlquist, 1975). As they evolved xylem with greater conductive efficiency, they were able to move from mesic highlands to hot tropical lowlands with wide fluctuations in soil moisture. Further evolution enabled these early flowering plants to invade more stressful climates and temperate woodlands and forests. The fossil record indicates that they adapted very early to extreme habitats by assuming varied growth forms.

The cooperative efforts of plant anatomists, palynologists, paleobotanists, chemo-taxonomists and molecular-taxonomists, and many other botanists have supplied phylogenists with much reliable information about ancestral features of the angiosperms and established objectively and independently many largely unidirectional trends of specialization through comparative studies and outgroup comparison in stem, leaf, root, and floral anatomy, pollen grains, embryos, seeds, fruits, and chemistry (Barthlott and Froelich, 1983; Behnke, 1976; Bendz and Santesson, 1974; Carlquist, 1961; Corner, 1976; Dickison, 1976, 1992; Ehrendorfer et al., 1968; Eyde, 1976; Fairbrothers et al., 1976; Froelich and Barthlott, 1988; Frost, 1930a, 1930b, 1931; Harborne and Swain, 1969; Hegnauer, 1988; Hickey and Wolfe, 1976; Jensen et al., 1975; Johri, 1984; Martin, 1946; Metcalfe, 1987; Metcalfe and Chalk, 1950, 1983; Palser, 1976; Raven, 1976; Roth, 1977; Teichman and van Wyk, 1991; Walker and Doyle, 1976; Walker and Walker, 1984; Wolfe et al., 1976; Young and Seigler, 1981). Other trends of specialization have been recognized and generally accepted through correlation studies (Sporne, 1948, 1956) with these irreversible or seldom reversible trends. Thus, using data from the above-listed sources, combined with the evidence from fossils and the relict flowering plants mentioned above, we can with reasonable safety describe the probable characteristics of the ancestral angiosperms (Taylor and Hickey, 1990b, 1992; Thorne, 1976, 1992a, Chapter 9; Wolfe et al., 1976).

The common ancestors of our living flowering plants were probably small trees, shrubs, or woody perennial herbs with leaves simple, alternate, exstipulate, entire, petiolate, glabrous, evergreen, with poorly organized reticulate venation and paracytic or anomocytic stomata, and with two leaf traces from a single leaf gap (Bailey, 1956; Hickey and Wolfe, 1976). The wood anatomy (Bailey, 1944a; Dickison, 1976; Frost, 1930a, 1930b, 1931; Metcalfe, 1987; Metcalfe and Chalk, 1950, 1983) was very unspecialized with vessels absent or tracheid-like with elements characterized by small diameter, great length, angular cross-sectional appearance, thin walls, tapering ends with many-barred scalariform perforation plates, scalariform intervascular pitting, and solitary, diffuse arrangement; tracheids moderately long with conspicuous bordered pits; wood parenchyma cells absent or diffuse; rays of primitive heterocellular type with numerous uniseriates. The flowers (Eyde, 1976; Friis, 1989) were probably borne separately in the axils of subtending leaf-like bracts, and they were bisexual, radially symmetrical, with perianth parts poorly differentiated and all floral parts indefinite in number, distinct from one another, and spirally arranged [but with early development of trimerous flowers with parts whorled and definite in number (Drinnan et al., 1990)]; stamens were basifixed, broad, unmodified into filament, anther, and connective, with microsporangia marginal and opening at maturity by linear slits. Pollen grains (Brenner, 1987; Walker and Doyle, 1976; Walker and Walker, 1984; Ward et al., 1988) were distally monoaperturate (anasulcate) and binucleate; however, according to Brenner (personal communication, Chapter 5), the earliest angiosperm pollen from the Israeli Negev is "small, inaperturate and round, like *Gnetum* and Piperaceae—not monosulcate. Sulcus is derived in Hauterivian from this grain" (see Brenner and Bickoff, 1992; Brenner, Chapter 5). Carpels (Bailey and Swamy, 1951; Nishida, 1985) were styleless with broad, convolute or involute lamina bearing an indefinite number of marginal ovules, margins stigmatic and not at all or only partly sealed [although Taylor [1991a] prefers ascidiate ancestral carpels], with the ovules (Palser, 1976) anatropous [Taylor (1991a) prefers them orthotropous], bitegmic, and crassinucellate, with embryo sac development the normal monosporic *Polygonum* type, and endosperm development cellular. Fruits were follicular (Crane and Dilcher, 1984; Dilcher and Crane, 1984) with many seeds, each with a rudimentary embryo embedded in abundant endosperm (Martin, 1946). All of these primitive, that is, ancestral, attributes can be found in the Magnoliales (sensu latissimo) and many of them in the least specialized family of all, the Winteraceae (Bailey, 1944a; Bailey and Nast, 1943a, 1943b, 1944a, 1944b, 1945; Nast, 1944; Swamy, 1952).

Although our knowledge of the ancestral features and trends of specialization in the flowering plants and my classification of the Angiospermae (Magnoliopsida) (Thorne, 1992a) were largely developed before the advent of the currently very popular molecular approach to angiosperm phylogeny, it is most comforting to observe that *rbc*L sequence variation (Chase et al., 1993; Qiu et al., 1993) generally shows rather close congruence with our most recent taxonomic classifications of the flowering plants. For example, their basalmost clades within the flowering plants are the very ones treated in this chapter as the least specialized angiosperms, Magnolianae, Nymphaeanae, and more primitive members of Rosanae and Lilianae (List in Table 11.1).

MAGNOLIANAE

Magnoliales

Winterineae.   The family Winteraceae of the suborder Winterineae of the order Magnoliales and superorder Magnolianae (Thorne, 1974) retains more primitive features in the vegetative and reproductive structures of its members than any other angiosperm family known to me. Depending on the expert (Vink, 1993) followed, there are 4–8 widely accepted genera with perhaps 65–100 species of small trees or shrubs found primarily in areas fringing the Pacific Basin from Fuegia and the Juan Fernandez Islands north to eastern Brazil,

Guyana, and Mexico [*Drimys* Forst. & Forst. (Fig. 11.1)] and from Tasmania and New Zealand north to the Philippines and Borneo (most of the other genera) with the sole species of *Takhtajania* M. Baranova & J. Leroy collected once in the rain forests of northeastern Madagascar. The earliest known fossils of the family are pollen grains of Early Cretaceous time, from strata 100–140 million years old, from Gabon and Israel (Brenner, 1987). Other winteraceous fossils are reported from South Africa and California (Coetzee and Praglowski, 1988; Page, 1979). Only the Chloranthaceae, of the Laurineae in the same order Magnoliales, include recognized fossils as old or older than those of the Winteraceae (Friis et al., 1986).

Drimys winteri J. R. & G. Forst. (Fig. 11.1) illustrates many of the primitive features listed above. Among the more distinctive winteraceous features shown are the entirely vesselless wood (Fig. 11.1i,l); alternate, entire, glabrous, exstipulate leaves (Fig. 11.1a); and pollen grains in permanent tetrahedral tetrads, each grain with a distal germination pore (Fig. 11.1m). Among the more archaic features of this small tree from temperate South America are, besides the vesselless xylem, the long cambial initials and tracheids (Fig. 11.1i); bordered tracheary pitting (Fig. 11.1j,k); heterogeneous rays; diffuse to tangentially banded wood parenchyma; paracytic stomatal apparatus (having one or more subsidiary cells parallel to the stoma) in evergreen leaves with their occluded stomata and poorly organized pinnate venation; intercalary (Fig. 11.1a) to subterminal cymose inflorescences; relatively small, actinomorphic, bisexual flowers (Fig. 11.1b) with a varied number of mostly separate, spirally arranged parts; single-veined, relatively short microsporophylls with protuberant, lateral to apical microsporangia (Fig. 11.1c); carpels styleless, megasporophylls with several anatropous ovules on submarginal placentae (Fig. 11.1b,d); and seeds with rudimentary embryo and abundant endosperm. Like all extant flowering plants, this species has a mixture of primitive and specialized characteristics. Other members of the family (Thorne, 1974; Vink, 1993) bear less specialized features such as metaxylem tracheids with scalariform pitting; perianth less well differentiated; stamens appearing as broad microsporophylls with lateral microsporangia; anaulcerate pollen grains monocolpate and single; carpels conduplicate, often stipitate, with partly to wholly "unsealed" stigmatic margins; and fruit baccate follicles rather than the more common fleshy berries.

Many trends of morphological specialization (Bailey, 1944a; Bailey and Nast, 1943a, 1943b, 1944a, 1944b, 1945; Thorne, 1974; Vink, 1993) are displayed among the many species of the several winteraceous genera. Among the more prominent trends are dwarfing of shrubs and leaves; production of single, terminal flowers; calyptrate calyxes; petals well differentiated from the sepals, few in number, or absent; functional unisexuality; differentiated and few stamens; carpels reduced in number, sometime to one, with closure of the megasporophyll and restriction of the stigmatic crests to a subapical projection

Figure 11.1. *Drimys winteri* R. G. Forst., Winteraceae. (**a**) Flowering branch with alternate, evergreen leaves and intercalary, cymose inflorescences of cream-colored flowers, x0.4. (**b**) Flower with much of perianth and some stamens removed and one carpel in longisection to show ovules, x4.2. (**c**) Stamens, with lateral microsporangia, x7.7. (**d**) Whorl of separate carpels, 3 in cross section to show ovule placement and vascular bundles, x3.5. (**e**) Microscopic section of carpel wall especially to show spherical secretory (essential) oil cells in the parenchymatous tissues, x26. (**f**) Developing cluster of carpels in old flower, x1.7. (**g**) Carpel, x2.4. (**h**) Seed, x7. (**i**) Tracheids, 10 μm in diameter, from stem in radial view showing bordered pits. (**j**) Bordered pit between 2 tracheids. (**k**) Bordered pit, 2-3 μm in diam. (**l**) Cross section of xylem showing tracheids with bordered pits mostly limited to radial faces. (**m**) Permanent tetrahedral tetrads, 50 μm in diameter, of distally porate and distally reticulate pollen grains.

(Fig. 11.1b); and syncarpy. One characteristic found essentially throughout the Magnoliales, hence synapomorphic for the order, is the presence of spherical, essential-oil cells in parenchymatous tissues, as illustrated here in the microscopic section of a carpel wall (Fig. 11.1e). Synapomorphic for most of the superorder Magnolianae is the prevalence of benzyl-isoquinoline or apophine alkaloids and the absence of ellagic acid, myricetin, and often of proanthocyanins. The evolutionary significance of these latter anatomical and chemical features, so important in classification, remains unknown or controversial.

Illicineae.   In some features, other members of the Magnoliales are even less specialized than members of the Winteraceae. The closely related, sister family Illiciaceae (Bailey and Nast, 1948; Keng, 1993a; Smith, 1947) of the adjacent suborder Illicineae [containing also the Schisandraceae (Keng, 1993b) of similar eastern Asian–eastern North American distribution], for example, has more primitive follicular fruit produced in a whorled follicetum (Fig. 11.2a,c). The illustrated *Illicium floridanum* Ellis (Fig. 11.2), due to the absence of Winteraceae north of the Mexican border, is by default probably the least specialized of flowering plants indigenous in the United States. The Illiciaceae, however, are more specialized in their tricolp(oid)ate pollen grains (Fig. 11.2f) and the presence of very primitive vessels, the elements of which are long, slender, angular, thin walled with greatly overlapping end walls with many-barred scalariform perforation plates, scalariform or transitional intervascular pitting, and which are diffusely scattered through the xylem. Thus, although vessels are present, they are barely more specialized than primitive tracheids.

Magnoliineae.   Better represented in the United States are members of the Magnoliineae, which include the Magnoliaceae, Annonaceae, Aristolochiaceae, and Canellaceae, this last family represented only in Florida. The archaic Magnoliaceae (Canright, 1952, 1953, 1955, 1960; Nooteboom, 1993) are especially noteworthy for their many-barred scalariform, primitive vessel elements and other primitive xylem features; relatively large flowers (perhaps a specialization for beetle pollination) with mostly spirally arranged, numerous, separate floral parts, the perianth undifferentiated and spiral, to well differentiated, trimerous, and whorled; microsporangia ribbon-like and producing mostly distally monosulcate, boat-shaped pollen grains; carpels maturing into follicles and producing mostly two, large seeds with a showy sarcotesta, much endosperm, and a very small embryo, often suspended from the follicle on a long, slender funiculus.

The large tropical family Annonaceae (Kessler, 1993; Walker, 1971, 1972), as represented by the temperate American *Asimina triloba* (L.) Dunal (Fig. 11.3), has, like the related Myristicaceae (Kühn and Kubitzki, 1993) and Canellaceae (Kubitzki, 1993b), its least specialized members in tropical America and Africa, formerly West Gondwanaland. The Annonaceae are noteworthy

Figure 11.2. *Illicium floridanum* Ellis., Illiciaceae. (**a**) Fruiting branch bearing ever-green, alternate leaves and single follicetum of one-seeded, ventrally dehiscent follicles in whorl, x0.4. (**b**). Single flower consisting of undifferentiated perianth of many purplish red tepals, stamens similarly numerous and many-seriate, and single whorl of separate carpels, x0.8. (**c**) Floral axis with all parts removed except 2 carpels in longisection to show in each a single, basally attached ovule, x3. (**d**) Developing young fruit, x1.5. (**e**). Stamens showing introrse dehiscence by longitudinal slits, x1.7. (**f**) Apparently syncolpoidate pollen grain, 28 μm. (**g**) Two glossy, flattened seeds in lateral view, x2.7, and 1 from hilum view, x4.

for their often nodding flowers with well-differentiated, trimerous perianth in three whorls (Fig. 11.3b,d), many spirally arranged "peltate" stamens (Fig. 11.3g), producing tetrads of catasulcate pollen grains (Fig. 11.3h), few separate carpels (Fig. 11.3c,e), and seeds with ruminate endosperm (Fig. 11.3j).

The related Aristolochiaceae (Huber, 1993) are so similar karyomorphologically (Morawetz, 1985) and in other features to the Annonaceae that it is unwise to treat the Aristolochiaceae as a separate order or suborder. Those who have done so have been too strongly influenced by the highly specialized and large genus *Aristolochia* L., which often has large, bizarre, bilaterally symmetrical and apetalous flowers. The western American *Asarum caudatum* Lindl. (Fig. 11.4), however, offers adequate clues to the family's relationship: spherical secretory oil cells; simple, entire, alternate, exstipulate leaves (Fig. 11.4a); trimerous bisexual flowers (Fig. 11.4a,c); inaperturate pollen grains (Fig. 11.4f); and seeds with rudimentary embryo embedded in abundant endosperm (Fig. 11.4g).

There is even available a nonmissing link, the relict *Saruma henryi* Oliv. of Hupeh, China (Dickison, 1992; Thorne, 1963). That perennial caulescent herb has radially symmetrical flowers with three conspicuous, separate, reniform petals, as well as three green sepals, a dozen separate stamens, six largely separate, semisuperior carpels, anasulcate pollen grains, and follicular fruit, plus the other primitive features listed above for the rest of the family. The swallowtail butterflies of the Papilioninae (Munroe, 1953) indicate pragmatically the biochemical affinities of the family. The larvae of the tribe Troidini feed primarily on foliage of members of Aristolochiaceae, and the larvae of the other two tribes, Graphiini and Papilionini, feed mostly on Annonaceae and other members of the Magnoliales. The three families, Degeneriaceae (Kubitzki, 1993c), Eupomatiaceae (Endress, 1993c), and Himantandraceae (Endress, 1993f), of the southwestern Pacific, area are closely related to the Magnoliaceae and share many primitive characteristics with that family.

Austrobaileyineae.   The monotypic *Austrobaileya scandens* C. T. White (Bailey and Swamy, 1949; Endress, 1993a; Srivastava, 1970) of northeastern Queensland has primitive vessels and other unspecialized stem features; bisexual flowers with many, spirally arranged, undifferentiated tepals; anasulcate pollen grains; carpels with unsealed, conduplicate styles, and numerous marginal ovules, but with a specialized habit as a scandent to high-climbing liana with opposite leaves. It is so intermediate between the Magnoliinae and the Laurineae that it is best treated as a monotypic suborder, Austrobaileyineae (Thorne, 1983). For this and other reasons, the next group of Magnoliales, the Laurineae, should not be treated as a separate order, Laurales. The gap between the Magnoliineae and putative Laurales is much too meager to be considered of ordinal size.

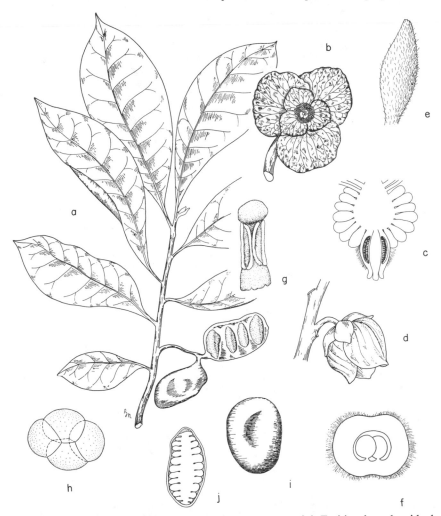

Figure 11.3. *Asimina triloba* (L.) Dunal, Annonaceae. (**a**) Fruiting branch with deciduous, alternate leaves and 2 baccate fruits, 1 in longisection to show the several flattened seeds, x0.4. (**b**) Flower from above showing 6 purplish-brown petals in 2 whorls, the outer much larger, numerous stamens, and few distinct carpels, x0.8. (**c**) Vertical section of flower, less perianth, showing spiral arrangement of many stamens and few separate carpels, 2 in longisection to show many marginal ovules, x1.5. (**d**) Nodding flower bud showing whorl of 3 valvate sepals and outer whorl of 3 imbricated petals, X 2. e. Pubescent carpel, x4. (**f**) Cross section of carpel showing 1 ovule in each of 2 rows in ovary, x6. (**g**) Stamen with 2 pollen sacs dehiscing by longitudinal slits and overtopped by enlarged "peltate" connective, x6. (**h**) Tetragonal tetrad of heteropolar, bilateral, apparently catasulcate (proximally aperturate) pollen grains, each with longest axis ca. 100 μm. i. Seed in longisection to show ruminate endosperm, x1.

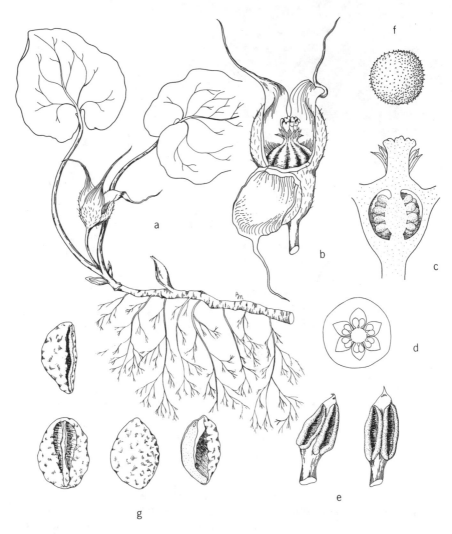

Figure 11.4. *Asarum caudatum* Lindl., Aristolochiaceae. (**a**) Flowering plant showing acaulescent, rhizomatous, herbaceous habit, basally attached, cordate leaves, and solitary, axillary flower, x0.4. (**b**) Flower with 1 of brownish-purple, petaloid sepals cut and bent back to show top of inferior ovary, 12 stamens, and 6 stigmas, x1. (**c**) Longisection through compound ovary showing numerous ovules on axile placentae in 2 of 6 locules, x1.7. (**d**) Cross section of pistil showing ovules attached in 2 rows to axis in each of 6 locules, x1.5. (**e**) Stamens dehiscent extrorsely by longitudinal slits and terminated by apiculate connective, x5.6. (**f**) Inaperturate, spherical, spinulose pollen grain, ca. 50 μm in diam. (**g**) Seeds from dorsal, ventral, and side views, and 1 in longisection to show large raphe and rudimentary embryo in abundant endosperm, x4.

Laurineae. The suborder Laurineae is mostly more specialized than the closely related suborders Magnoliineae and Austrobaileyineae; however, it does retain many primitive features. *Amborella trichopoda* Baill. (Bailey, 1957; Bailey and Swamy, 1948; Philipson, 1993a) of New Caledonia, for example, retains primitively vesselless wood, alternate leaves, largely hypogynous flowers, several separate carpels, an abundance of endosperm, and stamens dehiscent by longitudinal slits.

The six other families of the Laurineae are well supplied with the magnolialian, spherical, essential-oil cells in parenchymatous tissues and often with an abundance of benzyl-isoquinoline or aporphine alkaloids so characteristic of the superorder. Unilacunar nodes, opposite leaves, and anthers often dehiscing by valves are general features of the Laurineae, but the suborder is also known for numerous trends of specialization (Thorne, 1974). Among these trends are the usual specializations of the xylem tissues; modification of the basic two leaf traces; attainment of a definite number of floral parts in whorls; connation and adnation in the flower leading to gamosepaly, syncarpy, and epigyny; specializations of the stamens and carpels; reduction in the flowers, including perianth loss and reduction in number of stamens, carpels, and ovules to one of each or even to unisexuality; pollen grains evolving from the anasulcate to the acolpate, dicolpate, polycolpate, or even polyporate condition; basic follicles replaced by indehiscent berries, drupes, or nutlets; and exendospermous seeds with food storage taken over by the well-developed cotyledons. In addition, in several families, the arborescent habit has been replaced by shrubs, lianas, suffruticose herbs, and even by slender, twining, parasitic vines in the lauraceous *Cassytha* L.

The small pantropical family Chloranthaceae (Endress, 1987a; Todzia, 1993) is the oldest of the Laurineae in the fossil record, reaching back to the Barremian of the Early Cretaceous as the fossil pollen *Clavatipollenites* (Friis et al., 1986; Walker and Walker, 1984; see also Brenner, Chapter 5), rather comparable to the extant genus *Ascarina* Forster & Forster. Chloranthaceous xylem is very unspecialized (Carlquist, 1992b, 1992c; Swamy 1953a, 1953c, 1953d; Swamy and Bailey, 1950; Vijayaraghavan, 1964), the stems of *Sarcandra* Gardner are primitively vesselless (Editors' Note: vessels are now reported, see Carlquist, Chapter 4), and the vessels in the other three genera long with much tapered, many-barred scalariform perforation plates. Yet the flowers of the family are greatly specialized, being highly reduced, sometimes to a single, naked, uniovulate carpel with a single stamen, adnate to its side. It has been suggested that the family may have been the earliest in which wind pollination can be recorded. Its closest relative seems to be *Trimenia* Seeman, the only genus in the southern Pacific Trimeniaceae (Money et al., 1950; Philipson, 1993c; Rodenburg, 1971).

Rather characteristic of the Laurineae is the small family Calycanthaceae (Blake, 1972; Kubitzki, 1993a; Nicely, 1965) with four genera of temperate

North American and eastern Asian and tropical Queensland forests. Except for the exendospermous seeds with well-developed embryo and the relatively large, conspicuous flowers (Fig. 11.5a,c), the family rather closely relates to the tropical Monimiaceae. The Californian *Calycanthus occidentalis* H. & A. (Fig. 11.5) displays the shrubby habit; opposite leaves (Fig. 5a); poorly differentiated, numerous tepals (Fig. 11.5a,b); perigynous, cup-shaped receptacle with many separate stamens on the rim, enclosing separate, two- or one-seeded carpels (Fig. 11.5c); dicolpate pollen grains (fig. 11.5g); and indehiscent achenes (Fig. 11.5f). Food bodies terminating the stamen connectives (Fig. 11.5d) indicate beetle pollination. The well-developed cotyledons contain a strychninelike substance toxic to ruminants. Other families in the Laurineae are the Lauraceae (Rohwer, 1993), Gomortegaceae (Kubitzki, 1993d), and Hernandiaceae (Kubitzki, 1993e), all also rather closely related to the Monimiaceae (Philipson, 1993b), especially in the valvate dehiscence of the microsporangia.

Piperineae. The Piperineae, with the large tropical family Piperaceae, the monotypic Juan Fernandez Lactoridaceae, and largely temperate and species-few Saururaceae, are often treated as a separate order Piperales. Again this is quite unnecessary because of the abundance in members of the spherical secretory oil cells in parenchymatous tissues, numerous plesiomorphic features, such as the basically trimerous although naked flowers, mostly anasulcate pollen grains, crassinucellate, usually bitegmentary ovule, and rudimentary embryo in the seed, also characteristic of the Magnoliales, and the probable close relationship of the group to the Chloranthaceae of the Laurineae. The Piperineae are, however, distinctive in their mostly naked, highly reduced flowers with usually syncarpous pistil and orthotropous ovules, frequent herbaceous habit, and shared perisperm in the seeds of Piperaceae and Saururaceae.

The Saururaceae (Holm, 1926; Quibell, 1941; Raju, 1961; Steenis, 1948; Wood, 1971; Wu and Kubitzki, 1993c), illustrated here by the western American *Anemopsis californica* Hook. (Fig. 11.6), are less specialized than the Piperaceae, at least in their gynoecia, which vary from nearly apocarpous to completely syncarpous with axile placentation and separate locules or parietal placentation in a single, epigynous locule (Fig. 11.6b,d). *Anemopsis californica* is a stoloniferous perennial (Fig. 11.6a; Tucker and Douglas, Chapter 7) with small, bracteate flowers crowded into terminal conical spikes subtended by an involucre of white, reddish-maculate bracts (Fig. 11.6a,c); individual flowers without perianth and with six stamens, three conduplicate styles, and a compound inferior ovary with numerous ovules on parietal placentae (Fig. 11.6b); anasulcate, boat-shaped pollen grains (Fig. 11.6f); and globose seeds (Fig. 11.6g) with rudimentary embryo in copious perisperm.

The closely related Piperaceae (Tebbs, 1993), with highly varied growth habit and foliage, are generally more specialized in their xylem features; tiny,

Figure 11.5.   *Calycanthus occidentalis* H. & A., Calycanthaceae. (**a**) Flowering twig with opposite, deciduous leaves and solitary, terminal reddish-brown flower, x0.4. (**b**) Flower with some of the strap-shaped tepals and stamens cut away, x1. (**c**) Longisection through flower showing numerous separate carpels enclosed in flask-shaped hollow of pubescent cup-shaped receptacle but with filiform styles exserted, and some of the numerous stamens inserted on receptacle, x2. (**d**). Stamens, each terminated by succulent food body on connective, x11. (**e**) Ovoid, conspicuously veined, indehiscent pseudocarp from accrescent receptacle, x0.7. (**f**) Velvety villous achene, x1.5. (**g**). Disulculate pollen grains, 50 μm in diameter.

Figure 11.6. *Anemopsis californica* Hook., Saururaceae. (**a**). Flowering perennial herb with 1 leafy stolon, mostly basal leaves, and 2 flowering stems terminated by *Anemone*-like conical spike subtended by involucre of white, often reddish-maculate bracts, each of the small flowers in spike also subtended by small, white bract, x0.4. (**b**) Individual flower dissected from rachis of spike to show absence of perianth, the 6 stamens, 3 styles, and inferior, compound ovary with many ovules on parietal placentae, x6. (**c**) Fruiting spike with persistent bracts, x0.6. (**d**) Cross section of ovary with ovules attached in 2 lines to each of the 3 parietal placentae, x6. (**e**). Stamen, x12. (**f**) Anasulcate pollen grain, ca. 13.5 μm long. (**g**) Globose seed, containing copious perisperm and rudimentary embryo, x24.

highly reduced, bilaterally symmetrical, often unisexual flowers; solitary, erect, basal, orthotropous ovule; and baccate or drupaceous fruits. *Lactoris fernandeziana* Phil. (Carlquist, 1964, 1990; Kubitzki, 1993f) of Masatierra is more primitive in several respects than the rest of the Piperineae, especially in its trimerous flowers with three sepals, six stamens, and three separate carpels, anatropous ovules, beaked follicular fruits, and seeds with abundant endosperm (Tucker and Douglas, Chapter 7).

Nelumbonales

Probably the closest relatives to the Magnoliales are the Nelumbonales with only two extant species in the genus Nelumbo Adans. (Ito, 1986, Moseley and Uhl, 1985; Williamson and Schneider, 1993b) of the Nelumbonaceae. Formerly included in the Nymphaeaceae, it has nothing to do with that family other than convergence in habitat and habit. The two species are rich in benzyl-isoquinoline and aporphine alkaloids so characteristic of the Magnolianae. Unlike the Nymphaeaceae, there is no perisperm nor gallic and ellagic acids, the roots have vessels with scalariform perforation plates with numerous cross-bars, and the pollen is tricolpate. Serological and biochemical studies support the relationship to the Magnoliales (Hegnauer, 1971; Seigler, 1977; Simon, 1970, 1971) though *rbc*L studies (Chase et al., 1993; Qiu et al., 1993) appear to place *Nelumbo* near *Platanus* L. and *Lambertia* Sm, a highly unlikely relationship among all three taxa. Numerous fossils of *Nelumbo* and *Nelumbites* have been reported from the Lower to Upper Cretaceous. The showy, large flowers are bisexual and cantharophilous, with the tepals and stamens numerous and spirally arranged and the many separate carpels embedded in a top-shaped, spongy receptacle. The anatropous, bitegmic, crassinucellar ovule is solitary and pendulous and matures into a nut filled with the edible embyo and very little endosperm.

Ceratophyllales

Possibly related to the Nelumbonales are the Ceratophyllales, also with a single genus, *Ceratophyllum* L., with six species in the Ceratophyllaceae (Les, 1993). This ancient genus is so highly specialized for its submersed aquatic existence that its relationships are largely obscured, yet the molecular taxonomists regard it as the sister genus to all the other angiosperms (Chase et al., 1993; Qiu et al., 1993). The species lack roots, normal vessels in the xylem, and perianth in the unisexual flowers; the exine is much reduced in the inaperturate, hydrophilous pollen grains. The solitary carpel contains a single ovule, pendulous, orthotropous, crassinucellar, and unitegmic, and matures into a nut-like achene with the single seed lacking both endosperm and perisperm. *Ceratophyllum* is reported in the fossil record from the early Aptian (Dilcher, 1989).

Because of its great age and retention of some primitive characteristics, I have retained the order in my Magnolianae for want of a better place to put it. Possibly it does rate a separate superorder Ceratophyllanae.

Paeoniales

Two monogeneric families, Paeoniaceae and Glaucidiaceae, represented by *Paeonia* L. (Keefe and Moseley, 1978) and *Glaucidium* Sieb. & Zucc. (Tamura, 1972), respectively, have long been misplaced either in the Ranunculaceae or near the Dilleniaceae (Cronquist, 1981). Their retention of numerous primitive features, both vegetative and reproductive, including relatively unspecialized xylem, spiral arrangement and indefinite number of separate floral parts in *Paeonia*, bitegmic, crassinucellar ovules, follicular fruit, and seeds with rudimentary embryo, and the scattered secretory oil cells in *Paeonia* parenchymatous tissues suggest a relationship to the Magnoliales. Recently, Ditsch and Barthlott (1994) have proposed the incorporation of Paeoniaceae into Magnoliidae because the *Aristolochia*-type epicuticular wax crystalloids so typical of Magnoliidae predominate in *Paeonia*.

On the other hand, the perennial herb or shrubby habit, large, often deeply lobed or dissected leaves, centrifugal initiation and development of the stamens, low chromosome count ($x = 5$), follicular fruit, and other distinctive features common to the two families suggest recognition of them as a distinct order, perhaps best placed between Magnoliales and Berberidales (Thorne, 1981). *Paeonia californica* Nutt. (Fig. 11.7) is one of two western American *Paeonia* species, all the others being found only in Eurasia or Mediterranean North Africa. This perennial herb with fascicled fleshy roots (Fig. 11.7a,g) bears biternate, glaucous leaves and a solitary, large, terminal flower (Fig. 11.7a–c) in which the leaf-like bracts, poorly differentiated perianth parts, stamen trunks, and several separate carpels are arranged in a continuous spiral. The numerous stamens (Fig. 11.7d) develop centrifugally and bear tricolporate pollen grains, and the three to five coarse carpels (Fig. 11.7b) each produce a marginal row of anatropous, bitegmic, crassinucellar ovules that mature in the follicles as large seeds (Fig. 11.7f) with funicular aril, abundant endosperm, and minute embryo.

Glaucidium palmatum Sieb. & Zucc. (Tamura, 1972), endemic to Japan, is a rhizomatous perennial herb that differs from *Paeonia* primarily in its four petaloid tepals and two carpels connate at the base that ripen into quadriform follicles that dehisce along the dorsal and ventral sutures and produce up to 20 broadly winged seeds. Embryological, cytological, and other differences also suggest separate familial treatment for this putative *Paeonia* relative.

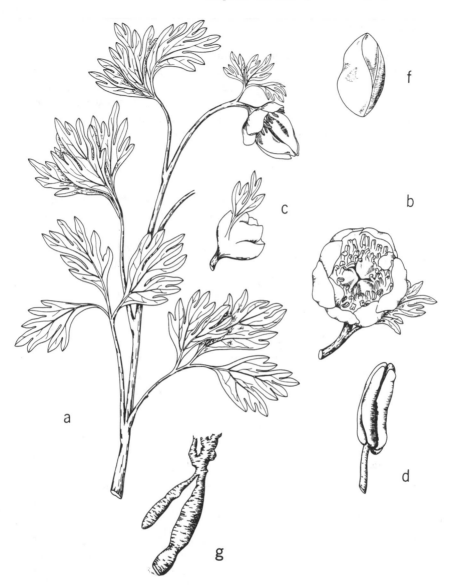

Figure 11.7.   *Paeonia californica* Torr. & A. Gray, Paeoniaceae. (**a**) perennial herb with biternate, glaucous leaves and single flower developing into cluster of follicles, x 0.4. (**b**). Large, terminal flower subtended by leaflike bract and with poorly differentiated perianth parts, stamen trunks (stamens developing centrifugally from the trunks), and 3 separate carpels arranged in continuous spiral, x1. (**c**) Single terminal flower with only leafy bract and calyx shown, x0.4. (**d**) Stamen with 4 microsporangia, x8. (**f**) Seed, x1.6. (**g**) Fascicled fleshy roots, x0.4.

Berberidales

The Berberidales, including the Papaverineae, contain, other than *Ceratophyllum*, the most specialized members of the Magnolianae. They lack the spherical secretory oil cells of the Magnoliales, but, like them, usually possess the benzyl-isoquinoline and aporphine alkaloids in abundance and numerous plesiomorphic features. They seem to be primarily herbaceous with a very varied habit of growth ranging from annual herbs to secondarily woody shrubs, lianas, and small trees. Their xylem is more specialized than that of magnolialian species; leaves are commonly compound or much dissected; flowers, bisexual or unisexual, are mostly trimerous and sometimes much reduced, or bilaterally symmetrical; perianth parts, when petals are present, are well differentiated; microsporophylls are "normal" stamens, and pollen grains are mostly triaperturate or multiaperturate; carpels are one to many, separate or variously united; and placentation, ovules, and fruits are highly varied. I include (Thorne, 1992a) eight families in the order, with Lardizabalaceae (Wu and Kubitzki, 1993b) of eastern Asia and central Chile the least specialized, and the widely distributed Papaveraceae (including Fumariaceae) (Kadereit, 1993; Líden, 1986, 1993) containing some of the most specialized members.

The fairly large, mostly temperate Berberidaceae (Loconte, 1993) illustrated here by *Berberis nervosa* Pursh (Fig. 11.8), are moderately specialized. They are best defined by their bisexual, trimerous flowers with a single carpel (Fig. 11.8b,e); stamens commonly equal to and opposite the petals (Fig. 11.8b); anthers opening by flap-like valves (Fig. 11.8d); dry or fleshy (Fig. 11.8c), mostly indehiscent fruits; and seeds (Fig. 11.8g) containing abundant endosperm and a small embryo. Probably only the Papaveraceae and the Chinese endemic *Circaeaster* Maxim. (Nowicke and Skvarla, 1982; Wu and Kubitzki, 1993a) are more specialized in the order. Lardizabalaceae, including *Sargentodoxa* Rehder & E. Wilson (Nowicke and Skvarla, 1982; Wu and Kubitzki, 1993b), Menispermaceae (Kessler, 1993), Ranunculaceae (Tamura, 1993), and Hydrastidaceae (Tobe and Keating, 1985; Keener, 1993), have, in general, larger assortments of unspecialized features.

NYMPHAEANAE

Nymphaeales

The aquatic families Cabombaceae (Williamson and Schneider, 1993a) and Nymphaeaceae (Schneider and Williamson, 1993), despite their long specialization for life in the aquatic habitat, retain many unspecialized features. Some botanists (Cronquist, 1981) have placed them in the Magnolianae, usually

Figure 11.8.   *Berberis nervosa* Pursh, Berberidaceae. (**a**) Flowering stem of low, rhizomatous shrub with stem covered by persistent bud scales and bearing 2 alternate, pinnate, glossy, evergreen leaves and 2 erect racemes with buds, flowers, and young berries, x0.4. (**b**) Flower with 6 sepals in 2 series, 6 petals in 2 series, 6 stamens opposite the petals, and single carpel, x4. (**c**) Pendulous, fruiting raceme bearing one-seeded blue berries with gray bloom, x1. (**d**) Stamen with anthers opening by 2 uplifting, flaplike valves, x13. (**e**) Carpel with several largely basal ovules and broad, sessile stigma, x9. (**f**) Spiraperturate pollen grain, 50 μm in diam. (**g**) Seeds, each containing a small embryo in abundant endosperm, x4.

adjacent to or including *Nelumbo* and *Ceratophyllum*. That they have no close relationship to those taxa is indicated especially by their chemistry (Hegnauer, 1971, 1988; Seigler, 1977), for they lack benzyl-isoquinoline alkaloids and essential oils, but have, instead, sesquiterpene alkaloids, ellagitannins, and myricetin. There are also serological differences (Simon, 1970, 1971), lack of vessels, perhaps secondary, in the xylem, pollen grains monosulcate or mono-sulcate derived, and seeds operculate and rich in perisperm. Perhaps least specialized in the Nymphaeales is the cabombaceous monotypic *Brasenia schreberi* Gmel. (Fig. 11.9) of sporadic subcosmopolitan distribution. Arising from slender rhizomes (Fig. 11.9g) are caulescent floating stems, floating elliptic, peltate leaves, and emersed small purplish flowers (Fig. 11.9a), the submersed parts of the plant covered with a thick layer of mucilaginous jelly (Fig. 11.9b,f). Also characteristic are the trimerous, biseriate perianth and numerous stamens (Fig. 11.9b,c); anasulcate pollen grains (Fig. 11.9h); several to 18 separate carpels (Fig. 11.9b,c); and indehiscent nutlets with one to three operculate, perisperm-rich seeds (Fig. 11.9d,e). The other genus *Cabomba* Aublet, with seven New World species, differs from *Brasenia* mainly in its opposite, much dissected submersed leaves and fewer floral parts, as three to six stamens and usually three carpels.

The second family in the order, Nymphaeaceae (Schneider and Williamson, 1993), with six genera, is, like *Brasenia,* subcosmopolitan in fresh, quiet waters. The members are somewhat more specialized in their usually larger flowers, often epigynous and syncarpous, with the petals, apparently of staminodial origin, sometimes absent; the ovary multilocular and maturing into a many-seeded, spongy berry; and the operculate seeds often arillate. Fossilized remains of this family or similar aquatic types indicate that it had a wide distribution early, possibly extending back to the Albian of the Early Cretaceous.

RAFFLESIANAE

The small superorder Rafflesianae consists of two families, Hydnoraceae and Rafflesiaceae (including *Mitrastemma* Makino), of achlorophyllous parasites that are perhaps the most specialized of all terrestrial flowering plants. They must be mentioned here, however, for the pollen grains of the root parasites of the two hynoraceous genera, *Hydnora* Thunb. of Africa and *Prosopanche* Bary of tropical America (Meijer, 1993a), are monosulcate and disulcate to trisulcate or trichotomosulcate, respectively, indicating the archaic origin of the group and possible ancient magnolialian affinities. The pollen grains of species in several genera of Rafflesiaceae are described as unisulcate ulcerate (Meijer, 1993b), also very primitive. Parasitic angiosperms often do seem to retain very primitive features, for reasons unknown to the phylogenists. The widely disjunct and often

Figure 11.9.   *Brasenia schreberi* Gmel., Cabombaceae. (**a**) Flowering caulescent aquatic with floating, alternate, elliptic, centrally peltate leaves and emersed, small, purplish flowers on branches arising from slender rhizomes that root at nodes, x0.3. (**b**) Flower with 6 rather similar sepals and petals, 18-36 stamens, and 6 distinct carpels, x1.3. (**c**) Flower maturing into essentially ripe fruits, x1.3. Note that, as shown also in (**b**) and (**f**), the submersed parts of the plant are covered by a thick layer of mucilaginous jelly. (**d**) Slenderly panduriform, two-seeded, indehiscent, nutlike fruit, x2. (**e**) Globose, operculate, grayish seed, x3. (**f**) Young submersed branch with buds, developing leaves, and stems all thickly covered with mucilaginous jelly, x0.5. (**g**) Node of rhizome from bottom mud producing branch, leaves, and numerous fibrous roots, x1. (**h**) Anasulcate pollen grains seen from different angles, 58 μm long.

highly sporadic nature of the distribution of members of the order also indicate the great age of the group. The Rafflesiaceae are especially fascinating because their vegetative bodies are contained entirely within the tissues of the host plants and are largely filamentous like fungus mycelia. The peculiar flowers arise endogenously from the roots or stems of the host and have an enormous range in size. The flowers of the Sumatran *Rafflesia arnoldii* R. Br., parasitic in the roots and stems of species of the vitaceous *Tetrastigma* (Miq.) Planch., have been measured at 1 meter across and are considered the largest flowers in the world.

ROSANAE

Hamamelidales

The huge superorder Rosanae, including my former Hamamelidiflorae, contains many members with numerous primitive features, particularly in the Hamamelidales (including Trochodendrales). The Trochodendraceae (Endress, 1986, 1993i), with two oriental species *Trochodendron aralioides* Sieb. & Zucc. and *Tetracentron sinense* Oliv., retain primitively vesselless wood and corresponding other primitive xylem features, as well as numerous other unspecialized characteristics of the vegetative and reproductive organs that are very reminiscent of the Magnoliales. The secretory, simple or branched idioblasts in the parenchymatous tissues of *Tetracentron* Oliv. possibly are homologous with the spherical secretory oil cells of the Magnoliales. However, the pollen grains are tricolpate and the overall character of the two species is hamamelidalian. It is likely that these archaic genera, along with the related Cercidiphyllaceae (Endress, 1993b), Eupteleaceae (Endress, 1993d; Nast and Bailey, 1946), and Platanaceae (Kubitzki, 1993g), had at least distant common ancestry with the Magnolianae.

Besides the primitively vesselless condition of the Trochodendraceae, the Hamamelidales (Endress, 1993i, 1993b, 1993d, 1993e) display numerous other primitive features in their stems, leaves, and reproductive organs. In the xylem, the tracheids are often long and scalariform-pitted in the early wood; vessel elements, when present, tend to be small, angular, with slanting endplates with often many-barred scalariform perforations; imperforate tracheary elements have bordered pits; wood-rays are heterocellular and mixed uniseriate and pluriseriate; and wood-parenchyma are diffuse or in small aggregates. Leaves are simple, with margins often serrate or crenate, pinnately or palmately veined, and frequently borne on short shoots. Flowers are small, perfect or imperfect, hypogynous, often without petals and some also lacking sepals; stamens and carpels indefinite or definite in number, separate or some carpels laterally

Table 11.1.   Classification of the Least Specialized Taxa Discussed in This Chapter.

Class: Angiospermae (Magnoliopsida)
Subclass: Dicotyledoneae (Magnoliidae)

Superorder: Magnolianae (Annonanae)
  Order: Magnoliales
    Suborder: Winterineae
        Family: Winteraceae (*Drimys, Takhtajania,* etc.)
    Suborder: Illiciineae
        Family: Illiciaceae (*Illicium*)
        Family: Schisandraceae (*Kadsura, Schisandra*)
    Suborder: Magnoliineae (Annonineae)
        Family: Magnoliaceae (*Magnolia,* etc.)
        Family: Degeneriaceae (*Degeneria*)
        Family: Himantandraceae (*Galbulimima*)
        Family: Eupomatiaceae (*Eupomatia*)
      Family: Annonaceae (*Asimina, etc.*)
        Family: Aristolochiaceae (*Aristolochia, Asarum, Saruma,* etc.)
        Family: Myristicaceae (*Myristica,* etc.)
        Family: Canellaceae (*Canella,* etc.)
    Suborder: Austrobaileyineae
        Family: Austrobaileyaceae (*Austrobaileya*)
    Suborder: Laurineae
        Family: Amborellaceae (*Amborella*)
        Family: Trimeniaceae (*Trimenia*)
        Family: Chloranthaceae: (*Ascarina, Clavatipollenites, Sarcandra,* etc.)
        Family: Monimiaceae (*Monimia,* etc.)
        Family: Gomortegaceae (*Gomortega*)
        Family: Calycanthaceae (*Calycanthus, Idiospermum,* etc.)
        Family: Lauraceae (*Cassytha, Laurus,* etc.)
        Family: Hernandiaceae (*Gyrocarpus, Hernandia,* etc.)
    Suborder: Piperineae
        Family: Lactoridaceae (*Lactoris*)
        Family: Saururaceae (*Anemopsis,* etc.)
        Family: Piperaceae (*Piper,* etc.)
  Order: Ceratophyllales
        Family: Ceratophyllaceae (*Ceratophyllum*)
  Order: Nelumbonales
        Family: Nelumbonaceae (*Nelumbo*)
  Order: Paeoniales
        Family: Paeoniaceae (*Paeonia*)
        Family: Glaucidiaceae (*Glaucidium*)
  Order: Berberidales
        Family: Lardizabalaceae (*Lardizabala, Sargentodoxa,* etc.)

Table 11.1.   (*cont.*)

        Family: Menispermaceae (*Menispermum,* etc.)
        Family: Berberidaceae (*Berberis [Mahonia],* etc.)
        Family: Hydrastidaceae (*Hydrastis*)
        Family: Ranunculaceae (*Ranunculus,* etc.)
        Family: Circaeasteraceae (*Circaeaster*)
Superorder: Nymphaeanae
  Order: Nymphaeales
        Family: Cabombaceae (*Brasenia, Cabomba*)
        Family: Nymphaeaceae (*Nymphaea,* etc.)
Superorder: Rafflesianae
  Order: Rafflesiales
        Family: Hydnoraceae (*Hydnora, Prosopanche*)
        Family: Rafflesiaceae (*Mitrastemma, Rafflesia,* etc.)
Superorder: Rosanae
  Order: Hamamelidales
  Suborder: Trochodendrineae
        Family: Trochodendraceae (*Tetracentron, Trochodendron*)
        Family: Eupteleaceae (*Euptelea*)
        Family: Cercidiphyllaceae (*Cercidiphyllum*)
  Suborder: Hamamelidineae
        Family: Platanaceae (*Platanus*)
        Family: Hamamelidaceae (*Hamamelis,* etc.)
  Order: Rosales
  Order: Saxifragales
  Order: Cunoniliales
Superorder: Arlianae (Cornanae)
  Order: Brexiales (Hydrangeales)
  Order: Cornales
  Order: Pittosporales
    Subclass: Monocotyledoneae (Liliidae)
Superorder: Acorinae
  Order: Acorales
        Family: Acoraceae (*Acorus*)
Superorder: Lilianae
  Order: Liliales
    Suborder Liliineae
        Family: Melanthiaceae (*Harperocallis, Petrosavia,* etc.)
Superorder: Alismatanae
  Order: Alismatales
        Family: Butomaceae (*Butomus*)
        Family: Limnocharitaceae (*Limnocharis,* etc.)
        Family: Alismataceae (*Damasonium,* etc.)
Superorder: Aranae
Superorder: Arecanae
  Order: Areales
        Family: Arecaceae (Thrinacinae, *Nypa,* etc.)

---

connate, the free carpels often conduplicate and some with stigmatic margins partly unsealed. Pollen grains are mostly triaperturate, and pollination ranges from entomophilous to anemophilous. Ovules are mostly pendulous from marginal placentae, anatropous, bitegmic, crassinucellar, with endosperm development cellular. Fruits are follicles, capsules, samaras, achenes, or nutlets producing many to few seeds or a single seed with tiny to large embryo with much or scanty endosperm.

The fossil record of these archaic families is extensive, indicating a wide distribution of all of them in the northern hemisphere, and at least some of them with fossils reported back to the Late Cretaceous. Today *Trochodendron* Sieb. & Zucc., *Tetracentron, Cercidiphyllum* Sieb. & Zucc., and *Euptelea* Sieb. & Zucc. are restricted in range to eastern Asia, and *Platanus* L. has a widely disjunct distribution in the Northern Hemisphere. More specialized generally in their flowers, pollen, fruits, and seeds are the members of the Hamamelidaceae (Endress, 1993e), with about 30 genera and 100 species. They too, however, have a fossil record back at least to the Late Cretaceous and, presently, a very disjunct distribution pattern.

LILIANAE

The monocotyledons appear to be a very early offshoot of the most primitive dicotyledons. Because they retain many of the same primitive features as do the Magnoliales and Nymphaeales, such as uniaperturate (monosulcate or monosulcate-derived) pollen grains, trimerous flowers with separate floral parts, and so on, some of us consider them to have common ancestry with the Magnoliales and Nymphaeanae. In their *rbc*L sequence investigations, Chase et al. (1993) and Qiu et al. (1993) found the monocots to be monophyletic and derived from within the monosulcate Magnoliidae [Magnoliales sensu Thorne (1992a)]. There is considerable argument as to which monocotyledons are the least specialized. Duvall, Clegg et al. (1993) in their *rbc*L sequence studies, using Magnoliales and Nymphaeales ("paleoherbs") as outgroups, found *Acorus* L., formerly misplaced in the Araceae, to be "the most basal extant lineage of monocotyledons." This specialized aquatic genus, like the highly specialized aquatic *Ceratophyllum*, likewise considered by the molecular taxonomists (Chase et al., 1993; Qiu et al., 1993) as the most basal extant lineage of dicots, has no close relatives and is surely of ancient origin. Their many tens of millions of years of adapting to the aquatic habitat have resulted in the extreme specialization of both genera.

Some of us consider the Melanthiaceae (Dahlgren et al., 1985) of the Liliales, although probably polyphyletic, to be the least specialized monocots. Duvall, Clegg et al. (1993) treat the Melanthiales along with Aranae and Alismatanae, next to *Acorus*, at the bottom of their consensus trees. Members of the family are mostly perennial, rhizomatous herbs, sometimes rising from a corm-like or bulb-like base. Vessels are limited to the fibrous roots and the vessel elements generally have scalariform perforation plates. The mostly alternate leaves are sheathing at the base, have parallel venation, and vary from linear to lanceolate or ovate in shape. Flowers are borne in racemes, panicles, or spikes and are mostly bisexual, hypogynous, actinomorphic, and trimerous. The flower generally consists of three plus three similar tepals, separate or slightly connate at the base; three plus three separate stamens dehiscing longitudinally and producing mostly monosulcate or bisulculate pollen grains with a reticulate exine and a binucleate condition when shed; and a tricarpellate, triloculate pistil with separate stylodia. Two genera, *Harperocallis* McDaniel of Florida and the achlorophyllous, mycophytic *Petrosavia* Becc. of southeastern Asia and Borneo, have their three carpels free nearly to the base and maturing into follicles. *Petrosavia* also has septal nectaries and produces campylotropous ovules. The three locules of melanthiaceous pistils bear numerous to few, usually anatropous, bitegmic, crassinucellate ovules on axile placentae. Embryo-sac formation is according to the *Polygonum* type, and endosperm formation is helobial. The fruits are usually loculicidal or septicidal capsules with rounded seeds that have small, ovoid or globose embryos and abundant endosperm.

Other Unspecialized Monocotyledons

Few monocots come close to the Melanthiaceae in number of retained primitive features. Some of the unspecialized members of the Alismatales, Arales, and Arecales do retain many primitive characteristics; *Butomus umbellatus* L. (Dahlgren et al., 1985) of the Butomaceae, for example, bears flowers with three green outer tepals, three inner similar but lavender tepals, six plus three stamens producing monosulcate pollen grains, and six essentially separate carpels, connate only at the very base, and maturing into many-seeded follicles. The various genera of the Limnocharitaceae and Alismataceae (Dahlgren et al., 1985) have essentially free carpels, with those of the three genera of Limnocharitaceae and *Damasonium* Miller of the Alismataceae maturing into follicles. Also 14 genera of coryphoid palms (Thrinacinae) plus *Nypa fruticans* Wurmb. (Uhl and Dransfield, 1987) have flowers with separate or solitary carpels.

OTHER UNSPECIALIZED DICOTYLEDONS

Space precludes extended discussion of the many other dicotyledons that retain numerous primitive features but some should at least be mentioned. Among the

Rosanae, the Balanopales (Buxales), Bruniales, Rosales, Saxifragales, and Cunoniales have members with significant numbers of retained primitive features in both their vegetative and reproductive organs. All seem to have arisen from common ancestry with the Hamamelidales, and I include them along with the Hamamelidales in my Rosanae. Related to the Rosales are the Aralianae (Cornanae) with such orders as the Brexiales (Hydrangeales), Cornales, and Pittosporales with numerous unspecialized members, especially in regard to their stem anatomy. Finally, we must remind ourselves again that no extant angiosperm is wholly unspecialized or wholly specialized.

CONCLUSIONS

The least specialized angiosperms belong to four major groups: Magnolianae, Nymphaeanae, Rosanae (including Hamamelidanae), and Lilianae. The more archaic members of these four groups are defined largely by such primitive features as woody or herbaceous perennial habit; simple, alternate, pinnately veined, exstipulate leaves; vesselless or otherwise primitive xylem; perianth undifferentiated; stamens unmodified microsporophylls producing *Gnetum*-like inaperturate or cycad-like, anasulcate pollen grains; styleless carpels with "unsealed," stigmatic margins and anatropous ovules on marginal or submarginal placentae maturing into endospermous seeds, each with rudimentary embryo. Magnolianae, especially Magnoliales, Nelumbonales, and Paeoniales, retain the largest number of unspecialized features. Having nearly common ancestry with them are Nymphaeanae, Rafflesianae, and Lilianae. Less closely related are the Rosanae, including especially Hamamelidales, Bruniales, Rosales, Saxifragales, and Cunoniales. Related to the Rosanae are the Aralianae (Cornanae) with the relatively unspecialized Brexiales (Hydrangeales), Cornales, and Pittosporales. Among the monocots, linked by rather distant common ancestry to the Lilianae, are unspecialized members of Alismatales, Araceae, and Arecaceae.

ACKNOWLEDGMENTS

I wish to thank the editor of *Aliso*, Dr. Richard K. Benjamin, for permission to use seven of the plates included here that were originally published in one of my papers (Thorne, 1974). The original drawings were made by Michael Cole and Paula Nicholas, and I am most grateful to them for their patience with my demands and for their artistic skill.

Molecular Phylogenies and the Diversification of the Angiosperms

Kenneth J. Sytsma and David A. Baum

The angiosperms, with approximately 440 families and nearly 250,000 species (Thorne, 1992b), represent the most significant radiation in the plant kingdom. As a result, understanding the manner and tempo of angiosperm evolution has preoccupied botanists for well over 100 years. Earlier chapters of this book have reviewed the various sources of data that have been brought to bear on this problem, which range from fossil evidence, biogeographical distributions, comparative morphology and anatomy, and, more recently, protein and nucleotide sequences. Phylogeny-based analysis of the origin and radiation of angiosperms is well underway and is utilizing both morphological and molecular data (Crane, 1985; Loconte and Stevenson, 1991; Chase et al., 1993; Qiu et al., 1993; Doyle and Donoghue, 1993; Albert et al., 1994; Donoghue, 1994; Meacham, 1994; Nixon et al., 1994; Sanderson and Donoghue, 1994; Nickrent and Soltis, 1995; reviewed in Sytsma and Hahn, 1994, and Crane et al., 1995). The questions eliciting the most interest include: Are the angiosperms a monophyletic group? What is/are the closest gymnosperm(s) to the angiosperms? What are the major angiosperm lineages? What are the basal angiosperm lineages? What did the earliest angiosperms look like? When did the angiosperms arise? Why are the angiosperms so diverse? Answering these questions depends on resolving the phylogeny of the early angiosperm radiation.

The fossil record suggests that the angiosperms may have separated from their closest gymnosperm sister groups perhaps as far back as the Triassic (Doyle and Donoghue, 1986b, 1993; Crane et al., 1995) but that the major angiosperm diversification started approximately 140 million years ago (mya; Brenner, Chapter 5) and occurred relatively quickly in the course of a few million years (Hickey and Doyle, 1977; Doyle, 1978; Lidgard and Crane, 1988; Crane and Lidgard, 1989; Taylor and Hickey, 1990a; Crane et al., 1995). This timing poses severe challenges for phylogenetic analysis using either morphological or molecular data. When the time intervals separating basal branches are short, phylogenetic analyses are vulnerable to even minor biases in data collection or analysis. One factor that can potentially confound phylogeny reconstruction arises when taxa are selected to represent the estimated quarter of a million angiosperm species. Particularly important is the possibility of spurious relationships among ancient and isolated lineages, or "long branch attraction," due to the statistical properties of the evolutionary process (Felsenstein, 1978). Indeed, we present preliminary evidence below to show that taxon sampling, perhaps acting via long branch attraction, can bias the inference of molecular phylogenies. Whether such problems also apply to morphological data is unknown but quite possible (see Stevenson and Davis, 1994).

The problem of taxon sampling can be compounded by other biases such as concerted homoplasy, wherein unrelated groups undergo convergent/parallel evolution under strong directional selection. This is likely to be especially problematic for morphological characters (Kocher et al., 1993; Givnish et al., 1994, 1995) but has also been reported for molecular data (Stewart et al., 1987). Considering the wide breadth of habits, habitats, and reproductive modes of early angiosperms, the impact of concerted homoplasy on morphological analyses could be large. For example, multiple switches to a woody (or herbaceous) habit or to an aquatic environment could have led to convergence in whole suites of morphological characters. Evaluating the role of concerted convergence in angiosperm phylogeny is difficult, and will not be discussed further. However, we would like to stress that this phenomenon should be borne in mind when the characters supporting a clade appear to be tightly associated with the clade's selective regime.

Superimposed on these directional biases is the problem of the low signal-to-noise ratio of ancient divergences. In such situations, determining whether a pattern could be due to noise, rather than signal, is critically important. This entails the evaluation of internal support, for example using bootstrap analysis, decay indices and similar statistics, as well as exploration of sensitivity to different modes of analysis such as neighbor-joining, maximum-likelihood, and compatibility methods (Saitou and Nei, 1987; Nei, 1991; Clegg, 1993; Hillis et al., 1993, 1994; Clegg and Zurawski, 1992; Stewart, 1993; Baum et al., 1994; Meacham, 1994). However, the molecular datasets that pertain to questions of angiosperm origin and radiation have generally not been subjected to rigorous studies of internal support.

Our aim here is to review the available molecular datasets for angiosperms and evaluate the success of these data in resolving their radiation. This chapter is not intended to provide a resolution of these issues but rather to summarize the molecular data with their strengths and weaknesses and suggest directions for further work. We will begin by reviewing the molecular data bearing on the issue of the time of origin of angiosperms. The vast difference in time of origin for angiosperms inferred from the fossil record and from sequence data using a "molecular clock" is one of the most interesting and puzzling aspects of "Darwin's Abominable Mystery." We will then consider the topology of the basal angiosperm radiation, and compare the phylogenies derived from sequences of nuclear rRNA and the chloroplast gene *rbc*L. We will not survey the extensive protein sequence data that have been accumulated for angiosperms over the last two decades (e.g., Boulter et al., 1972; Martin and Dowd, 1990, 1991) because of the incomplete sampling, especially of critical "primitive" families, the small numbers of characters [e.g., the first 40 amino acids of *rbc*S (Martin and Dowd, 1991)], the inherent and still unresolved computational problems associated with protein sequence data, and their relatively high homoplasy (Bremer, 1988). A more thorough critique of protein sequence data is needed, however, before they are laid aside perhaps unjustly. For example, we note the similarity of broad-scale relationships within monocots based on protein and DNA sequence data (Martin and Dowd, 1991; Duvall, Clegg et al., 1993; Duvall, Learn et al., 1993). Likewise, the results of nuclear 5S rRNA will not be surveyed because this gene is clearly too small and evolving too fast to satisfactorily address issues of early angiosperm evolution (Sytsma, 1990; Steele et al., 1991).

We shall conclude by reexamining in detail *rbc*L data for over 100 angiosperm taxa and various gymnosperm outgroups to test the robustness of certain results that are critical for an understanding of early angiosperm evolution. We have selected the *rbc*L dataset for this reanalysis because it is the only molecular dataset that has sufficient representation of critical taxa (i.e., potential gymnosperm outgroups and putatively basal angiosperm lines) for the issue of the basal topology of angiosperms to be addressed. The nuclear rRNA data is approaching the level of coverage necessary with about 200 seed plant 18S sequences available (D. Nickrent, personal communication 1995), whereas coverage for other genes (e.g., *gap*C) is presently too scattered to be useful.

MOLECULAR CLOCKS AND DATING ANGIOSPERM ORIGINS

One of the most controversial aspects in the study of seed plant evolution is dating the rise of the angiosperms. Evidence pertinent to this issue comes from the fossil record (e.g., Lidgard and Crane, 1988, 1990; Crane and Lidgard, 1989; Crane et al., 1995; Brenner, Chapter 5), seed plant cladograms with branching points dated using the fossil record (Troitsky et al., 1991; Doyle and Donoghue, 1986, 1992), and molecular clocks (Martin et al., 1989, 1993; Troitsky et al., 1991; Wolfe et al., 1989). Troitsky et al. (1991, p. 257) and, later in more detail, Doyle and Donoghue (1993) and Crane et al. (1995) noted that two dates are frequently confounded, creating confusion (Fig. 12.1). Node A dates the splitting of the stem lineage of extant angiosperms from its sister groups among extant or fossil seed plants. Node A thus defines the origin of the stem angiosperms: "angiophytes" sensu Doyle and Donoghue (1993), or "proangiosperms" sensu Troitsky et al. (1991). Node B dates the angiosperm diversification, that is the splitting of the crown group into extant clades. For example, the date of earliest angiosperm pollen fossils (140 mya) in the early Cretaceous provides a minimum (i.e., could be earlier but not later) estimate for node B, whereas the late Triassic is a minimum estimate for node A because the sister clade of angiosperms extends at least this far back [Doyle and Donoghue (1986b, 1992); but see the works of Troitsky et al. (1991) and Crane (1985) for different sister clade compositions and thus ages of node A.] Accurately determining the dates for both node A and node B would help resolve important questions such as whether angiosperms radiated explosively in the mid-Cretaceous soon after their origin, or arose considerably earlier but remained rare and/or restricted to habitats not conducive to fossilization.

There have been a number of recent attempts to estimate these divergence times by applying a molecular clock (the assumption of approximate constancy of evolutionary rate over time) to nucleotide sequence data. The first detailed attempt to date the radiation of angiosperms (node B) with molecules (Martin et al., 1989) used nine angiosperm sequences from *gapC*, the nuclear gene encoding cytosolic glyceraldehyde-3-phosphate dehydrogenase (GAPDH). The observed number of nonsynonymous substitutions between each pair of species (K_a) was compared to estimated rates of K_a (substitutions per site per year) inferred from various "known" divergence times (e.g., plants–animals, plants–yeast, mammal–chicken, human–rat). The results implied a separation of monocots and dicots at $319 + 35$ mya, a dicot radiation at $276 + 33$ mya, Solanaceae divergence (*Petunia* versus *Nicotiana*) at $199 + 32$ mya, and cereal grass divergence (*Hordeum* versus *Zea*) at $103 + 22$ mya. However, these results were questioned by Wolfe et al. (1989), Crane et al. (1989), Clegg (1990), and Troitsky et al. (1991) because only a single gene was studied, few

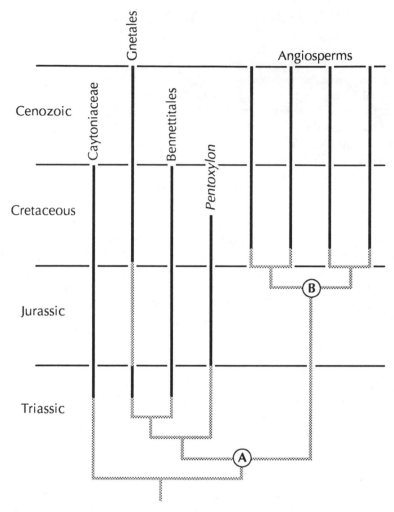

Figure 12.1.   Distinction between the two ages of the angiosperm lineage (redrawn with permission from Doyle and Donoghue, 1993, Fig. 8). Node A represents the date that the angiosperm stem lineage split from its extant sister group. Node B represents the date that the angiosperm crown group split into extant subgroups. Note that alternative ideas of the relationships of gymnosperm outgroups to each other and to angiosperms will change the position and date of node A (e.g., Crane, 1985). Dotted lines represent times when fossil evidence is lacking.

angiosperms and no gymnosperms were included, and the organisms used to calibrate the *gap*C clock showed a twofold variation in rate. Using the lowest and highest inferred rates of *gap*C evolution, the monocot–dicot split could have occurred between 251 and 511 mya. Additionally, because the presence of duplicated *gap*C was not ruled out the possibility existed that not all comparisons were strictly homologous (i.e., orthologous) and, hence, that the inferred dates are gene-duplication events not lineage-branching events.

Wolfe et al. (1989) attempted to date the monocot–dicot split using a large number of genes in the chloroplast genome. They took a three-tiered approach: *Zea* versus *Triticum* versus *Nicotiana* using synonymous substitutions (K_s) as saturation of such sites has not occurred at this level; angiosperms versus bryophytes versus green algae using K_a of protein coding genes in the chloroplast genome and substitutions in chloroplast rRNA; and plants versus fungi versus animals using nuclear rRNA sequences. Wolfe et al. (1989) considered 200 mya (late Triassic) as a likely estimate of the monocot–dicot split based on these data, although the actual dates obtained range from as low as 140 to as high as 300 mya depending on sequence and outgroup selected. Troitsky et al. (1991) argued that the few taxa examined (only three angiosperms, for example) call into question the generality of the results. It is interesting to note, however, that Martin and Dowd (1990) using a protein clock also supported a 200 mya split for monocots and dicots.

Martin et al. (1993) provided new data and arguments in support of their earlier hypothesis of a Carboniferous ($\sim$ 300 mya) origin of angiosperms with a rate analysis of both *rbc*L (cpDNA) and *gap*C (nuclear DNA) sequences. In this study the substitution rates were calibrated using various bryophytes and gymnosperms, and attempts were made to counter the criticisms leveled against their previous study (Martin et al., 1989). The monocot–dicot split was again found to be the basal division in angiosperms. They criticized the study of Wolfe et al. (1989) in that *Nicotiana*, with a significantly slower rate of nucleotide substitution than other angiosperms, was selected, thus decreasing the date of the monocot–dicot split. However, it should be noted concerning the study of Martin et al. (1993), that the monocot taxa chosen (cereal grasses) have some of the highest rates of sequence divergence known in angiosperms (see below).

How should the results of these calibrated molecular clock studies be viewed? Based on mammalian nuclear sequences, rates of synonymous substitutions are better correlated with generation time than with clock time, whereas rates of nonsynonymous substitutions fall between these two time scales (Li et al., 1987). With respect to plants, recent reviews have argued that a strict molecular clock cannot be invoked for *rbc*L (Clegg, 1990, 1993; Clegg et al., 1995), primarily because the choice of taxa will have a major impact on the dates obtained (Wilson et al., 1990; Bosquet et al., 1992; Clegg, 1993; Frascaria et al., 1993; Gaut et al., 1992, 1993; Hasebe et al., 1993). Of particular importance was the survey of 35 *rbc*L sequences within monocots in which a fivefold difference in the substitution rate was seen (Gaut et al., 1992). The pattern of rate variation in this case was consistent with the generation time hypothesis, with grasses and palms exhibiting the fastest and slowest rates, respectively.

Albert et al. (1994, p.535) argued that whereas *rbc*L data cannot be considered perfectly ultrametric (i.e., satisfying a clock assumption), the small range of absolute variation (mean of 2 X 10^{-10} total substitutions per site per million years) suggests that some predictions of the clock hypothesis still apply (e.g., linearity in the relationship between time and the accumulation of nucleotide substitutions). Although the range of absolute variation in *rbc*L substitution rates is indeed small, a fivefold difference in rates (although small in absolute sense) in selected monocots (cf. Gaut et al., 1992) could nonetheless have a significant impact on dating the origin of angiosperms.

These results strongly caution against using the molecular clock for dating unless extensive sampling of taxa and genes with quite different rates of molecular evolution is completed. Similar relative rate studies to those carried out with *rbc*L are needed for rDNA but are not presently available. However, the use of nuclear genes for molecular clock analyses also raises the specter of gene duplications and, hence, inadvertent paralogous comparisons as well as the possible confounding effects of concerted evolution.

MOLECULAR DATASETS FOR ANGIOSPERM DIVERSIFICATION

Two major molecular datasets exist for reconstructing the pattern of early angiosperm evolution, nuclear ribosomal RNA/DNA (Zimmer et al., 1989; Troitsky et al., 1991; Hamby and Zimmer, 1992; Nickrent and Soltis, 1995), and the chloroplast gene *rbc*L (Chase et al., 1993; Qiu et al., 1993). The surveys of rRNA were initiated largely to address basal angiosperm relationships and evolution within the monocots (Zimmer et al., 1989). As a result, sampling of the "primitive" dicots and of monocots was good, but overall sampling of dicots was poor. A more thorough survey of 18S rDNA has since been initiated to correct these deficiencies (D. Soltis and D. Nickrent, personal communication 1995), and a preliminary analysis of an expanded 18S dataset has recently been published (Nickrent and Soltis, 1995). The *rbc*L dataset arose out of a number of labs independently addressing questions of seed plant relationships at a number of different levels (family to division). More taxa have been sampled for *rbc*L than for rDNA and, although concentrations are seen in particular groups (e.g., Asteridae), the sampling within the angiosperms is surprisingly even.

Nuclear ribosomal RNA sequences

Two research programs have attempted to uncover relationships of the angiosperms (and gymnosperms) using nuclear rRNA sequencing. The first of these (Troitsky et al., 1991, and references therein) presented a limited survey of seed plant relationships (14 angiosperms and 6 gymnosperms in the largest study) based on various rRNA sequences (18S, 5.8S, and 5S from the cytoplasm). Two findings were noteworthy: the gymnosperms were monophyletic with no one extant group sister to the angiosperms, and the monocots were paraphyletic and located at the base of the angiosperms with a lilioid (*Narcissus*) most basal, and *Acorus* (Acoraceae) as the sister to the dicots. These results are not congruent with those of other rRNA studies and may reflect the low taxon density (see below).

The second of the rRNA sequencing projects was initiated by E. Zimmer [the more comprehensive 18S sequencing study initiated by D. Nickrent, D. Soltis and P. Soltis is still underway and results were not available for inclusion in this chapter, but see Nickrent and Soltis (1995)]. It involved a number of researchers and had as its goals the elucidation of the phylogeny of the major lineages of angiosperms and their relationships to various gymnosperm groups (Zimmer et al., 1989; Knaak et al., 1990; Hamby and Zimmer, 1992; Doyle et al., 1994). Specifics of the research methods are provided in Hamby and Zimmer (1992). One of two most parsimonious Fitch cladograms, based on 60 taxa, 1701 nucleotides, and 13 gaps (insertions/deletions) from the most detailed analysis, is shown in Fig. 12.2 (Hamby and Zimmer, 1992). An expanded conifer + *Ginkgo* + cycad lineage is shown as the sister group to angiosperms with the Gnetales even more basal. However, Gnetales is sister to angiosperms in trees one step longer and in maximum-likelihood and neighbor-joining trees, as found in earlier versions of the rRNA dataset (Zimmer et al., 1989) and by the most recent analyses (Doyle et al., 1994). In any case, the gymnosperms appear to be paraphyletic, in contrast with the result of Troitsky et al. (1991).

A consistent result of the accumulating rRNA evidence [excluding the more limited study of Troitsky et al. (1991)] is the placement of the order Nymphaeales at the base of the angiosperm radiation, followed next by members of the order Piperales. However, this basal group, comprising a large portion of the "paleoherb" alliance sensu Donoghue and Doyle (1989), does not include *Nelumbo* or *Ceratophyllum*. The distinctiveness of the latter two relative to the other paleoherbs is a consistent result of most modern morphological and molecular studies (e.g., Ito, 1987; Les, 1988; Les et al., 1991; Chase et al., 1993). After the paleoherb clades, the monocots appear as sister to the remaining dicots. Several aquatic taxa, including *Nelumbo* and *Ceratophyllum*, form a paraphyletic assemblage at the base of the monocots, suggestive of an aquatic

rRNA

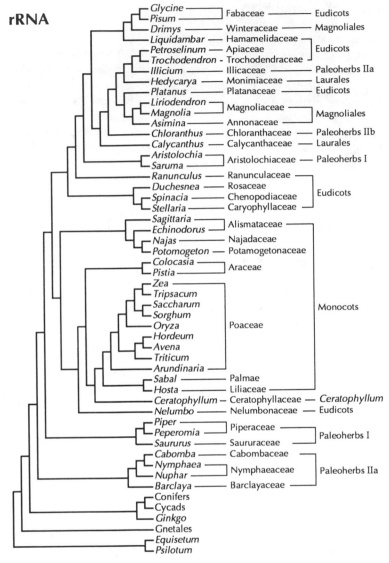

Figure 12.2. Relationships of major lineages of angiosperms and other seed plants based on parsimony analysis of nuclear rRNA sequences (redrawn with permission from Hamby and Zimmer, 1992, Fig. 4.6). The cladogram includes 60 taxa and has a consistency index of 0.39 (excluding autapomorphies). Lineage designations at right have been added to permit comparisons with other trees. Maximum likelihood and neighbor-joining analyses of the same dataset place the Gnetales closest to angiosperms (Hamby and Zimmer, 1992).

ancestry for the clade. The magnoliids (woody members of subclass Magnoliidae) form a clade together with those dicots with 3-colpate or derived pollen [= "eudicots," a term hereafter used sensu Doyle and Hotton (1991)], plus Aristolochiales, Chloranthales, and Illiciales.

Keeping in mind the lack of extensive sampling in higher dicots and the lack of rigorous data analysis (e.g., bootstrapping on all taxa, decay analysis for all branches), the results of rRNA sequencing are not incompatible with analyses of morphology (Doyle et al., 1994) and *rbcL* (see below). Although several morphological datasets have consistently placed the woody magnoliids at the base of angiosperms (e.g., Donoghue and Doyle, 1989b), the placement of Nymphaeales at the base was obtained in trees one step longer. Similarly, in a recent morphological dataset (Doyle et al., 1994), Nymphaeales + monocots are favored as the basal lineage, but other arrangements were found in trees one step longer. A paleoherb root for angiosperms was also suggested by the phylogenetic analysis of Taylor and Hickey (1992), whereas the recent rDNA analysis of Nickrent and Soltis (1995) is unresolved at the base of the angiosperms.

Chloroplast rbcL Sequences—Previous Analyses

Coordinated sequencing of the chloroplast gene *rbcL* from over a dozen laboratories culminated in the publication of two broad analyses of gymnosperm and angiosperm relationships involving 475 taxa and 499 taxa, respectively (Chase et al., 1993). A cladogram summarizing the results of the latter study is presented in Fig. 12.3. A related study (Qiu et al., 1993) examined in more detail the early radiation of angiosperms by analyzing *rbcL* from 82 gymnosperms, magnoliids, monocots, and eudicots. Most of the trees are basically similar to that in Fig. 12.3 except that the monocot lineage diverges first, before the Laurales. The very limited study of Hasebe et al. (1992b) examined the inferred protein sequences of *rbcL* for only six gymnosperms and four angiosperms. Based on the results we present below, this study included too few taxa to provide reliable information about angiosperm phylogeny and must, therefore, be treated with considerable caution.

Seven primary lineages of flowering plants are identified in Chase et al. (1993): Ceratophyllaceae, Magnoliales, Laurales, Nymphaeales + Chloranthaceae, Piperales (including other "paleoherbs"), monocots, and eudicots (see Fig. 12.3). Nymphaeales, however, are modified to include four woody families lacking vessels: Austrobaileyaceae, Amborellaceae, Illiciaceae, and Schisandraceae and excludes *Ceratophyllum* and *Nelumbo*. Ceratophyllaceae are placed as the sister group to all other angiosperms, a result contrary to that obtained from nuclear rRNA sequence studies discussed above. Interestingly, the only global morphological study of seed plants that included *Ceratophyllum* (Nixon

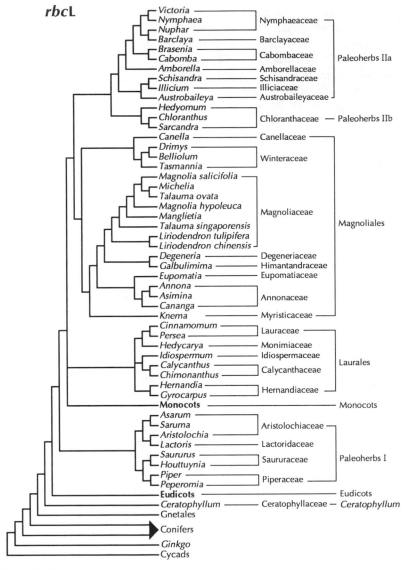

Figure 12.3. Relationships of major lineages of angiosperms and gymnosperms based on Fitch parsimony analysis of *rbc*L sequences (redrawn with permission from Chase et al., 1993, Figs. 2b and 4b). The cladogram contains 499 taxa and has a consistency index of 0.102 (excluding autapomorphies). An unresolved polytomy leads to the lineages comprising the monocots, the Laurales, and the Magnoliales + Paleoherbs II. Lineage designations at right differ from those presented in Chase et al. (1993) only in the recognition of two Paleoherb II subgroups.

et al., 1994) found the genus as sister to the other angiosperms in some but not all of the shortest trees. Similarly, the limited morphological analysis of Les (1988) is consistent with *Ceratophyllum* being in this position. Unfortunately, most of the other morphological analyses of seed plants have omitted *Ceratophyllum* (e.g., Loconte and Stevenson, 1991; Doyle and Donoghue, 1992), so its divergence near the base of angiosperms is neither rejected nor supported.

The monocots form a strongly supported monophyletic clade and divide into a number of major lineages (see Chase et al., 1993 and Duvall, Clegg et al., 1993), many of these previously advocated by Dahlgren et al. (1985). *Acorus*, now in its own family Acoraceae (Grayum, 1987), is the most basal monocot (Duvall, Learn et al., 1993) followed by the Aranae/Alismatanae. The Lilianae next diverge and form a paraphyletic grade of three lineages roughly corresponding to Dioscoreales, Liliales, and Asparagales. The remainder of the monocots, loosely the Commelineae + Arecanae, constitute a strongly supported clade.

The eudicots form an early diverging clade and are one of the best supported of all angiosperm groups (Qiu et al., 1993). This diverse clade comprises all the dicots with triaperturate or triaperturate-derived pollen, with Ranunculales + Papaverales (traditionally Magnoliidae) at its base. Within the eudicots, two large clades are identified that approximately correspond to Rosidae and Asteridae. Various lineages, including the "lower" Hamamelidae (Trochodendraceae, Tetracentraceae, Platanaceae), Proteaceae, Buxaceae, Sabiaceae, Gunneraceae, Nelumbonaceae, and Caryophyllidae appear as a paraphyletic assemblage below the Rosidae + Asteridae.

Synthesis of Molecular Results for Basal Angiosperm Relationships

How different are the results of the *rbc*L and rRNA datasets? The *rbc*L- and rRNA- based trees presented in the various published articles unquestionably differ in several substantial ways: (1) the placement of *Ceratophyllum* versus the paleoherbs at the base of the angiosperm radiation with *rbc*L and rRNA, respectively; (2) the splitting of paleoherbs into two divergent groups (Paleoherbs I and II) based on *rbc*L data; (3) the early separation of eudicots relative to all monosulcate (or monosulcate-derived) taxa in the *rbc*L tree, in contrast to the nesting of eudicots within the Magnoliidae in the rRNA tree; (4) and the differential placement of the monocots. However, none of these four results is strongly supported by either of the two datasets. This is evident in the instability of angiosperm relationships as more rRNA sequences have been added (e.g., Nickrent and Soltis, 1995), in the results of limited decay analyses on the rRNA data (Hamby and Zimmer, 1992; Doyle et al., 1994) and subsets of the *rbc*L data (see Qiu et al., 1993, and below), and in ongoing studies of the 499 taxon *rbc*L dataset where alternative basal relationships have been uncovered (Rice et al., 1995). We provide additional evidence below that a number of problems exist or potentially exist in the *rbc*L dataset so as to call into doubt most of the basal relationships implied by the Chase et al. (1993) trees (the rRNA and morphology datasets have not been analyzed in the same fashion).

Not only do these two molecular datasets have points of conflict with each other but they also differ from the morphological datasets (Crane, 1985; Loconte and Stevenson, 1991; Donoghue, 1994; Doyle and Donoghue, 1986b, 1993; Doyle et al., 1994; Nixon et al., 1994). Whether the incongruence among these datasets are the result of a failure to resolve relationships by one or another type of data, as suggested by Donoghue and Doyle (1989b), or are genuine conflicts, remains to be seen. It is noteworthy that the studies that have made pairwise comparisons of different datasets with respect to angiosperm basal relationships [morphology and nuclear rRNA (Doyle et al., 1994), morphology and *rbc*L (Albert et al., 1994), and *rbc*L and rDNA (Nickrent and Soltis, 1995)] have failed to find conflicts that are well supported by both types of data.

EARLY ANGIOSPERM EVOLUTION: EXAMINATION OF rbcL FROM SELECTED TAXA

The sheer size of the *rbc*L dataset (over 1 kb of sequence and almost 500 taxa) analyzed by Chase et al. (1993) prevented rigorous analysis of the data and, therefore, left a number of questions unanswered concerning angiosperm origins and early diversification. Many of these analytical problems were pointed out and discussed by Chase et al. (1993) and later by Baum (1993). Importantly, the length of time necessary for computation and the lack of memory to store large numbers of trees did not permit efficient sampling in tree "space" raising the possibility that "islands" (*sensu* Maddison, 1991) were missed (see also Olmstead et al., 1993; Page, 1993). Indeed, at least one island of trees five steps shorter than those described by Chase et al. (1993) has been discovered (Rice et al., 1995). In order to assess what can and cannot be learned about early angiosperm evolution from *rbc*L sequences, we have performed a number of analyses on a large subset of taxa. In some respects, our analyses are similar to those reported by Qiu et al. (1993), but they differ significantly as described below (see Table 12.1). Our main aim in this section is to highlight the danger of taking any topology based on the *rbc*L dataset as "real" when the issue of early angiosperm diversification is considered. The problem we emphasize here is taxon sampling for phylogenetic analysis, whether due to lineage extinction or through the choice of exemplar taxa ("placeholders").

Table 12.1.   Manipulations of a 109 Taxon, 1400 bp *rbc*L Data Subset, and Resulting Tree Information.

Manipulation	Number of taxa	Number of trees	Length	Length baseline topology	CI
1. Baseline	109	27	4222 (4017)	NA	.277 (.240)
2. Outgroups removed	97	27	3432 (3202)	+1	.312 (.263)
3. Conifers only as outgroups	103	90	3849 (3640)	+1	.293 (.252)
4. *Ginkgo* only as outgroup	98	27	3532 (3296)	+0	.308 (.259)
5. Gnetales only as outgroups	102	36	3799 (3587)	+7	.296 (.254)
6. *Gnetum gemon* only as outgroup	98	261	3598 (3366)	+7	.305 (.257)
7. *Welwitschia* only as outgroup	98	27	3582 (3348)	+10	.306 (.257)
8. *Ceratophyllum* removed	108	36	4169 (3964)	+1	.280 (.243)
9. Monocots removed	93	210	3352 (3156)	+0	.315 (.272)
10. Paleoherbs I removed	101	54	3909 (3701)	+1	.292 (.252)
11. Paleoherbs IIa removed	98	45	3974 (3770)	+3	.288 (.249)
12. Paleoherbs IIb removed	106	27	4139 (3940)	+0	.279 (.243)
13. Magnoliales removed	87	74	3775 (3593)	+0	.291 (.255)
14. Laurales removed	100	30	4025 (3820)	+2	.286 (.248)
15. Eudicots removed	82	24	3018 (2831)	+2	.339 (.295)
16. Ranunculales removed	98	9	3825 (3612)	+0	.295 (.253)
17. Lower Hamamelids removed	102	18	4007 (3811)	+1	.286 (.248)
18. Higher dicots removed	93	12	3433 (3248)	+3	.313 (.273)
19. *Acorus* placeholder	94	81	3409 (3210)	+2	.313 (.271)
20. *Alisma* placeholder	94	378	3422 (3228)	+18	.311 (.270)
21. *Oryza* placeholder	94	195	3483 (3280)	+27	.310 (.267)
22. *Peperomia* placeholder	102	54	4005 (3798)	+7	.287 (.248)
23. *Nelumbo* placeholder	83	39	3048 (2861)	+0	.336 (.293)
24. *Papaver* placeholder	83	3	3078 (2892)	+2	.334 (.291)
25. *Dianthus* placeholder	83	6	3121 (2930)	+6	.331 (.288)
26. *Geranium* placeholder	83	3	3127 (2933)	+1	.332 (.287)
27. *Lactuca* placeholder	83	6	3109 (2921)	+7	.332 (.289)

Note: "Length baseline topology" gives the minimum number of extra steps required to find one of the 27 baseline trees. The 27 most-parsimonious trees were pruned to the appropriate subset of taxa and their lengths compared to that of the most-parsimonious trees for that condition. This statistic gives a rough indication as to the degree of difference between the most-parsimonious topologies under a given condition and the baseline topologies. Length and consistency index (CI) are given based on all variable characters with values minus autapomorphies given in parentheses.

The Baseline Study

A total of 109 taxa were examined, which is still a larger sample of taxa than any non-*rbcL* angiosperm study (e.g., Hamby and Zimmer, 1992; Doyle et al., 1994; Nickrent and Soltis, 1995). Of these taxa, 12 were gymnosperms representing the outgroups *Ginkgo*, conifers, and the Gnetales, and 97 were angiosperms. Sequences of *Gnetum parvifolium* and *Ephedra sinica* were obtained through GenBank; all other sequences were used by Chase et al. (1993). The angiosperms were subsampled in a systematic fashion so as to have good representation of the seven major lineages identified by Chase et al. (1993): (1) *Ceratophyllum* (1 taxon); (2) Magnoliales (22 taxa); (3) Laurales (9 taxa); (4) monocots (16 taxa); (5) Paleoherbs I (= Aristolochiales + Piperales) (8 taxa); (6a) Paleoherbs IIa (= Nymphaeales + Illiciales)(11 taxa); (6b) Paleoherbs IIb (= Chloranthaceae) (3 taxa); and (7) eudicots (= tricolpate dicots) (27 taxa representing Ranunculales, "lower hamamelids," Caryophyllidae, Rosidae, and Asteridae; the latter 3 subclasses to be referred to hereafter as the "higher dicots.") Phylogenetic analysis was performed with PAUP 3.1.1 (Swofford, 1993) on the entire *rbcL* sequence except the first 28 bases (5' priming site). The search strategy followed that of Olmstead et al. (1993) using a four-step heuristic approach on a Mac IIsi starting with 360 rounds of random sequence addition with mulpars off and NNI branch swapping and ending with TBR branch swapping with mulpars on (steepest descent always on). Trees were rooted with *Ginkgo*, as most phylogenetic studies [morphological and molecular, including Chase et al. (1993)] place *Ginkgo* as more remote from angiosperms than conifers or Gnetales when gymnosperms are assumed to be paraphyletic with respect to angiosperms. Bootstrap support for different clades was determined using 100 replicates (simple addition sequences, TBR branch swapping, maxtrees = 10, steepest descent off). Decay values for each branch (Bremer, 1988) were obtained using converse topological constraints as described in Swofford (1991b) and implemented by Baum et al. (1994). A variation on the baseline approach to correct for transition/transversion bias at each codon position was performed using the weighting method of Albert et al. (1993) as implemented by Chase et al. (1993), Conti et al. (1993), and others.

The baseline analysis took over a week to complete and found 27 most parsimonious trees from a single island (see Table 12.1, row 1 for tree specifics). When the same dataset was run using a "simple" addition sequence, TBR and mulpars from the start, the same 27 trees were found within 12 hours. In several of the other tests described below, where both random and simple additions were employed, we found that both methods recovered the same trees. In view of the speed of the simple addition sequence and the lack of evidence for it being significantly more prone to become trapped on suboptimal islands for these data, all other permutations of the dataset were based on the shorter method. The bootstrap analysis required nearly 250 hours of computation on a Macintosh Quadra 650. The parsimony analysis weighted using the step-matrix of Alberts et al. (1993) found 3 shortest trees (not shown) that were equivalent to three of the 27 trees found in the baseline study.

The consensus of the 27 Fitch parsimony trees, indicating major lineages, is shown in Fig. 12.4 and one of the 27 trees is depicted as a phylogram in Fig. 12.5. The Gnetales are shown to be the closest extant gymnosperms to the angiosperms, supporting the results of all previous analyses based on *rbc*L (Chase et al., 1993; Qiu et al., 1993), some analyses based on rRNA (Hamby and Zimmer, 1992; Doyle et al., 1994), and the most recent studies of morphology (Loconte and Stevenson, 1990; Doyle and Donoghue, 1993; Doyle et al., 1994, Nixon et al., 1994). The angiosperms hold together with 61 site changes, and the branch leading to all angiosperms has 100% bootstrap support and decays only after relaxing parsimony 22 steps. The placement of *Ceratophyllum* as the sister taxon to the rest of the angiosperms followed by the separation of eudicots from the remainder of dicots and monocots is similar to that found in most *rbc*L studies (Les et al., 1991; Chase et al., 1993; Qiu et al., 1993), although one of the two islands of trees uncovered by Qiu et al. (1993) suggests a different position for *Ceratophyllum*. Following the early divergence of *Ceratophyllum* and the eudicots, our tree shows exactly the reverse sequence of splits from tree B of Chase et al. (1993) and of Qiu et al. (1993) (compare Figs. 12.3 and 12.4). Whereas the angiosperms minus the eudicots and *Ceratophyllum* are rooted on the Paleoherb I branch in Chase et al. (1993), ours is rooted in Paleoherb IIb. Both the bootstrap and decay analyses, however, indicate that relationships among the main lineages of angiosperms are tenuous (Fig. 12.4). The branches defining many of the eight major lineages are well supported, whereas all of the internal branches uniting any of these major lineages decay in trees one step longer and have bootstrap values well below 50% (Fig. 12.4). For example, trees one step longer place genera of the Alismataceae (monocots) below *Ceratophyllum* at the base of angiosperms. These decay results are similar to those found in Qiu et al. (1993), although the latter performed decay on an *rbc*L dataset excluding all gymnosperms and, thus, their results are not strictly comparable.

Neighbor-joining (Saitou and Nei, 1987), implemented in PHYLIP 3.5 (Felsenstein, 1993), on Kimura 2-T divergence values for all pairwise comparisons in the 109 taxon dataset gave quite different results relative to the baseline parsimony studies. If the distance tree is rooted on the branch leading to *Ginkgo*, the angiosperm root (branch leading from Gnetales) is within a paraphyletic monocot lineage (tree not shown). Most of the monocots (with Alismataceae being the earliest split in this clade) represent the sister group to the rest of the angiosperms. In consecutive fashion, the following lineages separate from the remaining angiosperms: Araceae, *Acorus*, Piperales, Aristolochiales, *Ce-*

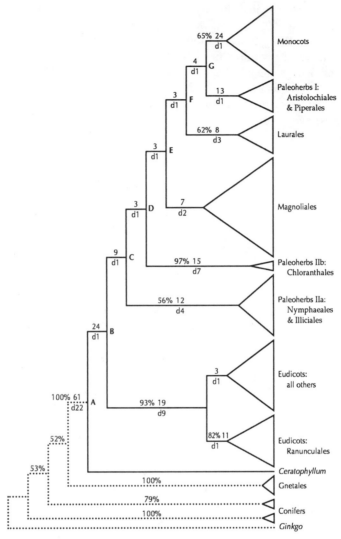

Figure 12.4.   Consensus of the 27 most-parsimonious cladograms generated from the baseline, 109 taxon, *rbc*L data subset of angiosperms and gymnosperms. The seven major angiosperm lineages are represented for clarity. Branches containing two or more lineages are identified by the letters A–G. Lengths of internal branches and branches leading to each of the angiosperm lineages are given above the branch. Decay values for each of these branches are given underneath (for example, d22 at node A indicates that the monophyly of all angiosperms is lost only when examining trees 22 steps longer than the most parsimonious). Bootstrap support is given above a branch only when it exceeded 50%. Unresolved nodes in the consensus tree are found within the eudicots and Laurales.

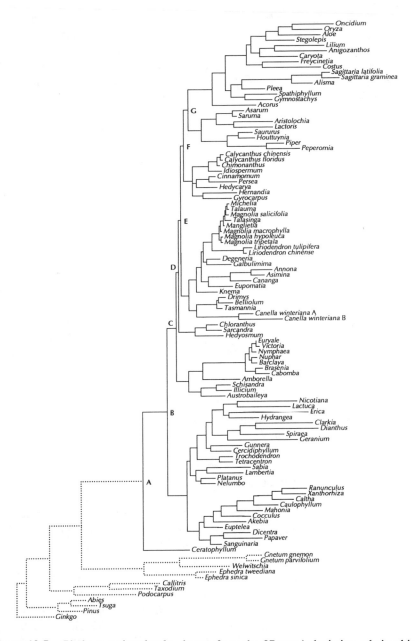

Figure 12.5. Phylogram (randomly chosen from the 27 trees) depicting relationships within angiosperms and outgroup gymnosperms using the baseline *rbc*L sequences. Nodes indicated by letter are identical to those in Fig. 12.4.

ratophyllum + Eudicots, Paleoherbs IIa + IIb, and Laurales + Magnoliales. Thus, the results of the neighbor-joining analysis of *rbc*L sequence divergence are much more similar to those of the rRNA study of Hamby and Zimmer (1992) than to the results of Fitch parsimony on *rbc*L sequences. However, this fact should not lead one to necessarily support this topology, as branch-length inequalities could distort both data-sets simultaneously.

Effects of Outgroups on Angiosperm Relationships

Among the taxon-sampling issues, the importance of the choice of outgroups has been long recognized (Maddison et al., 1984). It has already been demonstrated with morphological data that outgroup choice can effect phylogenetic inference in the angiosperms (Doyle and Donoghue, 1993). In molecular data, one specific concern is high levels of sequence divergence between outgroups and the ingroup and the subsequent possibility of spurious long-branch attraction (cf. Felsenstein, 1978; see Albert et al., 1994). Studies are now underway using randomly generated outgroup sequences to test the robustness of various angiosperm roots indicated in the broad *rbc*L studies (Donoghue, personal communication 1994). We have explored the effects both of excluding outgroups (cf. Qui et al., 1993) and of choosing different outgroups on angiosperm topology. Removal of all outgroups generated 27 shortest unrooted trees topologically equivalent to those found in the baseline study except for a minor change within the eudicot lineage (Table 12.1, row 2). It is important to note that the unstable positioning of *Ceratophyllum* when only angiosperms are compared, as seen in the study of Qiu et al. (1993) and discussed by Doyle et al. (1994), is not apparent in this dataset.

Using *Ginkgo* as the only outgroup resulted in trees with identical relationships among the main angiosperm lineages to that seen in the baseline study (Table 12.1, row 3). When conifers were used as the only outgroup the consensus tree was less resolved, with nodes C through F collapsed (Table 12.1, row 4; Fig. 12.4). The most drastic changes seen within the angiosperm clade occurred when the Gnetales were used as the only outgroup. These differences are reflected in the relatively large numbers of extra steps required to obtain any of the baseline topologies (Table 12.1, rows 5–7), and may reflect the "attraction" of Gnetales for specific lineages of the angiosperms. Using all five gnetalean outgroups, the rooting of the angiosperms occurs between the Ranunculales and the higher dicots (Fig. 12.6a), and *Ceratophyllum* becomes the sister group to all the angiosperms minus the eudicots. If only *Gnetum gemon* is used, the root of angiosperms is placed at the base of a paraphyletic monocot lineage; specifically, Alismataceae is basal and *Ceratophyllum* is sister to the eudicots (Fig. 12.6b). When *Welwitschia mirabilis* is used, similar trees are found except that *Oryza* is basal in a paraphyletic monocot lineage (Fig. 12.6c). These results

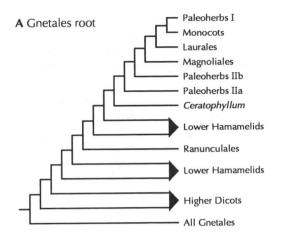

A Gnetales root

- Paleoherbs I
- Monocots
- Laurales
- Magnoliales
- Paleoherbs IIb
- Paleoherbs IIa
- *Ceratophyllum*
- Lower Hamamelids
- Ranunculales
- Lower Hamamelids
- Higher Dicots
- All Gnetales

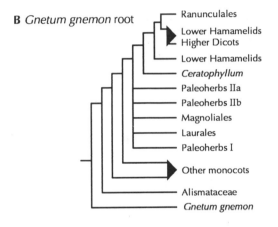

B *Gnetum gnemon* root

- Ranunculales
- Lower Hamamelids
- Higher Dicots
- Lower Hamamelids
- *Ceratophyllum*
- Paleoherbs IIa
- Paleoherbs IIb
- Magnoliales
- Laurales
- Paleoherbs I
- Other monocots
- Alismataceae
- *Gnetum gnemon*

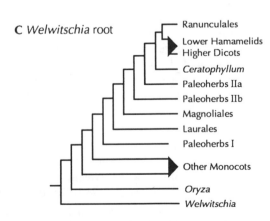

C *Welwitschia* root

- Ranunculales
- Lower Hamamelids
- Higher Dicots
- *Ceratophyllum*
- Paleoherbs IIa
- Paleoherbs IIb
- Magnoliales
- Laurales
- Paleoherbs I
- Other Monocots
- *Oryza*
- *Welwitschia*

Figure 12.6. Differential rooting of angiosperms and changes in angiosperm relationships when all gymnosperms except Gnetales are removed from the baseline *rbc*L dataset; outgroups: **a**, all Gnetales; **b**, *Gnetum gemon*, and; **c**, *Welwitschia mirabilis*. Paraphyletic lineages are indicated by arrowheads.

may well be due to long-branch attraction, as it appears from the most parsimonious baseline phylograms (e.g., Fig. 12.5) that the longest branches are
(a) the Gnetales, (b) the monocots, including *Oryza* and Alismataceae, and (c)
the eudicots. The lineage leading to *Ceratophyllum* is not as long and apparently
is not attracted to the gnetalean lineage in these analyses (see below).

Clearly, we do not know that the baseline rooting is "true," but these
manipulations show great instability when single outgroups are used. These
results suggest that sole reliance on a single outgroup taxon, even if all evidence
points to its sister group relationship with the ingroup, should be avoided
(Maddison et al., 1984). However, sampling further out not only increases the
number of taxa to sample (and thus the computational load) but also increases
the likelihood that sequence divergence is greater than that appropriate for
phylogenetic analysis of the given gene (Felsenstein, 1978; Sytsma, 1990;
Albert et al., 1994). It is noteworthy that distance approaches are apparently not
immune to problems observed using parsimony. Removing all gymnosperm
sequences and running neighbor-joining only on the angiosperm sequence distance matrix generated quite different trees. For example, the Canellaceae +
Winteraceae clade moves from a basal position in Magnoliales to a position as
sister to Magnoliales + Lauraceae + Paleoherbs II.

Effects of Lineage Removal on Angiosperm Relationships

Lineage extinction and failure to sample critical lineages are inherent problems in phylogenetic analyses that may generate distorted relationships. Lineage
extinction is always going to be an issue and is likely to be more severe for
studies based on molecules than on morphology because in the latter one can
potentially include fossil taxa (Donoghue et al., 1989). Not sampling certain
extant lineages has been a prevalent problem in both molecular and morphological analyses and can be attributed to a lack of data (e.g., in preliminary studies)
and to *a priori* notions of what lineages should be sampled or included (e.g.,
the lack of sampling of *Ceratophyllum* in most but not all morphological
studies). We explored the removal of various angiosperm taxa on the stability
of the baseline trees as a way to gauge the impact of lineage extinction or failure
to sample extant lineages. Each of the eight major angiosperm lineages were
removed one at a time and the matrix was analyzed with PAUP (Table 12.1,
rows 8–15). The removal of *Ceratophyllum*, Paleoherbs IIb, and Magnoliales
had no effect on the remaining angiosperm topology, and the removal of
monocots and Paleoherbs I gave topologically equivalent but less resolved trees
(Fig. 12.7a-e). Zero or one extra step was required in order to find the baseline
trees (with appropriate taxa pruned) under the new conditions. The effects of
deleting Paleoherbs IIa, Laurales, or eudicots were more substantial (Figs

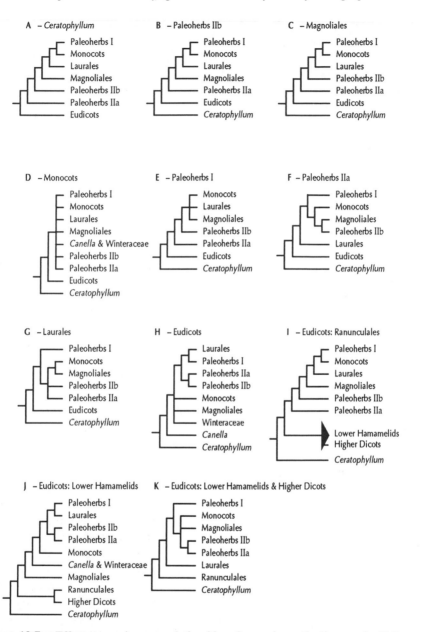

Figure 12.7. Effects on angiosperm relationships of removing entire lineages (**a–h**) from the angiosperm baseline *rbc*L dataset or removal of portions of the eudicot lineage (**i–k**). Note that none of the lineage removals shifted the placement of *Ceratophyllum*, although many other relationships are altered. Paraphyletic lineages are indicated by arrowheads.

12.7f–k). It is also noteworthy that removal of subsets of the dicots resulted in significant instability, perhaps reflecting the fact that the eudicots represent a clade with some of the longest branches in the angiosperm tree (see Fig. 12.5) and that the branches tying together the basal angiosperm lineages are already weak (see Fig. 12.4). These results should be sobering. They suggest that limited sampling of major angiosperms clades when using *rbc*L, and perhaps other, sequence data can lead to spurious results. The absence or low representation of "higher dicots" and *Ceratophyllum* in most morphological analyses of angiosperm or seed plant relationships (e.g., Doyle and Donoghue, 1993, and previous studies) is, thus, a possible source of error.

Effects of Using "Exemplars" on Angiosperm Relationships

One of the most controversial aspects of attempting to reconstruct origins and diversification of the angiosperms is the choice of taxon sampling within lineages. Chase et al. (1993) frankly point out that extensive taxon sampling has the major drawback of requiring computational efforts that can become unrealistic, at least using present tree building strategies [see Olmstead et al. (1993), for alternative strategies for analyzing the *rbc*L dataset]. The use of exemplars or "placeholders" to represent larger lineages is one method to reduce this computational load. Exemplars have also been used when initially surveying larger groups, where sampling is necessarily limited to one or few taxa per lineage, and when reducing numbers of taxa in analyses in which different datasets are compared and/or merged in order to sample equivalent taxa (e.g., Doyle et al., 1994; Smith and Sytsma, 1994).

We explored the outcomes on inferred angiosperm relationships when different exemplars are selected as placeholders for three major lineages: monocots, Paleoherbs I, and eudicots. These lineages were selected because they each contain a range of branch lengths and, thus, were most likely to affect the angiosperm tree. Three taxa were used as alternate exemplars for the monocots: *Acorus*, *Alisma*, and *Oryza* (Table 12.1, rows 19–21). Reducing the monocot lineage to its basal taxon, *Acorus*, had no effect on angiosperm relationships except for an unresolved node (Fig. 12.8a). However, replacing the monocot lineage with either *Alisma* or *Oryza* had a major impact. In both, cases the monocot lineage became the sister group to all other angiosperms, including *Ceratophyllum* (Fig. 12.8b). This position of the monocot clade was maintained even when all three (or various pairs of these) monocot taxa were included together (data not shown). The movement of the monocot lineage is perhaps not unexpected, considering that at least the Alismataceae subclade shifts to the basal position in trees that are one step longer in the baseline study (see above). However, the topology of these baseline trees is quite different from the

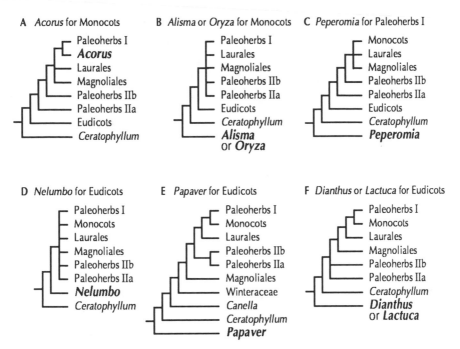

Figure 12.8. Effects of the use of exemplars to represent major lineages in the *rbc*L dataset. Genera in bold represent the placeholders: (**a–b**) monocot placeholders, **c**) Paleoherb I placeholder, and (**d–f**) eudicot placeholders.

topology obtained when *Alisma* or *Oryza* are used as exemplars (see Table 12.1, rows 20 and 21). Similarly, the placement of the Paleoherb I lineage as sister to the other angiosperms occurs when *Peperomia* is its exemplar (Table 12.1, row 22; Fig. 12.8c). Of the four taxa used as placeholders for eudicots only one (*Nelumbo*) maintained *Ceratophyllum* in the basal position (Table 12.1, row 23; Fig. 12.8d). The other three eudicot placeholders forced the shift of the eudicot lineage to the basal position within angiosperms (Table 12.1, rows 24–27; Fig. 12.8e–g).

Finally, we subsampled taxa within the entire *rbc*L dataset to give a reduced number of taxa comparable to prior analyses of nuclear rRNA and morphology. Each pseudorandom subset of the 109 taxa had roughly equal representation of each major lineage. Figures 12.9a–c give examples of the results obtained with 37, 22, and 11 taxa, respectively (two in each data subset were gymnosperms). The relative ease with which clades shift around in the angiosperm topology when the number of taxa sampled for each lineage is reduced suggests that there could be significant error introduced by the use of exemplars. Furthermore, our studies show that the effect of exemplars is notably nonrandom. Reducing the number of representatives of a clade is most likely to result in that clade

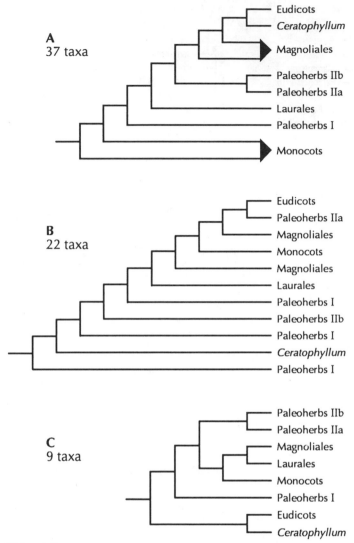

Figure 12.9.   Effects on angiosperm relationships and gymnosperm rooting by strat-ified-random (maintaining relatively equal numbers of taxa per major lineage, whenever possible) subsampling of the baseline 109 taxa *rbc*L dataset. Three representative results are shown for (**a**) 37 taxa, (**b**) 22 taxa, and (**c**) 9 taxa. The numbers of taxa surveyed represent the range seen in studies of other molecular and morphological angiosperm datasets. Paraphyletic lineages are indicated by arrowheads. Note that the major lineages identified by Chase et al. (1993) often appear non-monophyletic.

appearing at the base of the tree, presumably due to the attraction between the long branches of the exemplars and the outgroups. The apparent determinism of the exemplar effect offers both good and bad news. The good news is that we can be warned of possible artifacts when sparsely sampled lineages appear in basal positions and use that information to initiate analyses that are less prone to long-branch attraction (e.g., maximum-likelihood) or add further taxa to the dataset. The bad news is that if a basal clade is sparsely sampled due to extinction of its relatives, then it may prove very difficult to separate fact from artifact. These problems with the exemplar method should, however, not be taken as arguing that exemplars never be used. We often have to reduce data-sets to manageable size and it remains to be seen whether other approaches such as the "ancestral character state" method (see Donoghue, 1994) avoid these pitfalls. In any case, much more analysis is needed to test both the appropriateness and the efficacy of different methods for reducing the size of datasets.

Overall our studies suggest that the placement of *Ceratophyllum* at the base of angiosperms [also found by Les et al. (1991), Chase et al. (1993), Qiu et al. (1993), and some trees in Nixon et al. (1994)] should be treated with caution. Because only one sequence is available for the monogeneric Ceratophyllaceae (with 2–30 species), their position in the most parsimonious trees may be spurious due to the exemplar phenomenon we have just demonstrated. This instability of the genus's placement is also suggested by the rRNA analyses that place *Ceratophyllum* within monocots (Hamby and Zimmer, 1992; Fig. 12.2). Although additional sequences of *Ceratophyllum* are needed, there is no guarantee that they will be much different and thus they may not help solidify the placement of this interesting genus. This serves to highlight the fact although *rbc*L is certainly informative with respect to some questions in angiosperm phylogeny, it is of limited utility in resolving the arrangement of main angiosperm lineages. There is, however, room for optimism that additional sampling or modified methods of analysis will overcome these limitations. Also, one can hope that combining information from *rbc*L, other molecular data, and morphology may serve to improve the signal-to-noise ratio allowing clearer resolution of the angiosperm radiation.

CONCLUSIONS

Despite widespread interest and research, molecular answers to the timing and pattern of angiosperm diversification remain contradictory and ill-supported. The rate inconstancy of specific genes seen even within closely allied taxa calls into question attempts to use calibrated molecular clocks to date the origin of the angiosperms. Although the constitution of major angiosperm lineages is now becoming resolved, the basal branches and relationships among these lineages have inherently weak support. Furthermore, our studies suggest that taxon sampling issues may have large effects on the basal topology of angiosperms as inferred from molecular data. This fact should not be taken as arguing against the use of molecular data as we suspect that similar phenomena are occurring in the morphological analyses as evidenced by the continual changes in topology seen from year to year as taxa are added or the data are "tweaked." Thus, the already difficult problem of unraveling the ancient divisions within angiosperms is made even harder by the fact that we can only sample a small portion of extant angiosperm diversity and we lack the ability to extensively sample extinct lineages.

We are thus left with the troubling prospect that relationships among the basal lineages of angiosperms may be impossible to uncover with existing datasets analyzed independently. We believe that is likely, however, that those relationships can be resolved by additional data (more taxa and more sequences) and by combining sources of molecular and morphological data. Therefore, there is a need to coordinate the accumulation of all types of data to guarantee identical taxon coverage so as to facilitate combined analyses. It is reassuring that steps are now being taken to obtain 18S rRNA sequence information to match *rbc*L coverage (Soltis and Nickrent, 1995). However, before integrating multiple datasets we need to have a clear picture of how the information from different sorts of phylogenetic data can be combined most effectively. This questions is, however, still a topic of much discussion and controversy (e.g., Kluge, 1989; Hillis, 1987; Sytsma, 1990; Swofford, 1991b; Donoghue and Sanderson, 1992; Bull et al., 1993; de Queiroz, 1993; Doyle et al., 1994). Thus resolution of basal angiosperm relationships may have to await both the collection of additional molecular and morphological data as well as further theoretical advances in phylogenetic systematics.

REFERENCES

Adams, E. N. 1972. Consensus techniques and the comparison of taxonomic trees. *Systematic Zoology* **21**:390–397.

Aharoni, E. 1964. Litho-electric correlation of the Kurnub Group (Lower Cretaceous) in the northern Negev. *Israel Journal of Earth Sciences* **13**:63–81.

Albert, V. A., A. Backlund, K. Bremer, M. W. Chase, J. R. Manhart, B. D. Mishler, and K. C. Nixon. 1994. Functional constraints and *rbc*L evidence for land plant phylogeny. *Annals of the Missouri Botanical Garden* **81**:534–567.

Albert, V. A., M. W. Chase, and B. D. Mishler. 1993. Character-state weighting for cladistic analysis of protein-coding DNA sequences. *Annals of the Missouri Botanical Garden* **80**:752–766.

Alvin, K. and W. G. Chaloner. 1970. Parallel evolution in leaf venation: an alternative view of angiosperm origins. *Nature* **226**:662–663.

Anderson, J. M. and H. M. Anderson. 1983. *Palaeoflora of Southern Africa Molteno Formation (Triassic)*. A. A. Balkema, Rotterdam.

Anderson, J. M. and H. M. Anderson. 1985. *Palaeoflora of Southern Africa, Prodromus of South African Megafloras Devonian to Lower Cretaceous*. A. A. Balkema, Rotterdam.

Anderson, R. E. and R. A. Brink. 1952. Kernel pattern in variegated pericarp of maize and the frequency of self-colored offspring. *American Journal of Botany* **39**:637–644.

Andrews, H. N. 1961. *Studies in Paleobotany*. John Wiley & Sons, New York.

Andrews, H. N. 1963. Early seed plants. *Science* **142**:925–931.

Arber, A. 1937. The interpretation of the flower: a study of some aspects of morphological thought. *Cambridge Philosophical Society Biology Review* **12**:157–184.

Arber, E. A. N. and J. Parkin. 1907. On the origin of angiosperms. *Botanical Journal of the Linnean Society of London* **38**:29–80.

Arber, E. A. N. and J. Parkin. 1908. Studies on the evolution of the angiosperms. The relationship of the angiosperms to the Gnetales. *Annals of Botany (London)* **22**:489–515.

Archangelsky, S. and D. W. Brett. 1961. Studies on Triassic fossil plants from Argentina. I. *Rhexoxylon* from the Ishigualasto Formation. *Royal Society of London Philosophical Transactions, Series B* **244**:1–19.

Arnold, C. A. 1929. On the radial pitting in *Callixylon. American Journal of Botany* **16**:391–393.

Arnott, H. J. 1959. Anastomoses in the venation of *Ginkgo biloba. American Journal of Botany* **50**:821–830.

Asama, K. 1976. *Gigantopteris* flora in southeast Asia and its phytopalaeogeographic significance. *Tokyo National Science Museum Bulletin* 12.

Asama, K. 1984. *Gigantopteris* flora in China and southeast Asia. In *Geology and Palaeontology of Southeast Asia*, eds. T. Kobayashi, R. Toriyama, and W. Hashimoto, pp. 311–323. University of Tokyo Press, Tokyo.

Ash, S. R. 1970. *Dinophyton*, a problematical new plant genus from the Upper Triassic of the southwestern United States. *Palaeontology* 13:646–663.

Ash, S. R. 1972. Late Triassic plants from the Chinle Formation in north-eastern Arizona. *Palaeontology* 15:598–618.

Ash, S. R. 1987. Growth habit and systematics of the Upper Triassic plant *Pelourdea poleoensis*, southwestern U.S.A. *Review of Palaeobotany and Palynology* 51:37–49.

Ax, P. 1987. *The Phylogenetic System*. John Wiley & Sons, Chichester.

Axelrod, D. I. 1952. A theory of angiosperm evolution. *Evolution* 6:29–60.

Axelrod, D. I. 1961. How old are the angiosperms? *Americal Journal of Science.* 259:447–459.

Baas, P. and S. Carlquist 1985. A comparison of the ecological wood anatomy of the floras of southern California and Israel. *International Association of Wood Anatomists Bulletin, New Series* 6:349–353.

Bailey, I. W. 1923. The cambium and its derivative tissues. IV. The increase in girth of the cambium. *American Journal of Botany* 10:499–509.

Bailey, I. W. 1944a. The comparative morphology of the Winteraceae, III. Wood. *Journal of the Arnold Arboretum* 25:97–103.

Bailey, I. W. 1944b. The development of vessels in angiosperms and its significance in morphological research. *American Journal of Botany* 31:421–428.

Bailey, I. W. 1956. Nodal anatomy in retrospect. *Journal of the Arnold Arboretum* 37:269–287.

Bailey, I. W. 1957. Additional notes on the vesselless dicotyledon, *Amborella trichopoda* Baill. *Journal of the Arnold Arboretum* 38:374–380.

Bailey, I. W. 1966. The significance of the reduction of vessels in the Cactaceae. *Journal of the Arnold Arboretum* 47:288–292.

Bailey, I. W. and C. G. Nast. 1943a. The comparative morphology of the Winteraceae. I. Pollen and stamens. *Journal of the Arnold Arboretum* 24:340–346.

Bailey, I. W. and C. G. Nast. 1943b. The comparative morphology of the Winteraceae. II. Carpels. *Journal of the Arnold Arboretum* 24:472–481.

Bailey, I. W. and C. G. Nast. 1944a. The comparative morphology of the Winteraceae. IV. Anatomy of the node and vascularization of the leaf. *Journal of the Arnold Arboretum* 25:215–221.

Bailey, I. W. and C. G. Nast. 1944b. The comparative morphology of the Winteraceae. V. Foliar epidermis and sclerenchyma. *Journal of the Arnold Arboretum* 25:342–348.

Bailey, I. .W. and C. G. Nast. 1945. The comparative morphology of the Winteraceae. VII. Summary and conclusions. *Journal of the Arnold Arboretum* 26:37–47.

Bailey, I. W. and C. G. Nast. 1948. Morphology and relationships of *Illicium, Schisandra* and *Kadsura* I. Stem and leaf. *Journal of the Arnold Arboretum* 29:77–89.

Bailey, I. W.,C. G. Nast, and A. C. Smith. 1943. The family Himantandraceae. *Journal of the Arnold Arboretum* 24:190–206.

Bailey, I. W. and A. C. Smith. 1942. Degeneriaceae, a new family of flowering plants from Fiji. *Journal of the Arnold Arboretum* 23:356–365.

Bailey, I. W. and B. G. L. Swamy. 1948. *Amborella trichopoda* Baill., a new morphological type of vesselless dicotyledon. *Journal of the Arnold Arboretum* 29:245–254.

Bailey, I. W. and B. G. L. Swamy. 1949. The morphology and relationships of *Austrobaileya*. *Journal of the Arnold Arboretum* 30:211–236.

Bailey, I. W. and B. G. L. Swamy. 1951. The conduplicate carpel of dicotyledons and its initial trends of specialization. *American Journal of Botany* 38:373–379.

Bailey, I. W. and W. W. Tupper. 1918. Size variation in tracheary cells. I. A comparison between the secondary xylems of cryptogams, gymnosperms, and angiosperms. *Proceedings of the American Academy of Arts and Sciences* 54:149–204.

Baillon, H. 1871. *Natural History of Plants*, Vol. 3:80–82, Nymphaeaceae. L. Reeve & Co., London.

Bakker, R. T. 1978. Dinosaur feeding behaviour and the origin of flowering plants. *Nature* 274:661–663.

Bakker, R. T. 1986. How dinosaurs invented flowers. *Natural History* 86:30–38.

Barnard, C. 1961. The interpretation of the angiosperm flower. *Australian Journal of Science* 24:64–72.

Barthlott, W. and D. Froelich. 1983. Mikromorphologie und Orientierungsmuster epicuticularer Wachs-Kristalloide: Ein neues systematisches Merkmal bei Monocotylen. *Plant Systematics and Evolution* 142:171–185.

Battaglia, E. 1989. The evolution of the female gametophyte of angiosperms: an interpretative key. *Annali di Botanica* 47:7–144.

Batten, D. J. 1989. Systematic relationships between Normapolles pollen and the Hamamelidae. In *Evolution, Systematics, and Fossil History of the Hamamelidae*, eds. P. R. Crane and S. Blackmore, Vol. 2, pp.9–21. Clarendon Press, Oxford.

Baum, D. A., K. J. Sytsma, and P. C. Hoch. 1994. A phylogenetic analysis of *Epilobium* (Onagraceae) based on nuclear ribosomal DNA sequences. *Systematic Botany* 19:363–388.

Baum, H. 1949. Der einheitliche Bauplan der Angiospermengynözeen und die Homologie ihrer fertilen Abschnitte. *Botanische Jahrbücher für Systematik, Pflanzengeschichte und Pflanzengeographie* 96:64–82.

Baum, H. 1952. Über die "primitivste" Karpellform. *Botanische Jahrbücher für Systematik, Pflanzengeschichte und Pflanzengeographie* 99:632–634.

Baum-Leinfellner, H. 1953. Die Peltationsnomenklatur der Karpelle. *Botanische Jahrbücher für Systematik, Pflanzengeschichte und Pflanzengeographie* 100:424–426.

344   *References*

Beach, J. H. and W. J. Kress. 1980. Sprophyte versus gametophyte: a note on the origin of self-incompatibility in flowering plants. *Systematic Botany* 5:1–5.

Behnke, H.-D. 1976. The bases of angiosperm phylogeny: ultrastructure. *Annals of the Missouri Botanical Garden* 62:647–663.

Behnke, H.-D. 1989. Structure of Phloem. In *Transport of Photoassimilates*, eds. D. A. Baker and J. A. Milburn, pp. 79–137. Longman Scientific & Technical, Essex, United Kingdom.

Behnke, H.-D. 1991. Sieve-element characters of *Ticodendron*. *Annals of the Missouri Botanical Garden* 78: 131–134.

Behnke, H.-D., and R. Dahlgren. 1976. The distribution of characters within an angiosperm system. 2. Sieve-element plastids. *Botaniska Notiser* 129:287–295.

Bell, W. A. 1938. Fossil flora of Sydney Coalfield, Nova Scotia. *Memoirs of the Geological Survey of Canada* 215:1–334.

Bell, W. A. 1962. Flora of Pennsylvanian Pictou Group of New Brunswick. *Bulletin of the Geological Survey of Canada* 87:1–71.

Bendz, G. and J. Santesson (eds.). 1974. *Chemistry in Botanical Classification. Proceedings of the 25th Nobel Symposium, 1973, Sweden.* Academic Press, New York.

Bentor, Y. and A. Vroman 1960. The geologic map of the Negev. *Geological Survey of Israel, Series A, scale 1:100,000, sheet 16*: Mount Sedom:1–117.

Bernhardt, P. and L. B. Thien. 1987. Self-isolation and insect pollination in the primitive angiosperms: new evaluations of older hypotheses. *Plant Systematics and Evolution* 156:159–176.

Bertrand, C. E. 1898. Rémarques sur la structure des graines de pollen de Cordaites. *Compte Rendus Association Française pour l'avancement des sciences*, 1898:436–441.

Bessey, C. E. 1897. Phylogeny and taxonomy of the angiosperms. *Botanical Gazette* 24:145–178.

Bessey, C. E. 1915. The phylogenetic taxonomy of flowering plants. *Annals of the Missouri Botanical Garden* 2:109–164.

Bhandari, N. N. 1963. Embryology of *Pseudowintera colorata* – a vesselless dicotyledon. *Phytomorphology*: 303–316.

Bhandari, N. N. 1971. Embryology of the Magnoliales and comments on their relationships [I]. *Journal of the Arnold Arboretum* 52:1–39, 285–304.

Bierhorst, D. W. 1960. Observations on tracheary elements. *Phytomorphology* 10:249–305.

Bierhorst, D. W. 1971. *Morphology of Vascular Plants*. Macmillan, New York.

Bierhorst, D. W. and P. M. Zamora. 1965. Primary xylem elements and element associations of angiosperms. *American Journal of Botany* 52:657–710.

Blake, S. T. 1972. *Idiospermum* (Idiospermaceae), a new genus and family for *Calycanthus australiensis*. *Contribution to the Queensland Herbarium* 12:1–37.

Bliss, M. C. 1921. The vessel in seed plants. *Botanical Gazette* 71:314–326.

Bock, W. 1969. *The American Triassic Flora and Global Distribution*, Vols. **2** and **3**. Geology Center for Research Services, North Wales, Pennsylvania

Boersma, M. and L. M. Broekmeyer. 1979. *Index of Plant Megafossils. Carboniferous 1971–1975*. Laboratory of Paleobotany and Palynology, University of Utrecht, Utrecht.

Bold, H. C. 1973. *Morphology of Plants*, 3rd ed., Harper & Row, New York.

Bond, W. J. 1989. The tortoise and the hare: ecology of angiosperm domaniance and gynmosperm persistence. *Biological Journal of the Linnean Society* **36**:227–249.

Bose, M. N. 1953. *Bucklandia sahnii* sp. nov. from the Jurassic of the Rajmahal Hills, Bihar. *Paleobotanist* **2**:41–50.

Bosquet, J., S. H. Strauss, A. H. Doerksen, and R. A. Price. 1992. Extensive variation in evolutionary rate of *rbc*L gene sequences among seed plants. *Proceedings of the National Academy of Sciences of the United States of America* **89**:7844–7848.

Boulter, D., J. A. M. Ramshaw, E. W. Thompson, M. Richardson, and R. H. Brown. 1972. A phylogeny of higher plants based on the amino acid sequences of cytochrome c and its biological implications. *Proceedings of the Royal Society of London B* **181**:441–455.

Bouman, F. 1984. The ovule. In *Embryology of Angiosperms*, ed. A. M. Johri, pp. 123–157. Springer-Verlag, Berlin.

Boureau, E. and J. Doubinger. 1975. *Traité de Paléobotanique. IV. Pteridophyta*. Masson et Cie, Paris.

Bowman, J. L., D. R. Smyth, and E. M. Meyerowitz. 1989. Genes directing flower development in *Arabidopsis. The Plant Cell* **1**:37–52.

Bowman, J. L., D. R. Smyth, and E. M. Meyerowitz. 1991. Genetic interactions among floral homeotic genes of *Arabidopsis. Development* **112**:1–20.

Brashier, C. K. 1968. Vascularization of cycad leaflets. *Phytomorphology* **18**:35–43.

Bremer, K. 1988. The limits of amino acid sequence data in angiosperm phylogenetic reconstruction. *Evolution* **42**:795–803.

Brenner, G.J. 1963. The spores and pollen of the Potomac Group of Maryland. *Maryland Department of Geology, Mines, Water Resources Bulletin* **27**:1–215.

Brenner, G.J. 1976. Middle Cretaceous floral provinces and early migration of angiosperms. In *Origin and Early Evolution of Angiosperms*, ed. C. B. Beck, pp. 23–47, Columbia University Press, New York.

Brenner, G. J. 1987. Paleotropical evolution of the Magnoliidae in the Lower Cretaceous of Northern Gondwana. *American Journal of Botany* **74**:677–678.

Brenner, G. J. 1990. An evolutionary model of angiosperm pollen evolution based on fossil angiosperm pollen from the Hauterivian of Israel. *American Journal of Botany (Supplement)* **77**:82.

Brenner, G.J. and T. Crepet. 1986. Paleotropical evolution of angiosperms in the Lower Cretaceous of Northern Gondwana. *American Association of Stratigraphic Palynologists 19th Meeting Abstracts*, p. 3.

Brenner, G. H. and I. Bickoff. 1992. Palynology and age of the Lower Cretaceous basal Kurnub Group from the coastal plain to the northern Negev of Israel. *Palynology* 16:137–185.

Brongniart, A. 1822. Sur la classification et la distribution des végétaux fossiles en général, et sur ceux des terrains de sédiment supérieur en particulier. *Mémoires du Muséum d'Histoire naturelle [Paris]* 8:203–348.

Brown, R. W. 1956. Palmlike plants from the Delores Formation (Triassic) in southwestern Colorado. *U.S. Geological Survey Professional Paper* 274:205–209.

Buckley, G. P. 1984. The uses of herbaceous companion species in the establishment of woody species from seed. *Journal of Environmental Management* 18:309–322.

Bull, J. J., J. P. Huelsenbeck, C. W. Cunningham, D. L. Swofford, and P. J. Waddell. 1993. Partitioning and combining data in phylogenetic analysis. *Systematic Biology* 42:384–397.

Burger, W.C. 1977. The Piperales and the monocots. Alternate hypotheses for the origin of monocotyledonous flowers. *The Botanical Review* 43:345–393.

Burger, W. C. 1981. Heresy revived: the monocot theory of angiosperm origin. *Evolutionary Theory* 5:189–225.

Burger, W. C. 1982. Chloranthaceae are related to the Piperales, an ancient dicotyledonous alliance. *Botanical Society of America, Miscellaneous Series Publications* 162:87.

Burk, L. G., R. N. Stewart, and H. Dermen. 1964. Histogenesis and genetics of a plastid-controlled chlorophyll variegation in tobacco. *American Journal of Botany* 51:713–724.

Campbell, L. M., D. W. Stevenson, and H. Loconte. 1993. Floral morphology of *Sinocalycanthus chinensis* (Calycanthaceae). *American Journal of Botany (Supplement)* 80:136.

Canright, J. E. 1952. The comparative morphology and relationships of the Magnoliaceae. I. Trends of specialization in the stamens. *American Journal of Botany* 39:484–497.

Canright, J. E. 1953. The comparative morphology and relationships of the Magnoliaceae—II. Significance of the pollen. *Phytomorphology* 3:355–365.

Canright, J. E. 1955. The comparative morphology and relationships of the Magnoliaceae. IV. Wood and nodal anatomy. *Journal of the Arnold Arboretum* 36:119–140.

Canright, J. E. 1960. The comparative morphology and relationships of the Magnoliaceae. III. Carpels. *American Journal of Botany* 47:145–155.

Canright, J. E. 1963. Contributions of pollen morphology to the phylogeny of some ranalean families. *Grana Palynologica* 4:64–72.

Canright, J. E. 1965. Some phylogenetic implications of pollen morphology in the Ranales. *American Journal of Botany* 52:652 (abstract).

Carlquist, S. 1961. *Comparative Plant Anatomy. A Guide to Taxonomic and Evolutionary Application of Anatomical Data in Angiosperms.* Holt, Rinehart, and Winston, New York.

Carlquist, S. 1962. A theory of paedomorphosis in dicotyledonous woods. *Phytomorphology* **12**:30–45

Carlquist, S. 1964. Morphology and relationships of Lactoridaceae. *Aliso* **5**:421–435.

Carlquist, S. 1966. Wood anatomy of Compositae. A summary, with comments on factors influencing wood evolution. *Aliso* **6**:25–44.

Carlquist, S. 1975. *Ecological Strategies of Xylem Evolution.* University of California Press, Berkeley.

Carlquist, S. 1980. Further concepts in ecological wood anatomy, with comments on recent work in wood anatomy and evolution. *Aliso* **9**:499–553.

Carlquist, S. 1983. Wood anatomy of *Bubbia* (Winteraceae), with comments on origin of vessels in dicotyledons. *American Journal of Botany* **70**:578–590.

Carlquist, S. 1984. Vessel grouping in dicotyledon woods: significance and relationship to imperforate tracheary elements. *Aliso* **10**:505–525.

Carlquist, S. 1985. Vasicentric tracheids as a drought survival mechanism in the woody flora of southern California and similar regions; review of vasicentric tracheids. *Aliso* **11**:37–68.

Carlquist, S. 1987. Presence of vessels in *Sarcandra* (Chloranthaceae); comments on vessel origins in angiosperms. *American Journal of Botany* **64**:1765–1771.

Carlquist, S. 1988a. *Comparative Wood Anatomy.* Springer-Verlag, Berlin.

Carlquist, S. 1988b. Tracheid dimorphism: a new pathway in evolution of imperforate tracheary elements. *Aliso* **12**:102–118.

Carlquist, S. 1988c. Near-vessellessness in *Ephedra* and its significance. *American Journal of Botany* **75**: 598–601.

Carlquist, S. 1989. Wood and bark anatomy of the New World species of *Ephedra. Aliso* **12**: 441–483.

Carlquist, S. 1990. Wood anatomy and relationships of the Lactoridaceae. *American Journal of Botany* **77**:1498–1504.

Carlquist, S. 1991a. Anatomy of vine and liana stems: a review and synthesis. In *The Biology of Vines*, F. E. Putz and H. A. Mooney, eds., pp. 53–71. Cambridge University. Press, Cambridge.

Carlquist, S. 1991b. Wood, bark, and pith anatomy of the Old World species of *Ephedra* and summary for the genus. *Aliso* **13**:255–295.

Carlquist, S. 1991c. Wood and bark anatomy of *Ticodendron*: comments on relationships. *Annals of the Missouri Botanical Garden* **78**: 96–104.

Carlquist, S. 1992a. Pit membrane remnants in perforation-plates of primitive dicotyledons and their significance. *American Journal of Botany* **79**:660–672.

Carlquist, S. 1992b. Wood anatomy and stem of *Chloranthus*. summary of wood anatomy of Chloranthaceae, with comments on relationships, vessellessness, and the origin of monocotyledons. *International Association of Wood Anatomists Bulletin, New Series* **13**:3–16.

Carlquist, S. 1992c. Wood anatomy of *Hedyosmum* (Chloranthaceae) and the tracheid-vessel element transition. *Aliso* **13**:447–462.

Carlquist, S. 1995. Wood and bark anatomy of *Gnetum gnemon* L. *Botanical Journal of the Linnean Society* **116**:203–221.

Carmichael, J. S. and W. E. Friedman. 1994. Fertilization in *Gnetum*: its bearing on the evolution of sexual reproduction within the Gnetales. *American Journal of Botany (Supplement)* **81**:20.

Carpenter, R. and E. S. Coen. 1990. Floral homoeotic mutations produced by transposon-mutagenesis in *Antirrhinum majus*. *Genes and Development* **4**:1483–1493.

Chamberlain, C. J. 1935. *Gymnosperms: Structure and Evolution*. Johnson Reprint Corp., New York. [reprinted 1957]

Chamberlain, C. J. 1966. *Gymnosperms, Structure and Evolution*. Dover Publications, New York.

Chandra, S. and K. J. Singh. 1992. The genus *Glossopteris* from the late Permian beds of Handapa, Orissa, India. *Review of Palaeobotany and Palynology* **75**:183–218.

Chandra, S. and K. R. Surange. 1979. Revision of the Indian species of *Glossopteris*. *Birbal Sahni Institute of Palaeobotany, Monograph 2*. Lucknow, India.

Chasan, R. 1991. Modeling flowers. *The Plant Cell* **3**:845–847.

Chase, M. W., D. E. Soltis, R. G. Olmstead, D. Morgan, D. H. Les, B. D. Mishler, M. R. Duvall, R. A. Price, H. G. Hills, Y.-L. Qiu, K. A. Kron, J. H. Rettig, E. Conti, J. D. Palmer, J. R. Manhart, K. J. Sytsma, H. J. Michaels, W. J. Kress, K. G. Karol, W. D. Clark, M. Hedren, B. S. Gaut, R. K. Jansen, K.-J. Kim, C. R. Wimpee, J. F. Smith, G. R. Furnier, S. H. Strauss, Q.-Y. Xiang, G. M. Plunkett, P. S. Soltis, S. M. Swensen, S. E. Williams, P. A. Gadek, C. J. Quinn, L. E. Equiarte, E. Golenberg, J. Learn GH, S. W. Graham, S. C. H. Barrett, S. Dayanandan, and V. A. Albert. 1993. Phylogenetics of seed plants: an analysis of nucleotide sequences from the plastid gene *rbcL*. *Annals of the Missouri Botanical Garden* **80**:528–580.

Cheadle, V. I. 1943. The origin and trends of specialization of the vessel in the Monocotyledoneae. *American Journal of Botany* **30**:11–17.

Cleal, C. J. and E. L. Zodrow. 1989. Epidermal structure of some medullosan *Neuropteris* foliage from the Middle and Upper Carboniferous of Canada and Germany. *Palaeontology* **32**:837–882.

Clegg, M. T. 1990. Dating the monocot-dicot divergence. *Trends in Ecology and Evolution* **5**:1–2.

Clegg, M. T. 1993. Chloroplast gene sequences and the study of plant evolution. *Proceedings of the National Academy of Sciences of the United States* **90**:363–367.

Clegg, M. T., B. S. Gaut, G. H. Learn, Jr., and B. R. Martin. 1995. Rates and patterns of chloroplast DNA evolution. *Tempo and Mode in Evolution: Genetics and Paleontology 50 Years After Simpson*, eds. W. M. Fitch, F. J. Ayala, pp. 215–234. National Academy Press, Washington, D.C.

Clegg, M. T. and G. Zurawski. 1992. Chloroplast DNA and the study of plant phylogeny: present status and future prospects. *Molecular Systematics of Plants*, eds. D. E. Soltis, P. S. Soltis, and J. J. Doyle, pp. 1–13. Chapman & Hall, New York.

Clement-Westerhof, J. A. 1988. Morphology and phylogeny of Palaeozoic conifers. In *Origin and Evolution of Gymnosperms*, ed. C. B. Beck, pp. 163–175. Columbia University Press, New York.

Coen, E. S. and E. M. Myerowitz. 1991. The war of the whorls: genetic interactions controlling flower development. *Nature* **353**:31–37.

Coetzee, J. A. and J. Praglowski. 1988. Winteraceae pollen from the Miocene of the southwestern Cape (South Africa). *Grana* **27**:27–37.

Conti, E., A. Fischbach, and K. J. Sytsma. 1993. Tribal relationships in Onagraceae: implications from *rbcL* sequence data. *Annals of the Missouri Botanical Garden* **80**:672–685.

Cordemoy, J. C. de. 1863. Monographie du groupe des Chloranthacées. *Adansonia* **3**:280–310.

Corner, E. J. H. 1964. *The Life of Plants*, The University of Chicago Press, Chicago.

Corner, E. J. H. 1967. On thinking big. *Phytomorphology* **17**: 24–28.

Corner, E. J. H. 1976. *The Seeds of Dicotyledons*. Cambridge University Press, Cambridge.

Cornet, B. 1986. The reproductive structures and leaf venation of a Late Triassic angiosperm, *Sanmiguelia lewisii*. *Evolutionary Theory* **7**:231–309.

Cornet, B. 1989a. Late Triassic angiosperm-like pollen from the Richmond Rift Basin of Virginia, U.S.A. *Palaeontographica Beitrage zur Naturgeschichte der Vorzeit* **213B**:37–87.

Cornet, B. 1989b. The reproductive morphology and biology of *Sanmiguelia lewisii*, and its bearing on angiosperm evolution in the Late Triassic. *Evolutionary Trends in Plants* **3**:25–51.

Cornet, B. 1993. Dicot-like leaf and flowers from the Late Triassic tropical Newark Supergroup rift zone, U.S.A. *Modern Geology* **19**:81–99.

Cornet, B. and D. Habib. 1992. Angiosperm-like pollen from the ammonite-dated Oxfordian (Upper Jurassic) of France. *Review of Palaeobotany and Palynology* **71**:269–294.

Coulter, J. M. 1908. The embryo sac and embryo of *Gnetum gnemon*. *Botanical Gazette* **46**:43–49.

Crane, P. R. 1985. Phylogenetic analysis of seed plants and the origin of angiosperms. *Annals of the Missouri Botanical Garden* **72**:716–793.

Crane, P. R. 1986. The morphology and relationships of the Bennettitales. In *Systematic and Taxonomic Approaches in Palaeobotany*, eds. B. A. Thomas and R. A. Spicer, pp. 163–175. Clarendon Press, Oxford.

Crane, P. R. 1987. Vegetational consequences of the angiosperm diversification. In *The Origins of Angiosperms and their Biological Consequences*, eds. E. M. Friis, W. G. Chaloner, and P. R. Crane, pp. 107–144. Cambridge University Press, Cambridge.

Crane, P. R. 1988a. Major clades and relationships in the "higher" gymnosperms. In *Origin and Evolution of Gymnosperms*, ed. C.B. Beck, pp. 218–272. Columbia University Press, New York.

Crane, P.R. 1988b. Review of Cornet, B., The leaf venation and reproductive structures of a Late Triassic angiosperm, *Sanmiguelia lewisii*. *Taxon* **36**:778–779.

Crane, P. R. 1989. Paleobotanical evidence on the early radiation of nonmagnoliid dicotyledons. *Plant Systematics and Evolution* **162**:165–191.

Crane, P. R. 1993. Time for the angiosperms. *Nature* **366**:631–632.

Crane, P. R. and D. L. Dilcher. 1984. *Lesqueria*: an early angiosperm fruiting axis from the mid-Cretaceous. *Annals of the Missouri Botanical Garden* **71**:384–402.

Crane, P. R., M. M. Donoghue, J. A. Doyle, and E. M. Friis. 1989. Angiosperm origins. *Nature* **342**:131.

Crane, P. R., E. M. Friis, and K. R. Pedersen. 1995. The origin and early diversification of angiosperms. *Nature* **374**:27–33.

Crane, P. R. and S. Lidgard. 1989. Angiosperm diversification and paleolatitudianl gradients in Cretaceous floristic diversity. *Science* **246**:675–678.

Crane, P. R. and S. Lidgard. 1990. Angiosperm radiation and patterns of Cretaceous palynological diversity. In *Major Evolutionary Radiations*, eds. P. D. Taylor and G. P. Larwood, pp. 377–407. Systematics Association by Clarendon Press, Oxford.

Crane, P. R, and G. R. Upchurch, Jr. 1987. *Drewria potomacencis* gen. et sp. nov., an Early Cretaceous member of the Gnetales from the Potomac Group of Virginia. *American Journal of Botany* **74**:1722–1736.

Crepet, W. L. 1972. Investigations of North American cycadeoids: pollination mechanisms in *Cycadeoidea*. *American Journal of Botany* **59**:1048–1056.

Crepet, W. L. 1974. Investigations of North American cycadeoids: the reproductive biology of *Cycadeoidea*. **148B**:44–169.

Crepet, W. L. 1983. The role of insect pollination in the evolution of the angiosperms. In *Pollination Biology*, ed. L. Real, pp. 31–50. Academic Press, Inc, New York.

Crepet, W. L. and T. Delevoryas. 1972. Investigations of North American cycadeoids: early ovule ontogeny. *American Journal of Botany* **59**:209–215.

Crepet, W. L. and E. M. Friis. 1987. The evolution of insect pollination in angiosperms. In *The Origins of Angiosperms and their Biological Consequences*, eds. E. M. Friis, W. G. Chaloner, and P. R. Crane, pp. 181–201. Cambridge University Press, Cambridge.

Crepet, W. L., E. M. Friis, and K. C. Nixon. 1991. Fossil evidence for the evolution of biotic pollination. *Philosophical Transactions of the Royal Society of London, B* **333**:187–195.

Crepet, W. L., K. C. Nixon, D. W. Stevenson, and E. M. Friis. 1993. The relationships of seed plants in reference to angiosperm outgroups. *American Journal of Botany (Supplement)* **80**:123.

Cronquist, A. 1968. *The Evolution and Classification of Flowering Plants*, Houghton Mifflin, New York.

Cronquist, A. 1981. *An Integrated System of Classification of the Angiosperms*, Columbia University Press, New York.

Cronquist, A. 1987. A botanical critique of cladism. *The Botanical Review* **53**:1–52.

Cronquist, A. 1988. *The Evolution and Classification of Flowering Plants, 2nd ed.*, The New York Botanical Garden, Bronx, New York.

Crookall, R. 1959. Fossil plants of the Carboniferous rocks of Great Britain (Second section). *Memoirs of the Geological Survey of Great Britain, Palaeontology* **4**:85–216.

Cutter, E. G. 1957. Studies of morphogenesis in the Nymphaeaceae. II. Floral development in *Nuphar* and *Nymphaea*: bracts and calyx. *Phytomorphology* **7**:57–73.

Cutter, E. G. 1959. Studies of morphogenesis in the Nymphaeaceae. IV. Early floral development in species of *Nuphar*. *Phytomorphology* **9**:263–275.

Dahlgren, R. 1983. General aspects of angiosperm evolution and macrosystematics. *Nordic Journal of Botany* **3**:119–149.

Dahlgren, G. 1989. An updated angiosperm classification. *Botanical Journal of the Linnean Society of London*, **100**: 197–203.

Dahlgren, R. and K. Bremer. 1985. Major clades of the angiosperms. *Cladistics* **1**:349–368.

Dahlgren, R. M. T., H. T. Clifford, and P. F. Yeo. 1985. *The Families of Monocotyledons*, Springer-Verlag, Berlin.

Daugherty, L. H. 1941. *The Upper Triassic Flora of Arizona with a Discussion of its Geologic Occurrence*. Contributions to Paleontology 526, Carnegie Institution of Washington.

Davis, G. L. 1966. *Systematic Embryology of the Angiosperms*. Wiley, New York.

Davis, J. I. 1993. Character removal as a means for assessing stability of clades. *Cladistics* **9**:201–210.

De Mason, D. A., and K. W. Stolte. 1982. Floral development in *Phoenix dactylifera*. *Canadian Journal of Botany* **60**:1439–1446.

Delevoryas, T. 1953. A new male coraitean fructification from the Kansas Carboniferous. *American Journal of Botany* **40**:144–150.

Delevoryas, T. 1968. Investigations of North American cycadeoids: structure, ontogeny and phylogenetic considerations of cones of *Cycadeoidea*. *Palaeontogrophica Beitrage zur Naturgeschichte der Vorzeit*. **121B**:121–133.

Delevoryas, T. 1969. Glossopterid leaves from the Middle Jurassic of Oaxaca, Mexico. *Science* **165**:895–896.

Delevoryas, T. and C. P. Pearson. 1975. *Mexiglossa varia* gen. et sp. nov., a new genus of glossopteroid leaves from the Jurassic of Oaxaca, Mexico. *Palaeontographica Beitrage zur Naturgeschichte der Vorzeit* **154B**:114–120.

Dellaporta, S. L., M. A. Moreno, and A. Delong. 1991. Cell lineage of the gynoecium of maize using the transposable element *Ac*. *Development (Supplement)* **1**:1–7.

Delpino, F. 1890. Applicazione di nuovi criterii per la classificazione delle plante. Terza memoria. *Memorie della reale academia della scienze dell' istituto di Bologna*, Ser. IV, **10**:43–75.

Dermen, H. 1960. Nature of plant sports. *The American Horticultural Magazine* **39**:123–173.

Dickison, W. C. 1976. The bases of angiosperm phylogeny: vegetative anatomy. *Annals of the Missouri Botanical Garden* **62**:590–620.

Dickison, W. C. 1989. Comparisons of primitive Rosidae and Hamamelidae. In *Evolution, Systematics, and Fossil History of the Hamamelidae*, eds. P. R. Crane and S. Blackmore, Vol. 1, pp. 47–73. Clarendon Press, Oxford.

Dickison, W. C. 1992. Morphology and anatomy of the flower and pollen of *Saruma henryi* Oliv., a phylogenetic relict of the Aristolochiaceae. *Bulletin of the Torrey Botanical Club* **119**:392–400.

Dilcher, D. L. 1989. The occurrence of fruits with affinities to Ceratophyllaceae in lower and mid-Cretaceous sediments. *American Journal of Botany* **76**:162.

Dilcher, D. L. and P. R. Crane. 1984. *Archaeanthus*: an early angiosperm from the Cenomanian of the Western Interior of North America. *Annals of the Missouri Botanical Garden* **71**:351–383.

Ditsch, F. and W. Barthlott. 1994. Mikromorphologie der Epicuticularwachse und die Systematik der Dilleniales, Lecythidales, Malvales und Theales. *Tropische und Subtropische Pflanzenwelt* **88**:7–74.

Donoghue, M. J. 1994. Progress and prospects in reconstructing plant phylogeny. *Annals of the Missouri Botanical Garden* **81**:405–418.

Donoghue, M. J. and J. A. Doyle. 1989a. Phylogenetic analysis of angiosperms and the relationships of Hamamelidae. In *Evolution, Systematics, and Fossil History of the Hamamelidae*, eds. P. R. Crane and S. Blackmore, pp. 17–45. Clarendon Press, Oxford.

Donoghue, M. J. and J. A. Doyle. 1989b. Phylogenetic studies of seed plants and angiosperms based on morphological characters. In *The Hierarchy of Life: Molecules and Morphology in Phylogenetic Analysis*, ed. B. Fernholm, K. Bremer, and H. Jornvall, pp. 181–193. Elsevier Science, Amsterdam.

Donoghue, M. J., J. A. Doyle, J. Gauthier, A. G. Kluge, and T. Rowe. 1989. The importance of fossils in phylogeny reconstruction. *Annual Review of Ecology and Systematics* **20**:431–460.

Donoghue, M. J. and M. J. Sanderson. 1992. The suitability of molecular and morphological evidence in reconstructing plant phylogeny. *Molecular Systematics of Plants*, eds. D. E. Soltis, P. S. Soltis, and J. J. Doyle, pp. 340–368. Chapman & Hall, New York.

Donoghue, M. J. and S. M. Scheiner. 1992. The evolution of endosperm: A phylogenetic account. In *Ecology and Evolution of Plant Reproduction*, ed. R. Wyatt, pp. 356–389. Chapman and Hall, New York.

Doyle, J. A. 1969. Cretaceous angiosperm pollen of the Atlantic Coastal Plain and its evolutionary significance. *Journal of the Arnold Arboretum* **50**:1–35.

Doyle, J. A. 1978. Origin of angiosperms. *Annual Review of Ecology and Systematics* **9**:365–392.

Doyle, J. A. 1992. Revised palynological correlations of the Potomac Group (USA) and the Cocobeach sequence of Gabon (Barremian-Aptian). *Cretaceous Research* **13**:337–349.

Doyle, J. A. 1994. Origin of the angiosperm flower: a phylogenetic perspective. *Plant Systematics and Evolution (Supplement)* **8**:7–29.

Doyle, J. A. and M. J. Donoghue. 1986a. Relationships of angiosperms and Gnetales: a numerical cladistic analysis. In *Systematic and Taxonomic Approaches in Paleobotany*, eds. R. A. Spicer and B. A. Thomas, 177–198. Systematics Association, Special Vol. 31, by Cloverdon Press, Oxford.

Doyle, J. A. and M. J. Donoghue. 1986b. Seed plant phylogeny and the origin of angiosperms: an experimental cladistic approach. *The Botanical Review* 52:321–431.

Doyle, J. A. and M. J. Donoghue. 1987a. The origin of angiosperms: a cladistic approach. In *The Origins of Angiosperms and Their Biological Consequences*, eds. E. M. Friis, W. G. Chaloner, and P. R. Crane, pp. 17–49. Cambridge University Press, Cambridge, U. K.

Doyle, J. A. and M. J. Donoghue. 1992. Fossils and seed plant phylogeny reanalyzed. *Brittonia* 44:89–106.

Doyle, J. A. and M. J. Donoghue. 1993. Phylogenies and angiosperm diversification. *Paleobiology* 19:141–167.

Doyle, J. A., M. J. Donoghue, and E. A. Zimmer. 1994. Integration of morphological and ribosomal RNA data on the origin of angiosperms. *Annals of the Missouri Botanical Garden* 81:419–450.

Doyle, J. A. and L. J. Hickey. 1976. Pollen and leaves from the mid-Cretaceous Potomac Group and their bearing on early angiosperm evolution. In *Origin and Early Evolution of Angiosperms*, ed. C. B. Beck, pp. 139–206. Columbia University Press, New York.

Doyle, J. A. and C. L. Hotton. 1991. Diversification of early angiosperm pollen in a cladistic context. In *Pollen and Spores*, eds. S. Blackmore and S. H. Barnes, pp. 169–195. Systematic Association by Clarendon Press, Oxford.

Doyle, J. J. 1993. DNA, phylogeny, and the flowering of plant systematics. *Bioscience* 43:380–389.

Drinnan, A. N., P. R. Crane, E. M. Friis, and K. R. Pedersen. 1990. Lauraceous flowers from the Potomac Group (Mid-Cretaceous) of eastern North America. *Botanical Gazette* 151:370–384.

Dubertret, L. and H. Vautrin. 1937. Revision de la stratigraphie du Cretacé du Liban. *Notes et Memoires, Haut-Commissariat de la Republique Francaise en Syrie et au Liban* 2:43–73.

Dumortier, B. C. 1829. *Analyse des Familles des Plantes, Avec l'Indication des Princi-paux Generes qui s'y Rattachent.* J. Casterman aîné, Tournay, France.

Duvall, M. R., M. T. Clegg, M. W. Chase, W. D. Clark, W. J. Kress, H. G. Hills, L. E. Eguiarte, J. F. Smith, B. S. Gaut, E. A. Zimmer, and G. H. Learn, Jr. 1993. Phylogenetic hypotheses for the monocotyledons constructed from *rbcL* sequence data. *Annals of the Missouri Botanical Garden* 80:607–619.

Duvall, M. R., G. H. Learn Jr., L. E. Equiarte, and M. T. Clegg. 1993. Phylogenetic analyis of *rbcL* sequences identifies *Acorus calamus* as the primal extant monocoty-ledon. *Proceedings of the National Academy of Sciences of the United States of America* 90:4641–4644.

Eames, A. J. 1952. Relationships of the Ephedrales. *Phytomorphology* 2:79–100.

Eames, A.J. 1961. *Morphology of the Angiosperms*, McGraw-Hill Book Co., New York.

Edwards, J. G. 1920. Flower and seed of *Hedyosmum nutans*. *Botanical Gazette* 70:409–424.

Ehrendorfer, F. 1976. Evolutionary significance of chromosomal differentiation patterns in gymnosperms and primitive angiosperms. In *Origin and Early Evolution of Angios-perms*, ed. C. B. Beck, pp. 220–240. Columbia University Press, New York.

Ehrendofer, F., F. Krendl, E. Habeler, and W. Sauer. 1968. Chromosome numbers and evolution in primitive angiosperms. *Taxon* 17:337–353.

Eichler, A. W. 1875–1878. *Blütendiagramme*. Pts. 1, 2. Wilhelm Engelmann, Leipzig.

Endress, P. K. 1975. Nachbarliche Formbeziehungen mit Hüllfunktion im Infloreszenz- und Blütenbereich. *Botanische Jahrbücher für Systematik, Pflanzengeschichte und Pflanzengeographie* 96:1–44.

Endress, P. K. 1980. The reproductive structures and systematic position of the Austrobaileyaceae. *Botanische Jahrbücher für Systematik, Pflanzengeschichte und Pflanzengeographie* 101: 393–433.

Endress, P. K. 1983a. Reproductive structures of the flowering plants. *Progress in Botany* 45:54–67.

Endress, P. K. 1983b. Dispersal and distribution in some small archaic relic angiosperm families (Austrobaileyaceae, Eupomatiaceae, Himantandraceae, Idiospermaceae). *Sonderbande des Naturwissenschaftlichen Vereins in Hamburg* 7: 201–217

Endress, P. K. 1985. IV. Reproductive structures of the flowering plants. *Progress in Botany* 47:52–65.

Endress, P. K. 1986a. Reproductive structures and phylogenetic significance of extant primitive angiosperms. *Plant Systematics and Evolution* 152:1–28.

Endress, P. K. 1986b. Floral structure, systematics, and phylogeny in Trochodendrales. *Annals of the Missouri Botanical Garden* 73:297–324.

Endress, P. K. 1987a. The Chloranthaceae: reproductive structures and phylogenetic position. *Botanische Jahrbücher für Systematik, Pflanzengeschichte und Pflanzengeographie* 109:153–226.

Endress, P. K. 1987b. Floral phyllotaxis and floral evolution. *Botanische Jahrbucher für Systematik Pflantzengeschichte und Pflantzengeographie* 109:417–438.

Endress, P. K. 1989. Chaotic floral phyllotaxis and reduced perianth in *Achlys* (Berberidaceae). *Botanica Acta* 142:159–164.

Endress, P. K. 1990. Evolution of reproductive structures and functions in primitive angiosperms (Magnoliidae). *Memoirs of the New York Botanical Garden* 55:5–34.

Endress, P. K. 1993a. Austrobaileyaceae. In *The Families and Genera of Vascular Plants*, Vol. 2 *Flowering Plants. Dicotyledons: Magnoliid, Hamamelid and Caryophyllid Families*, eds. K. Kubitzki, J. G. Rohwer, and V. Bittrich, pp. 138–140. Springer-Verlag, Berlin.

Endress, P. K. 1993b. Cercidiphyllaceae. In *The Families and Genera of Vascular Plants*, Vol. 2 *Flowering Plants. Dicotyledons: Magnoliid, Hamamelid and Caryophyllid Families*, eds. K. Kubitzki, J. G. Rohwer, and V. Bittrich, pp. 250–252. Springer-Verlag, Berlin.

Endress, P. K. 1993c. Eupomatiaceae. In *The Families and Genera of Vascular Plants*, Vol. 2 *Flowering Plants. Dicotyledons: Magnoliid, Hamamelid and Caryophyllid Families*, eds. K. Kubitzki, J. G. Rohwer, and V. Bittrich, pp. 296–298. Springer-Verlag, Berlin.

Endress, P. K. 1993d. Eupteleaceae. In *The Families and Genera of Vascular Plants*, Vol. 2 *Flowering Plants. Dicotyledons: Magnoliid, Hamamelid and Caryophyllid*

Families, eds. K. Kubitzki, J. G. Rohwer, and V. Bittrich, pp. 299–300. Springer-Verlag, Berlin.

Endress, P. K. 1993e. Hamamelidaceae. In *The Families and Genera of Vascular Plants*, Vol. 2 *Flowering Plants. Dicotyledons: Magnoliid, Hamamelid and Caryophyllid Families*, eds. K. Kubitzki, J. G. Rohwer, and V. Bittrich, pp. 322–330. Springer-Verlag, Berlin.

Endress, P. K. 1993f. Himantandraceae. In *The Families and Genera of Vascular Plants*, Vol. 2 *Flowering Plants. Dicotyledons: Magnoliid, Hamamelid and Caryophyllid Families*, eds. K. Kubitzki, J. G. Rohwer, and V. Bittrich, pp. 338–341. Springer-Verlag, Berlin.

Endress, P. K. 1993g. Monimiaceae. In *The Families and Genera of Vascular Plants*, Vol. 2 *Flowering Plants. Dicotyledons: Magnoliid, Hamamelid and Caryophyllid Families*, eds. K. Kubitzki, J. G. Rohwer, and V. Bittrich, pp. 426–437. Spring-er–Verlag, Berlin.

Endress, P. K. 1993h. Trimeniaceae. In *The Families and Genera of Vascular Plants*, Vol. 2 *Flowering Plants. Dicotyledons: Magnoliid, Hamamelid and Caryophyllid Families*, eds. K. Kubitzki, J. G. Rohwer, and V. Bittrich, pp. 596–599. Springer-Verlag, Berlin.

Endress, P. K. 1993i. Trochodendraceae. In *The Families and Genera of Vascular Plants*, Vol. 2 *Flowering Plants. Dicotyledons: Magnoliid, Hamamelid and Caryophyllid Families*, eds. K. Kubitzki, J. G. Rohwer, and V. Bittrich, pp. 599–602. Springer-Verlag, Berlin.

Endress, P. K. 1994a. *Diversity and Evolutionary Biology of Tropical Flowers*, Cambridge University Press, Cambridge.

Endress, P. K. 1994b. Evolutionary aspects of the floral structure in *Ceratophyllum*. *Plant Systematics and Evolution (Supplement)* **8**:175–183.

Endress, P. K., and L. D. Hufford. 1989. The diversity of stamen structures and dehiscence patterns among Magnoliidae. *Botanical Journal of the Linnean Society of London* **100**:45–85.

Endress, P. K. and S. Stumpf. 1991. The diversity of stamen structures in lower Rosidae (Rosales, Fabales, Proteales, Sapindales). *Botanical Journal of the Linnean Society of London* **107**: 217–293.

Engler, A. 1887. Uber die Familie der Lactoridaceae. *Botanische Jahrbücher für Systematik, Pflanzengeschichte und Pflanzengeographie* **8**:53–56.

Engler, A. 1894. Piperaceae. In *Die natürlichen Pflanzenfamilien* 3(1), ed. A. Engler and K. Prantl, pp. 3–11. Wilhelm Engelmann, Leipzig.

Engler, A. and K. Prantl (eds.). 1887–1909. *Die natürlichen Pflanzenfamilien*. Wilhelm Englemann, Leipzig, 32 vols.

Erbar, C., and P. Leins. 1981. Zur Spirale in Magnolien–Blüten. *Beiträge zur Biologie der Pflanzen* **56**:225–241.

Erbar, C., and P. Leins. 1983. Zur Sequenz von Blütenorganen bei einigen Magnoliiden. *Botanische Jahrbücher für Systematik* **103**:433–449.

Erdtman, G. 1952. *Pollen Morphology and Plant Taxonomy. I. Angiosperms*, Almqvist and Wiksell, Stockholm.

Erdtman, G. 1957. *Pollen and Spore Morphology*. Almqvist and Wiksell, Stockholm.

Erdtman, G. 1965. *Pollen and Spore Morphology/Plant Taxonomy*. Vol. 3. *Gymnospermae, Bryophyta*, pp. 42–44. Almqvist and Wiksell, Stockholm.

Eriksson, O. 1993. Dynamics of genets in clonal plants. *Trends in Ecology and Evolution* **8**:313–316.

Eriksson, O. and B. Bremer. 1992. Pollination systems, dispersal modes, life forms, and diversification rates in angiosperm families. *Evolution* **46**:258–266.

Esau, K. 1965. *Plant Anatomy*, 2nd ed., John Wiley & Sons, Inc, New York.

Esau, K. 1979. Phloem. In *Anatomy of the Dicotyledons*, eds. C. R. Metcalfe and L. Chalk, pp. 181–189. Clarendon Press, Oxford.

Eyde, R. H. 1975. The foliar theory of the flower. *American Scientist* **63**:430–437.

Eyde, R. H. 1976. The bases of angiosperm phylogeny: floral anatomy. *Annals of the Missouri Botanical Garden* **62**:521–537.

Faegri, K. and J. Iversen. 1975. *Textbook of Pollen Analysis*. Hafner, New York.

Fairbrothers, D. E., T. J. Mabry, R. L. Scogin, and B. L. Turner. 1976. The bases of angiosperm phylogeny: chemotaxonomy. *Annals of the Missouri Botanical Garden* 62:765–800.

Felsenstein, J. 1978. Cases in which parsimony or compatibility methods will be positively misleading. *Systematic Zoology* **27**:401–410.

Felsenstein, J. 1993. *PHYLIP (Phylogeny Inference Package), Version 3.5*. Department of Genetics, University of Washington, Seattle, Washington.

Feuer, S. 1991. Pollen morphology and the systematic relationships of *Ticodendron incognitum*. *Annals of the Missouri Botanical Garden* **78**: 143–151.

Flores, E. M. and M. F. Moseley. 1982. The anatomy of the pistillate inflorescence and flower of *Casuarina verticillata*. *American Journal of Botany* **69**: 73–84.

Florin, R. 1939. The morphology of the female fructifications in cordaites and conifers of Palaeozoic age. *Botaniska Notiser* **36**:547–565.

Florin, R. 1950. On female reproductive organs in the *Cordaitinae*. *Acta Horti Bergiani* **15**:111–134.

Florin, R. 1951. Evolution in cordaites and conifers. *Acta Horti Bergiani* **15**:285–388.

Florin, R. 1954. The female reproductive organs of conifers and taxads. *Biological Review of the Cambridge Philosophical Society* **29**:367–389.

Foster, A. S. 1952. Foliar venation in angiosperms from an ontogenetic viewpoint. *American Journal of Botany* **39**:752–766.

Frascaria, N., L. Maggia, M. Michaud, and J. Bosquet. 1993. The *rbc*L gene sequence from chestnut indicates a slow rate of evolution in the Fagaceae. *Genome* **36**:668–671.

Freud, S. 1967. *Moses and monotheism*. Vintage Books, New York.

Frey-Wyssling, A. 1976. The plant cell wall. *Handbuch der Pflanzenanatomie* Vol. III(4), pp. 1–294. Gebrüder Borntraeger, Berlin.

Friedman, W.E. 1990a. Double fertilization in *Ephedra*, a nonflowering seed plant: its bearing on the origin of angiosperms. *Science* **247**:951–954.

Friedman, W. E. 1990b. Sexual reproduction in *Ephedra nevadensis* (Ephedraceae): further evidence of double fertilization in a nonflowering seed plant. *American Journal of Botany* 77:1582–1598.

Friedman, W. E. 1992. Evidence of a pre-angiosperm origin of endosperm: implications for the evolution of flowering plants. *Science* 255:336–339.

Friedman, W. E. 1992. Double fertilization in nonflowering seed plants and its relevance to the origin of flowering plants. *International Review of Cytology* 140:319–355.

Friedman. W. E. 1994. The evolution of embryogeny in seed plants and in the developmental origin and early history of endosperm. *American Journal of Botany* 81: 1468–1486.

Friis, E. M. 1989. Paleobotany. *Progress in Botany* 50:312–326.

Friis, E. M., P. R. Crane, and K. R. Pedersen. 1986. Floral evidence for Cretaceous chloranthoid angiosperms. *Nature* 320:163–164.

Friis, E.M., P.R. Crane, and K.R. Pedersen. 1991. Stamen diversity and in situ pollen of Cretaceous angiosperms. In *Pollen and spores: Patterns of Diversification*, eds. S. Blackmore and S.H Barnes, pp. 197–224. Clarendon Press, Oxford.

Friis, E. M. and W. L. Crepet. 1987. Time of appearance of floral features. In *The Origins of Angiosperms and their Biological Consequences*, eds. E. M. Friis, W. G. Chaloner, and P. R. Crane, pp. 145–179. Cambridge University Press, Cambridge.

Friis, E. M., H. Eklund, K. R. Pedersen, and P. R. Crane. 1994. *Virginianthus calycanthoides* gen. et sp. nov. – a calycanthaceous flower from the Potomac Group (Early Cretaceous) of eastern North America. *International Journal of Plant Science* 155: 772–785.

Friis, E. M. and P.K. Endress. 1990. Origin and evolution of angiosperm flowers. *Advances in Botanical Research* 17:99–162.

Friis, E. M., K. R. Pedersen, and P. R. Crane. 1994. Angiosperm floral structures from the Early Cretaceous of Portugal. *Plant Systematics and Evolution (Supplement)* 8:31–49.

Frolich, D., and W. Barthlott. 1988. Mikromorphologie der Epicuticularen Wachse und das System der Monokotylen. *Tropische und Subtropische Pflantzenwelt* 63:7–132.

Frost, F. H. 1930a. Specialization in secondary xylem in dicotyledons. I. Origin of vessels. *Botanical Gazette* 89:67–94.

Frost, F. H. 1930b. Specialization in secondary xylem of dicotyledons. II. Evolution of end-wall of vessel segment. *Botanical Gazette* 90:198–212.

Frost, F. H. 1931. Specialization in secondary xylem in dicotyledons. III. Evolution of lateral wall of vessel segment. *Botanical Gazette* 91:88–96.

Gaiser, J. C., K. Robinson-Beers, and C. S. Gasser. 1995. The *Arabidopsis SUPERMAN* gene mediates asymmetric growth of the outer integument of ovules. *The Plant Cell* 7:333–345.

Gasser, C. S. and K. Robinson-Beers. 1993. Pistil development. *The Plant Cell* 5:1231–1240.

Gaussen, H. 1946. *Les Gymnosperms actuelles et fossiles*. Pt. 3, Travaux du Laboratoire Forestier, Toulouse.

Gaut, B. S., S. V. Muse, W. D. Clark, and M. T. Clegg. 1992. Relative rates of nucleotide substitution at the *rbc*L locus of monocotyledonous plants. *Journal of Molecular Evolution* **35**:292–303.

Gaut, B. S., S. V. Muse, and M. T. Clegg. 1993. Relative rates of nucleotide substitution in the chloroplast genome. *Molecular Phylogenetics and Evolution* **2**:89–96.

Gerry, E. 1963. Ostracoda from the Lower Cretaceous of the Helez area. In *"Zone 1" in Heletz-Kokhav.* Pt. 1, pp. 1–4. Jerusalem Oil Company, Micropaleontological Laboratory, Jerusalem.

Giannasi, D. E. 1988. Flavonoids and evolution in dicotyledons. In *The Flavonoids, Advances in Research Since 1980*, ed. J. B. Harborne, pp. 479–504. Chapman & Hall, London.

Gibson, A. C. 1973. Wood anatomy of Cactoideae (Cactaceae). *Biotropica* **5**:29–65.

Gifford, E. M. 1943. The structure and development of the shoot apex of *Ephedra altissima* Desf. *Bulletin of the Torrey Botanical Club* **70**:15–25.

Gifford, E. M. and A. S. Foster. 1989. *Morphology and Evolution of Vascular Plants.* W. H. Freeman and Company, New York.

Givnish, T. J. 1976. Leaf form in relation to environment: a theoretical study. Ph.D. Thesis. Princeton University, Princeton, NJ.

Givnish, T. J. 1979. On the adaptive significance of leaf form. In *Topics in Plant Population Biology*, eds. O. T. Solbrig, S. Jain, G. B. Johnson, P. H. Raven, pp. 376–407. Columbia University Press, New York.

Givnish, T. J., K. J. Sytsma, J. F. Smith, and W. J. Hahn. 1994. Thorns and heterophylly in *Cyanea*: adaptations to extinct avian browsers on Hawaii? *Proceedings of the National Academy of Sciences of the United States of America* **91**:2810–2814.

Givnish, T. J., K. J. Sytsma, W. J. Hahn, and J. F. Smith. 1995. Molecular evolution, adaptive radiation, and geographic speciation in *Cyanea* (Campanulaceae), a species-rich genus of Hawaiian plants. In *Hawaiian Biogeography: Evolution on a Hot Spot Archipelago*, eds. W. L. Wagner, and V. L. Funk, pp. 288–337. Smithsonian Institution Press, Washington, D.C.

Gomez-Laurito, J. and L. D. Gomez P. 1989. *Ticodendron*: a new tree from Central America. *Annals of the Missouri Botanical Garden* **76**: 1148–1151.

Gomez-Laurito, J. and L. D. Gomez P. 1991. Ticodendraceae: a new family of flowering plants. *Annals of the Missouri Botanical Garden* **78**: 87–88.

Gothan, W. 1941. Paläobotanische Mitteilungen 5–7. Die Unterteilung der Karbonischen Neuropteriden. *Palaeontologische Zeitschrift* **22**:421–428.

Gothan, W. and W. Remy. 1957. *Steinkohlenpflanzen (Leitfaden zum Bestimmen der wichtigsten pflanzlichen Fossilen der paläozoikums im rheinisch-westfälischen Stein-kohlengebeit)*, Verlag Glück auf GMBH, Essen.

Goto, N., N. Katoh, and A. R. Kranz. 1991. Morphogenesis of floral organs in *Arabidopsis*: predominant carpel formation of the pin-formed mutant. *Japanese Journal of Genetics* **66**:551–567.

Gould, R. E. and T. Delevoryas. 1977. The biology of *Glossopteris*: evidence from petrified seed-bearing and pollen-bearing organs. *Alcheringa* **1**:387–399.

Grader, P., and Z. Reiss. 1958. On the Lower Cretaceous of the Helez Area. *Geological Survey of Israel* **16**:1–14.

Grayum, M. H. 1987. A summary of evidence and arguments supporting the removal of *Acorus* from the Araceae. *Taxon* **36**:723–729.

Grayum, M. H. 1990. Evolution and phylogeny of the Araceae. *Annals of the Missouri Botanical Garden* **77**:628–697.

Gregory, M. P. 1956. A phyletic rearrangement in the Aristolochiaceae. *American Journal of Botany* **43**:110–122.

Greguss, P. 1968. *Xylotomy of the Living Cycads*. Akademiai Kiadó, Budapest.

Greller, A. M., and E. B. Matzke. 1970. Organogenesis, aestivation, and anthers in the flower of *Lilium tigrinum*. *Botanical Gazette* **131**:304–311.

Gübeli, A. A., P. A. Hochuli, and W. Wildi. 1984. Lower Cretaceous turbiditic sediments from the Rift chain (Northern Morocco) palynology, stratigraphy and paleogeographic setting. *Geologische Rundschau* **73**:1081–1114.

Haberlandt, G. 1914. *Physiological Plant Anatomy* trans. M. Drummond, Macmillan, London.

Hagerup, O. 1934. Zur Abstammung einiger Angiospermen durch Gnetales und Coniferae. *Kongelige Danske Videnskabernes Selakab Biologiske Skrifter* **11**:3–83.

Hagerup, O. 1936. Zur Abstammung einiger Angiospermen durch Gnetales und Coniferae. *Kongelige Danske Videnskabernes Selakab Biologiske Skrifter* **13**:3–59.

Hagerup, O. 1938. On the origin of some angiosperms through the Gnetales and the Coniferae. *Kongelige Danske Videnskabernes Selakab Biologiske Skrifter* **14**:3–34.

Haig, D. 1990. New perspectives on the angiosperm female gametophyte. *The Botanical Review* **56**:236–274.

Haig, D. and M. Westoby. 1989. Parent-specific gene expression and the triploid endosperm. *The American Naturalist* **134**:147–155.

Haig, D. and M. Westoby. 1991. Seed size, pollination costs and angiosperm success. *Evolutionary Ecology* **5**:231–247.

Hallier, H. 1912. L'origne et la systeme phyletique des angiospermes exposes a l'aide de leru arbe genealogique. *Archives néerlandaises des sciences exactes et naturelles* Ser. 3B, 1:146–234.

Hamby, R. K. 1990. Ribosomal RNA and the Early Evolution of Flowering Plants. Ph.D. dissertation, Louisiana State University, Baton Rouge, Louisiana.

Hamby, R. K. and E. A. Zimmer. 1992. Ribosomal RNA as a phylogenetic tool in plant systematics. In *Molecular Systematics of Plants*, eds. P. S. Soltis, D. E. Soltis, and J. J. Doyle, pp. 50–91. Chapman and Hall, New York.

Harborne, J. B. and T. Swain (eds.). 1969. *Perspectives in Phytochemistry*. Academic Press, London.

Harper, J. L. 1977. *Population Biology of Plants*, Academic Press, London.

Harris, T. M. 1964. *The Yorkshire Jurassic Flora. II. Caytoniales, Cycadales and Pteridosperms*. British Museum (Natural History), London.

Harris, T. M. 1932a. The fossil flora of Scoresby Sound East Greenland. Part 2: description of seed plants *incertae sedis* together with a discussion of certain cycadophyte cuticles. *Meddelelser om Grønland* **85**(3):1–114.

Harris, T. M. 1932b. The fossil flora of Scoresby Sound East Greenland. Part 3: Caytoniales and Bennettitales. *Meddelelser om Grønland* **85**(5):1–133.

Harris, T. M. 1969. *The Yorkshire Jurassic Flora. III. Bennettitales*, British Museum (Natural History), London.

Harvey, P. H., R. M. May, and S. Nee. 1994. Phylogenies without fossils. *Evolution* **48**:523–529.

Hasebe, M., M. Ito, R. Kofuji, and K. Veda. 1992a. Phylogenetic relationships in Gnetopsida deduced from *rbcL* gene sequences. *Botanical Magazine (Tokyo)* **105**:385–391.

Hasebe, M., M. Ito, R. Kofuji, K. Iwatsuki, and K. Ueda. 1992b. Phylogeny of gymnosperms inferred from *rbcL* gene sequences. *Botanical Magazine (Tokyo)* **105**:673–679.

Hasebe, M., M. Ito, R. Kofuji, K. Ueda, and K. Iwatsuki. 1993. Phylogenetic relationships of ferns deduced from *rbcL* gene sequences. *Journal of Molecular Evolution* **37**:476–482.

Hasebe, M., R. Kofuji, M. Ito, M. Kato, K. Iwatsuki, and K. Ueda. 1992. *RbcL* genes indicate monophyly of gymnosperms. *American Journal of Botany (Supplement)* **79**:169.

Hegnauer, R. 1988. Biochemistry, distribution and taxonomic relevance of higher plant alkaloids. *Phytochemistry* **27**:2423–2427.

Hegnauer, R. 1971. Pflanzenstoffe und Pflanzensystematik. *Naturwissenschaften* **58**:585–598.

Henes, E. 1959. *Fossile Wandstrukturen*. Gebrüder Borntraeger, Berlin.

Herendeen, P. S. 1991. Charcoalified angiosperm wood from the Cretaceous of eastern North America and Europe. *Review of Palaeobotany and Palynology* **70**:225–239.

Herendeen, P. S., D. H. Les, and D. L. Dilcher. 1990. Fossil *Ceratophyllum* (Ceratophyllaceae) from the Tertiary. *American Journal of Botany* **77**: 7–16.

Herrera, C. M. 1989. Seed disperal by animals: a role in angiosperm diversification. *The American Naturalist* **133**:309-322.

Heslop-Harrison, J. 1971. Sporopollenin in the Biological Context, In *Sporopollenin*, eds. J. Brooks., P. R. Grant , M. Muir, P. van Gijzel and G. Shaw, pp. 1–30, Academic Press, New York.

Heslop-Harrison, Y. and K. R. Shivanna. 1977. The receptive surface of the angiosperm stigma. *Annals of Botany* **41**:1233–1258.

Hesse, M. 1984. Pollenkitt is lacking in Gnetatae: *Ephedra* and *Welwitschia*; further proof for its restriction to the angiosperms. *Plant Systematics and Evolution* **144**:9–16.

Hickey, L. J. 1973. Classification of the architecture of dicotyledonous leaves. *American Journal of Botany* **60**:17–33.

Hickey, L. J. 1977. Stratigraphy and paleobotany of the Golden Valley Formation (Early Tertiary) of Western North Dakota. *Geological Society of America Memoir* 150.

Hickey, L. J. 1978. Origin of the major features of angiospermous leaf architecture in the fossil record. *Courier Forschungs-Institute Senkenberg* **30**:27–34.

Hickey, L. J. 1979. A revised classification of the architecture of dicotyledonous leaves. In *Anatomy of the Dicotyledons*, eds. C. R. Metcalfe and L. Chalk, pp. 25–39. Oxford University Press, Oxford.

Hickey, L. J. 1986. Summary and implications of the fossil plant record of the Potomac Group. In *Land Plants: Notes for a Short Course*, ed. T. W. Broadhead, pp. 162–181. The Paleontological Society.

Hickey, L. J. and J. A. Doyle. 1977. Early Cretaceous fossil evidence for angiosperm evolution. *The Botanical Review* **43**:3–104.

Hickey, L. J. and R. K. Peterson. 1978. *Zingiberopsis*, a fossil genus of the ginger family from Late Cretaceous to early Eocene sediments of Western Interior North America. *Canadian Journal of Botany* **56**:1136–1152.

Hickey, L. J. and D. W. Taylor. 1991. The leaf architecture of *Ticodendron* and the application of foliar characters in discerning its relationships. *Annals of the Missouri Botanical Garden* **78**:105–130.

Hickey, L. J. and D. W. Taylor. 1992. Paleobiology of early angiosperms: evidence from sedimentological associations in the Early Cretaceous Potomac Group of eastern U.S.A. *The Paleontological Society Special Publications* **6**:128.

Hickey, L. J. and J. A. Wolfe. 1975. The bases of angiosperm phylogeny: vegetative morphology. *Annals of the Missouri Botanical Garden* **62**:538–589.

Hicks, G. S. and I. M. Sussex. 1970. Development in vitro of excised flower primordia of *Nicotiana tabacum*. *Canadian Journal of Botany* **48**:133–139.

Hicks, G. S. and I. M. Sussex. 1971. Organ regeneration in sterile culture after median bisection of the flower primordia of *Nicotiana tabacum*. *Botanical Gazette* **132**:350–363.

Hill, J. P. and E. M. Lord. 1989. Floral development in *Arabidopsis thaliana*: a comparison of the wild type and homeotic pistillata mutant. *Canadian Journal of Botany* **67**:2922–2936.

Hillis, D. M. 1987. Molecular versus morphological approaches to systematics. *Annual Review of Ecology and Systematics* **18**:23–42.

Hillis, D. M., M. W. Allard, and M. M. Miyamoto. 1993. Analysis of DNA sequence data: phylogenetic inference. *Methods in Enzymology* **224**:456–487.

Hillis, D. M., J. P. Huelsenbeck, and C. W. Cunningham. 1994. Application and accuracy of molecular phylogenies. *Science* **264**:671–677.

Holm, T. 1926. *Saururus cernuus* L., a morphological study. *American Journal of Science* **12**:162–168.

Holmes, W. B. K. 1977. A pinnate leaf with reticulate venation from the Permian of New South Wales. *Proceedings of the Linnean Society, New South Whales* **102**:52–57.

Hooker, J. D. 1886. *The Flora of British India*. Dehra Dun, Bischen Singh, Mahendra Pal Singh and Periodical Experts, Dehli (Reprinted 1973), 7 vols.

Huber, H. 1977. The treatment of the monocotyledons in an evolutionary system of classification. *Plant Systematics and Evolution (Supplement)* **1**:285–298.

Huber, H. 1993. Aristolochiaceae. In *The Families and Genera of Vascular Plants*. Vol. 2. *Flowering Plants. Dicotyledons: Magnoliid, Hamamelid and Caryophyllid Families*, eds. K. Kubitzki, J. G. Rohwer, and V. Bittrich, pp. 129–137. Springer-Verlag, Berlin.

Hufford, L. 1992. Rosidae and their relationships to other nonmagnoliid dicotyledons: a phylogenetic analysis using morphological and chemical data. *Annals of the Missouri Botanical Garden.* **79**: 218–248.

Hufford, L. and P. R. Crane. 1989. A preliminary phylogenetic analysis of the 'lower' Hamamelidae. In *Evolution, Systematics, and Fossil History of the Hamamelidae*, eds. P. R. Crane and S. Blackmore, Vol. 1, pp. 175–192. Clarendon Press, Oxford.

Hughes, N. F. 1976. *Palaeobiology of Angiosperm Origins*. Cambridge University Press, Cambridge.

Hughes, N. F. and A. B. McDougall 1987. Records of angiospermid pollen entry into the English Early Cretaceous succession. *Review of Palaeobotany and Palynology* **50**:255–272.

Hughes, N. F. and A. B. McDougall. 1990. New Wealden correlation for the Wessex Basin. *Proceedings of the Geologists Association* **101**:85–90.

Hughes, N. F., A. B. McDougall, and J. L. Chapman. 1991. Exceptional new record of Cretaceous Hauterivian angiospermid pollen from Southern England. *Journal of Micropalaeontology* **10**:75–82.

Humbert, H., and R. Capuron. 1955. Découvert d'une chloranthacée à Madagascar: *Ascarinopsis coursii*, gen. nov., sp. nov. Académie des Sciences (Paris): 3 January, 1955, pp. 28–30.

Humphries, C. J. and S. Blackmore. 1989. A review of the classification of the Moraceae. In *Evolution, Systematics, and Fossil History of the Hamamelidae*, eds. P. R. Crane and S. Blackmore, Vol. 1, pp. 267–277. Clarendon Press, Oxford.

Hutchinson, J. 1926. *The Families of Flowering Plants. I. Dicotyledons.* Macmillan, London.

Hutchinson, J. 1934. *The Families of Flowering Plants. II. Monocotyledons.* Macmillan, London.

Hutchinson, J. 1969. *Evolution and Phylogeny of Flowering Plants.* Academic Press, London.

International Association of Wood Anatomists Committee on Nomenclature. 1964. *Multilingual Glossary of Terms Used in Wood Anatomy*, Verlaganstalt Konkordia, Winterthur, Switzerland.

Irish, V. F. and I. M. Sussex. 1990. Function of the *apetala-1* gene during *Arabidopsis* floral development. *The Plant Cell* **2**:741–753.

Ito, M. 1986. Studies in the floral morphology and anatomy of Nymphaeales. IV. Floral anatomy of *Nelumbo nucifera*. *Acta Phytotaxonomica Geobotanica* **37**:82–96.

Ito, M. 1987. Phylogenetic systematics of the Nymphaeales. *Botanical Magazine (Tokyo)* **100**:17–35.

Jensen, S. R., B. J. Nielsen, and R. Dahlgren. 1975. Iridoid compounds, their occurrence and systematic importance in the angiosperms. *Botanisker Notiser* **128**:148–180.

Johansen, D. A. 1950. *Plant Embryology.* Chronica Botanica Publishers, Waltham, Massachusetts.

Johnson, D. S. 1902. On the development of certain Piperaceae. *Botanical Gazette* **34**:321–340.

Johnson, K. R. and L. J. Hickey. 1990. Megafloral change across the Cretaceous/Tertiary boundary in the northern Great Plains and Rocky Mountains, U.S.A. *Geological Society of America Special Paper* **247**:433–444.

Johnson, L. A. S. and K. L. Wilson. 1989. Casuarinaceae: a synopsis. In *Evolution, Systematics, and Fossil History of the Hamamelidae*, eds. P. R. Crane and S. Blackmore, Vol. 2, pp. 167–188. Clarendon Press, Oxford.

Johri, B. M. (ed.). 1984. *Embryology of Angiosperms.* Springer-Verlag, New York.

Johri, B. M., K. B. Ambegaokar, and P. S. Srivastava. 1992. *Comparative Embryology of Angiosperms.* Springer-Verlag, Berlin.

Josten, K. H. 1962. *Neuropteris semireticulata* eine neue Art als Bindeglied zwischen den Gattungen *Neuropteris* und *Reticulopteris*. *Palaeontology Zeitschrift* **36**:33–45.

Kadereit, J. W. 1993. Papaveraceae. In *The Families and Genera of Vascular Plants.* Vol. 2. *Flowering Plants. Dicotyledons. Magnoliid, Hamamelid and Caryophyllid Families*, eds. K. Kubitzki, J. G. Rohwer, and V. Bittrich pp.494–506. Springer-Verlag, Berlin.

Kaplan, D. R. 1973. The monocotyledons: their evolution and comparative biology. VII. The problem of leaf morphology and evolution in the monocotyledons. *The Quarterly Review of Biology* **48**:437–457.

Keefe, J. M. and M. F. Moseley, Jr. 1978. Wood anatomy and phylogeny of *Paeonia* section *Moutan. Journal of the Arnold Arboretum* **59**:274–297.

Keener, C. S. 1993. A review of the classification of the genus *Hydrastis* (Ranunculaceae). *Aliso* **13**:551–558.

Keng, H. 1993a. Illiciaceae. In *The Families and Genera of Vascular Plants.* Vol. 2. *Flowering Plants. Dicotyledons: Magnoliid, Hamamelid and Caryophyllid Families*, eds. K. Kubitzki, J. G. Rohwer, and V. Bittrich, pp. 344–347. Springer-Verlag, Berlin.

Keng, H. 1993b. Schisandraceae. In *The Families and Genera of Vascular Plants.* Vol. 2. *Flowering Plants. Dicotyledons: Magnoliid, Hamamelid and Caryophyllid Families*, eds. K. Kubitzki, J. G. Rohwer, and V. Bittrich, pp. 589–592. Springer-Verlag, Berlin.

Kessler, P. J. A. 1993. Annonaceae. In *The Families and Genera of Vascular Plants.* Vol. 2. *Flowering Plants. Dicotyledons: Magnoliid, Hamamelid and Caryophyllid Families*, eds. K. Kubitzki, J. G. Rohwer, and V. Bittrich, p.93–129. Springer-Verlag, Berlin.

Kiester, A. R., R. Lande, and D. W. Schemske. 1984. Models of coevolution and speciation in plants and their pollinators. *The American Naturalist* **124**:220–243.

Kirchner, G., C. J. Kinslow, G. C. Bloom, and D. W. Taylor. 1993. Nonlethal assay system of beta-glucuronidase acitivity in transgenic tobacco roots. *Plant Molecular Biology Reporter* **11**:320–325.

Kluge, A. G. 1989. A concern for evidence and a phylogenetic hypothesis of relationships among *Epicrates* (Boidae, Serpentes). *Systematic Zoology* **38**:7–25.

Knaak, C., R. K. Hamby, M. L. Arnold, M. D. LeBlanc, R. L. Chapman, and E. A. Zimmer. 1990. Ribosomal DNA variation and its use in plant biosystematics. In *Biological Approaches and Evolutionary Trends in Plants*, ed. S. Kawano, pp. 135–158. Academic Press, New York.

Knoll, M. A. and W. C. James. 1987. Effect of the advent and diversification of vascular land plants on mineral weathering through geologic time. *Geology* **15**:1099–1102.

Kocher, T. D., J. A. Conroy, K. R. McKaye, and J. R. Stauffer. 1993. Similar morphologies of cichlid fish in Lakes Tanganyika and Malawi are due to convergence. *Molecular Phylogenetics and Evolution* **2**:158–165.

Koidzumi, G. 1936. *Gigantopteris. Acta Phytotaxonomica Geobotanica* **5**:130–144.

Kosakai, H., M. F. Moseley, and V. I. Cheadle. 1970. Morphological studies in the Nymphaeaceae. V. Does *Nelumbo* have vessels? *American Journal of Botany* **57**:487–494.

Krassilov, V. A. 1977. Contributions to the knowledge of the Caytoniales. *Review of Paleobotany and Palynology* **24**:155–178.

Krassilov, V. A. 1984. New paleobotanical data on origin and early evolution of angiospermy. *Annals of the Missouri Botanical Garden* **71**:577–592.

Krassilov, V. A. and S. R. Ash. 1988. On *Dinophyton* – protognetalean Mesozoic plant. *Palaeontographica Beitrage zur Naturgeschichte der Vorzeit* **208B**:33–38.

Kribs, D. A. 1935. Salient lines of structural specialization in the wood rays of dicotyledons. *Botanical Gazette* **96**:547–557.

Kribs, D. A. 1937. Salient lines of structural specialization in the wood parenchyma of dicotyledons. *Bulletin of the Torrey Botanical Club* **64**:177–186.

Kubitzki, K. 1987. Origin and significance of trimerous flowers. *Taxon* 3621–28.

Kubitzki, K. 1990. Gnetatae with the single order Gnetales. *In The Families and Genera of Vascular Plants*. Vol. I. *Pteridophytes and Gymnosperms*, eds. K. U. Kramer and P. S. Green, p. 385. Springer–Verlag, Berlin.

Kubitzki, K. 1993a. Calycanthaceae. In *The Families and Genera of Vascular Plants*, Vol. 2 *Flowering Plants. Dicotyledons: Magnoliid, Hamamelid and Caryophyllid Families*, eds. K. Kubitzki, J. G. Rohwer, and V. Bittrich, pp. 138–140. Springer-Verlag, Berlin.

Kubitzki, K. 1993b. Canellaceae. In *The Families and Genera of Vascular Plants*, Vol. 2 *Flowering Plants. Dicotyledons: Magnoliid, Hamamelid and Caryophyllid Families*, eds. K. Kubitzki, J. G. Rohwer, and V. Bittrich, pp. 200–203. Springer-Verlag, Berlin.

Kubitzki, K. 1993c. Degeneriaceae. In *The Families and Genera of Vascular Plants*, Vol. 2 *Flowering Plants. Dicotyledons: Magnoliid, Hamamelid and Caryophyllid Families*, eds. K. Kubitzki, J. G. Rohwer, and V. Bittrich, pp. 290–291. Springer-Verlag, Berlin.

Kubitzki, K. 1993d. Gomortegaceae. In *The Families and Genera of Vascular Plants*, Vol. 2 *Flowering Plants. Dicotyledons: Magnoliid, Hamamelid and Caryophyllid Families*, eds. K. Kubitzki, J. G. Rohwer, and V. Bittrich, pp. 318–320. Springer-Verlag, Berlin.

Kubitzki, K. 1993e. Hernandiaceae. In *The Families and Genera of Vascular Plants,* Vol. 2 *Flowering Plants. Dicotyledons: Magnoliid, Hamamelid and Caryophyllid Families,* eds. K. Kubitzki, J. G. Rohwer, and V. Bittrich, pp. 334–348. Springer-Verlag, Berlin.

Kubitzki, K. 1993f. Lactoridaceae. In *The Families and Genera of Vascular Plants,* Vol. 2 *Flowering Plants. Dicotyledons: Magnoliid, Hamamelid and Caryophyllid Families,* eds. K. Kubitzki, J. G. Rohwer, and V. Bittrich, pp. 359–361. Springer-Verlag, Berlin.

Kubitzki, K. 1993g. Platanaceae. In *The Families and Genera of Vascular Plants,* Vol. 2 *Flowering Plants. Dicotyledons: Magnoliid, Hamamelid and Caryophyllid Families,* eds. K. Kubitzki, J. G. Rohwer, and V. Bittrich, pp. 521–522. Springer-Verlag, Berlin.

Kubitzki, K. and O. R. Gottlieb. 1984. Phytochemical aspects of angiosperm origin and evolution. *Acta Botanica Neerlandica* 33:457–468.

Kühn, U. and K. Kubitzki. 1993. Myristicaceae. In *The Families and Genera of Vascular Plants,* Vol. 2 *Flowering Plants. Dicotyledons: Magnoliid, Hamamelid and Caryophyllid Families,* eds. K. Kubitzki, J. G. Rohwer, and V. Bittrich, pp. 457–467. Springer-Verlag, Berlin.

Kuprianova, L. A. 1967. Palynological data for the history of the Chloranthaceae. *Pollen et Spores* 9:95–100.

Labandeira, C. C. and J. J. Sepkoski, Jr. 1993. Insect diversity in the fossil record. *Science* 261:310–315.

Lam, H. J. 1961. Reflections on angiosperm phylogeny. I and II. Facts and theories. *Koninklijke Akademie van Wetenschappen te Amsterdam, Afdruuken Natuurkunde, Procesverbaal* 64:251–276.

Lammers, T. G., T. F. Stuessy, and M. Silva O. 1986. Systematic relationships of the Lactoridaceae, an endemic family of the Juan Fernandez Islands, Chile. *Plant Systematics and Evolution* 152: 243–266.

Laveine, J.-P., S. Zhang, and Y. Lemoigne. 1989. Global paleobotany, as exemplified by some Upper Carboniferous pteridosperms. *Bulletin Societe Belgique Géologie* 98:115–125.

Leinfellner, W. 1950. Der Bauplan des synkarpen Gynozeums. *Oesterreichische Botanische Zeitschrift* 97:403–436.

Leins, P., and C. Erbar. 1985. Ein Beitrag zur Blütenentwicklung der Aristolochiaceen, einer Vermittlergruppe zu den Monokotylen. *Botanische Jahrbücher für Systematik, Pflanzengeschichte und Pflanzengeographie* 107:343–368.

Leins, P., C. Erbar, and W. A. van Heel. 1988. Notes on the floral development of *Thottea* (Aristolochiaceae). *Blumea* 33:357–370.

LeRoy, J.-P. 1983a. Interpretation nouvelle des appareils sexuels chez les Chloranthacee (Chloranthales, Magnoliidees). *Comptes rendus des seances de l'Academie des Sciences. Ser. III, Sciences de la vie* 296:747–752.

Leroy, J.-P. 1983b. The origin of angiosperms: an unrecognized ancestral dicotyledon, *Hedyosmum* (Chloranthales), with a strobiloid flower is living today. *Taxon* 32: 169–175.

Les, D. H. 1988. The origin and affinities of the Ceratophyllaceae. *Taxon* 37: 326–345.

Les, D. H. 1993. Ceratophyllaceae. In *The Families and Genera of Vascular Plants*. Vol. 2. *Flowering Plants. Dicotyledons: Magnoliid, Hamamelid and Caryophyllid Families*, eds. K. Kubitzki, J. G. Rohwer, and V. Bittrich, pp. 246–250. Springer-Verlag, Berlin.

Les, D. H., D. K. Garvin, and C. F. Wimpee. 1991. Molecular evolutionary history of ancient aquatic angiosperms. *Proceedings of the National Academy of Sciences of the United States of America* **88**:10119–10123.

Li, H., M. Tanimura, and P. M. Sharp. 1987. An evaluation of the molecular clock hypothesis using mammalian DNA sequences. *Journal of Molecular Evolution* **25**:330–342.

Li, H. and B. Tian. 1990. Anatomic study of foliage leaf of *Gigantonoclea guizhouensis* Gu et Zhi. *Acta Palaeontologica Sinica* **29**(2):216–227. (in Chinese)

Li. H., B. Tian. E. L. Taylor, and T. N. Taylor. 1994. Foliar anatomy of *Gigantoclea guizhoaensis* (Gigantopteridales) from the Upper Permian of Guizhou Province, China. *American Journal of Botany* **81**:678–689.

Liang, H.-X., and S. C. Tucker. 1989. Floral development in *Gymnotheca chinensis* (Saururaceae). *American Journal of Botany* **76**: 806–819.

Liang, H.-X., and S. C. Tucker. 1990. Comparative study of the floral vasculature in Saururaceae. *American Journal of Botany* **77**: 607–623.

Lidén, M. 1986. Synopsis of Fumarioideae (Papaveraceae) with a monograph of the tribe Fumarieae. *Opera Botanica* **88**:1–133.

Lidén, M. 1993. Fumariaceae. In *The Families and Genera of Vascular Plants*. Vol. 2. *Flowering Plants. Dicotyledons: Magnoliid, Hamamelid and Caryophyllid Families*, eds. K. Kubitzki, J. G. Rohwer, and V. Bittrich, pp. 310–318. Springer-Verlag, Berlin

Lidgard, S. and P. R. Crane. 1988. Quantitative analyses of the early angiosperm radiation. *Nature* **331**:344–346.

Lidgard, S. and P. R. Crane. 1990. Angiosperm diversification and Cretaceous floristic trends: a comparison of palynofloras and leaf macrofloras. *Paleobiology* **16**:77–93.

Lilligraven, J.A., Z. Kielan-Jawarowska, and W.A. Clemens (eds.). 1979. *Mesozoic Mammals, the First Two-Thirds of Mammalian History*. University of California Press, Berkeley.

Lloyd, D. G. and M. S. Wells. 1992. Reproductive biology of a primitive angiosperm, *Pseudowintera colorata* (Winteraceae), and the evolution of pollination systems in the Anthophyta. *Plant Systematics and Evolution* **181**:77–95.

Loconte, H. 1993. Berberidaceae. In *The Families and Genera of Vascular Plants*. Vol. 2. *Flowering Plants. Dicotyledons: Magnoliid, Hamamelid and Caryophyllid Families*, eds. K. Kubitzki, J. G. Rohwer, and V. Bittrich, pp. 147–152. Springer-Verlag, Berlin

Loconte, H., L. M. Campbell, and D. W. Stevenson. In press. Ordinal and familial relationships of ranunculid genera. In *Systematics and Evolution of the Ranunculiflorae*, eds. U. Jensen and J. W. Kadereit. Springer Verlag, Berlin.

Loconte, H. and J. R. Estes. 1989. Phylogenetic systematics of Berberidaceae and Ranunculales (Magnoliidae). *Systematic Botany* **14**:565–579.

Loconte, H. and D. W. Stevenson. 1990. Cladistics of the Spermatophyta. *Brittonia* 42:197–211.

Loconte, H. and D. W. Stevenson. 1991. Cladistics of the Magnoliidae. *Cladistics* 7:267–296.

Longman, K. A. and J. Jeník. 1987. *Tropical Forest and Its Enviroment*, 2nd ed., Longman Scientific & Technical, Essex, United Kingdom.

MacDuffie, R. C. 1921. Vessels of the gnetalean type in angiosperms. *Botanical Gazette* 71:438–445.

Maddison, D. R. 1991. The discovery and importance of multiple islands of most-parsimonious trees. *Systematic Zoology* 40:315–328.

Maddison, W. R., M. J. Donoghue, and D. R. Maddison. 1984. Outgroup analysis and parsimony. *Systematic Zoology* 33:83–103.

Maddison, W. P. and D. R. Maddison. 1993. MacClade. In *Analysis of Phylogeny and Character Evolution. Version 3.0.* Sinauer Associates, Inc., Sunderland, Massachusetts.

Maheshwari, P. 1935. Contributions to the morphology of *Ephedra foliata*. Proceedings of the Indian Academy of Sciences 1:586–606.

Maheshwari, P. 1950. *An Introduction to the Embryology of Angiosperms*. McGraw-Hill, New York.

Mamay, S. H. 1986. New species of Gigantopteridaceae from the Lower Permian of Texas. *Phytologia* 61:311–315.

Mamay, S. H. 1988. *Gigantonoclea* in the Lower Permian of Texas. *Phytologia* 64:330–332.

Mamay, S. H. 1989. *Evolsonia*, a new genus of Gigantopteridaceae from the Lower Permian Vale Formation, north-central Texas. *American Journal of Botany* 76:1299–1311.

Mamay, S. H., J. M. Miller, D. M. Rohr, and W. E. Stein, Jr. 1986. *Delnortea*, a genus of Permian plants from west Texas. *Phytologia* 60:345–346.

Mamay, S. H., J. M. Miller, D. M. Rohr, and W. E. Stein, Jr. 1988. Foliar morphology and anatomy of the gigantopterid plant *Delnortea abbottiae*, from the Lower Permian of West Texas. *American Journal of Botany* 75:1409–1433.

Mapes, G. and G. W. Rothwell. 1991. Structure and relationships of primitive conifers. *Neues Jahrbuch für Geologie Paläontologie* 183:269–287.

Mark, R. 1965. Tensile stress analysis of the cell walls of coniferous tracheids. In *Cellular Ultrastructure of Woody Plants*, W. A. Coté, ed., pp. 493–533. Syracuse University Press, Syracuse, New York.

Martens, P. 1971. *Les Gnétophytes*, Gerbrüder Borntraeger, Berlin.

Martens, P. 1977. *Welwitschia mirabilis* and neoteny. *American Journal of Botany* 64:916–920.

Martin, A. C. 1946. The comparative internal morphology of seeds. *American Naturalist* 36:513–660.

Martin, P. G. and J. M. Dowd. 1990. A protein sequence study of the phylogeny and origin of the dicotyledons. In *Biological Approaches and Evolutionary Trends in Plants*, ed. S. Kawano, pp. 171–181. Academic Press, New York.

Martin, P. G. and J. M. Dowd. 1991. Studies of angiosperm phylogenies using protein sequences. *Annals of the Missouri Botanical Garden* **78**:296–337.

Martin, W., A. Gierl and H. Saedler. 1989. Molecular evidence for pre-Cretaceous angiosperm origins. *Nature* **339**:46–48.

Martin, W., D. Lydiate, H. Brinkmann, G. Forkmann, H. Saedler, and R. Cerff. 1993. Molecular phylogenies in angiosperm evolution. *Molecular Biology and Evolution* **10**:140–162.

Masterson, J. 1994. Stomatal size in fossil plants: evidence for polyploidy in the majority of angiosperms. *Science* **264**:421–424.

Mauseth, J. D. 1988. *Plant Anatomy*. The Benjamin/Cummings Publishing Company, Menlo Park, California.

McLoughlin, S. 1990. Paleobotany and paleoenvironments of Permian strata, Bowen Basin, Queensland. Ph.D. dissertation, University of Queensland. St. Lucia, Queensland.

Meacham, C. A. 1994. Phylogenetic relationships at the basal radiation of angiosperms: further study by probability of character compatibility. *Systematic Botany* **19**:506–522.

Meeuse, A. J. D. 1963. The multiple origins of the angiosperms. *Advancing Frontiers of Plant Sciences* **1**:105–127.

Meeuse, A. D. J. 1972. Facts and fiction in floral morphology with special reference to the Polycarpicae. *Acta Botanica Neerlandica* **21**:113–127, 235–252, 351–365.

Meeuse, A. D. J. 1987. *All About Angiosperms*. Eburon, Delft.

Meeuse, A. J. D. 1990. *Flowers and Fossils*. Eburon, Delft.

Mehra, P. N. 1950. Occurrence of hermaphrodite flowers and the development of female gametophyte in *Ephedra intermedia* Shrenk et Mey. *Annals of Botany, New Series* **14**:165–180.

Meijer, W. 1993a. Hydnoraceae. In *The Families and Genera of Vascular Plants*. Vol. 2. *Flowering Plants. Dicotyledons: Magnoliid, Hamamelid and Caryophyllid Families*, eds. K. Kubitzki, J. G. Rohwer, and V. Bittrich, pp. 341–343. Springer-Verlag, Berlin.

Meijer, W. 1993b. Rafflesiaceae. In *The Families and Genera of Vascular Plants*. Vol. 2. *Flowering Plants. Dicotyledons: Magnoliid, Hamamelid and Caryophyllid Families*, eds. K. Kubitzki, J. G. Rohwer, and V. Bittrich, pp. 557–563. Springer-Verlag, Berlin.

Melchior, H. 1964. *Engler's Syllabus der Pflanzenfamilien*, 2nd ed., Vol. 2, *Angiospermen*. Gebrüder Borntraeger, Berlin.

Melville, R. 1962. A new theory of the angiosperm flower: I. The gynoecium. *Kew Bulletin* **16**:1–50.

Melville, R. 1963. A new theory of the angiosperm flower II. *Kew Bulletin* **17**:1–63.

Melville, R. 1969a. A new theory of the angiosperm flower III. *Kew Bulletin* **23**:133–180.

Melville, R. 1969b. Leaf venation patterns and the origins of angiosperms. *Nature* **224**:121–125.

Melville, R. 1970. Links between the Glossopteridae and the angiosperms. *Proceedings of the Second Gondwana Symposium, C.S.I.R. South Africa*, pp. 585–588.

Melville, R. 1983. Glossopteridae, Angiospermidae and the evidence for angiosperm origin. *Botanical Journal of the Linnean Society of London* **86**:279–323.

Merxmüller, H., and P. Leins. 1971. Zur Entwicklungsgeschichte männlicher Begonienblüten. *Flora* **160**:333–339.

Metcalfe, C. R. 1979. The stem. In *Anatomy of the Dicotyledons*, 2nd ed., Vol. I. *General Introduction*, eds. C. R. Metcalfe and L. Chalk, pp. 166–180. Clarendon Press, Oxford.

Metcalfe, C. R. (ed.) 1987. *Anatomy of the Dicotyledons*, 2nd ed., Vol. III, *Magnoliales, Illiciales, and Laurales*. Clarendon Press, Oxford.

Metcalfe, C. R., and L. Chalk (eds.). 1950. *Anatomy of the Dicotyledons; Leaves, Stem, and Wood in Relation to Taxonomy, with Notes on Economic Uses*. 2 vols. Clarendon Press, Oxford.

Metcalfe, C. R. and L. Chalk (eds.). 1983. *Anatomy of the Dicotyledons*, 2nd ed., Vol. II. *Wood Structure and Conclusion of the General Introduction*. Clarendon Press, Oxford.

Midgley, J. J. and W. J. Bond. 1991. Ecological aspects of the rise of angiosperms: a challenge to the reproductive superiority hypotheses. *Biological Journal of the Linnean Society* **44**:81–92.

Miller, H. J. 1975. Anatomical characteristics of some woody plants of the Angmassalik District of southeast Greenland. *Meddelelser om Grønland* **198**(6):1–30.

Modrusan, Z., L. Reiser, K. A. Feldmann, R. L. Fischer, and G. W. Haughn. 1994. Homeotic transformation of ovules into carpel-like structures in *Arabidopsis*. *The Plant Cell* **6**:333–349.

Moeliono, B. M. 1970. *Cauline or Carpellary Placentation Among Dicotyledons*, Van Gorcum and Co., Assen, Netherlands.

Moldowan, J. M., J. Dahl, B. J. Huizinga, F. J. Fago, L. J. Hickey, T. M. Peakman, and D. W. Taylor. 1994. The molecular fossil record of oleanane and its relationship to angiosperms. *Science* **265**:768–771.

Money, L. L., I. W. Bailey, and B. G. L. Swamy. 1950. The morphology and relationships of the Monimiaceae. *Journal of the Arnold Arboretum* **31**:372–404.

Mooney, H. A., and B. Gartner. 1991. Reserve economy of vines. In *The Biology of Vines*, eds., F. E. Putz and H. A. Mooney, pp. 161–179. Cambridge University. Press, Cambridge.

Moore, L. B. 1977. The Flowers of *Ascarina lucida* Hook. f. (Chloranthaceae). *New Zealand Journal of Botany* **15**:491–494.

Morawetz, W. 1985. Beitrage zur Karyologie und Systematik der Gattung *Thottea* (Aristolochiaceae). *Botanische Jahrbücher für Systematik, Pflanzengeschichte und Pfanzengeographie.* **107**:329–342.

Moseley, M. F. 1958. Morphological studies of the Nymphaeaceae. I. The nature of the stamens. *Phytomorphology* **8**:1–29.

Moseley, M. F. l974. Nymphaeales. In *Encyclopedia Britannica*, 15th ed., pp. 428–430. Encyclopedia Britannica, Inc., Chicago.

Moseley, M. F., I. J. Mehta, P. S. Williamson, and H. Kosakai. 1984. Morphological studies of the Nymphaeaceae (sensu lato). XIII. Contributions to the vegetative and floral structure of *Cabomba*. *American Journal of Botany* **71**:902–924.

Moseley, M. F., E. L. Schneider, and P. E. Williamson. 1993. Phylogenetic interpretations from selected floral vasculature characters in the Nymphaeaceae sensu lato. *Aquatic Botany* **44**:325–342.

Moseley, M. F. and N. W. Uhl. 1985. Morphological studies of the Nymphaeaceae sensu lato. XV. The anatomy of the flower of *Nelumbo*. *Botanische Jahrbücher für Systematik, Pflanzengeschichte und Pfanzegeographie.* **106**:61–98.

Muhammad, A. F. and R. Sattler. 1982. Vessel structure of *Gnetum* and the origin of angiosperms. *American Journal of Botany* **69**:1004–1021.

Mulcahy, D. L. 1979. The rise of the angiosperms: a genecological factor. *Science* **206**:20–23.

Mulcahy, D. 1983. Models of pollen tube competition in *Geranium maculatum*. In *Pollination Biology*, ed. L. Real, pp. 151–161. Academic Press, Inc, New York.

Mulcahy, D. L., G. B. Mulcahy, and K. B. Searcy. 1992. Evolutionary genetics of pollen competition. In *Ecology and Evolution of Plant Reproduction*, ed. R. Wyatt, pp. 25–36. Chapman and Hall, New York.

Muller, J. 1970. Palynological evidence on early differentiation of angiosperms. *Biological Review* **45**:417–450.

Muller, J. 1981. Fossil pollen records of extant angiosperms. *The Botanical Review* **47**:1–146.

Muller, J. 1984. Significance of fossil pollen for angiosperm history. *Annals of the Missouri Botanical Garden* **71**:419–443.

Munroe, E. 1953. The phylogeny of the Papilionidae. *Proceedings of the 7th Pacific Science Congress* **4**:83–87.

Murty, Y. S. 1960. Studies in the Order Piperales-VIII. A contribution to the morphology of *Houttuynia cordata* Thunb. *Phytomorphology* **10**:329–341.

Nast, C. G. 1944. The comparative morphology of the Winteraceae, VI. Vascular anatomy of the flowering shoot. *Journal of the Arnold Arboretum* **25**:454–466.

Nast, C. G. and I. W. Bailey. 1946. Morphology of *Euptelea* and comparison with *Trochodendron*. *Journal of the Arnold Arboretum* **27**:186–192.

Nei, M. 1991. Relative efficiencies of different tree-making methods for molecular data. In *Phylogenetic Analysis of DNA Sequences*, eds. M. M. Miyamoto and J. Cracraft, pp. 90–128. Oxford University Press, New York.

Neumayer, H. 1924. Die Geschichte der Blütte. *Abhandlung Zoologischen Botanische Gesellschaft Wien* **14**:1–110.

Nicely, K. A. 1965. A monographic study of the Calycanthaceae. *Castanea* **30**:38–81.

Nickrent, D. L. and D. E. Soltis. 1995. A comparison of angiosperm phylogenies from nuclear 18S rDNA and *rbc*L sequences. *Annals of the Missouri Botanical Garden* **82**: 208–234.

Nishida, H. 1985. A structurally preserved magnolialean fructification from the Mid-Cretaceous of Japan. *Nature* **318**:58–59.

Nishida, M. 1969. A petrified trunk of *Bucklandia choshiensis* sp. nov. from the Cretaceous of Choshi, Chiba Prefecture. *Phytomorphology* **19**:28–34.

Nixon, K. C., W. L. Crepet, D. Stevenson, and E. M. Friis. 1994. A reevaluation of seed plant phylogeny. *Annals of the Missouri Botanical Garden* **81**:484–533.

Nooteboom, H. P. 1993. Magnoliaceae. In *The Families and Genera of Vascular Plants*. Vol. 2. *Flowering Plants. Dicotyledons: Magnoliid, Hamamelid and Caryophyllid Families*, eds. K. Kubitzki, J. G. Rohwer, and V. Bittrich, pp. 391–401. Springer-Verlag, Berlin.

Nowicke, J. W., J. L. Bittner, and J. J. Skvarla. 1986. *Paeonia*, exine substructure and plasma ashing. In *Pollen and Spores: Form and Function*, eds. S. Blackmore and I. K. Ferguson, pp. 81–95. Academic Press, New York.

Nowicke, J. W., and J. J. Skvarla. 1982. Pollen morphology and the relationships of *Circaeaster*, of *Kingdonia*, and of *Sargentodoxa* to the Ranunculales. *American Journal of Botany* **69**:990–998.

Olmstead, R. G., B. Bremer, K. M. Scott, and J. D. Palmer. 1993. A parsimony analysis of the Asteridae sensu lato based on *rbc*L sequences. *Annals of the Missouri Botanical Garden* **80**:700–722.

Osborn, J.M., T.N. Taylor, and P.R. Crane. 1991. The ultrastructure of *Sahnia* pollen (Pentoxylales). *American Journal of Botany* **78**:1560–1569.

Osborn, J. M., T. N. Taylor, and M. R. de Lima. 1993. The ultrastructure of fossil ephedroid pollen with gnetalean affinities from the Lower Cretaceous of Brazil. *Review of Paleobotany and Palynology* **77**:171–184.

Ozima, M., I. Kaneoka and M. Yanagisawa. 1979. ^{40}Ar-^{39}Ar geochronological studies of drilled basement from Leg 51 and Leg 52. In *Initial Reports of the Deep Sea Drilling Project*, Vol. 51–53, pp. 1127–1128. U.S. Government Printing Office, Washington, D.C.

Padmanabhan, D., and M. V. Ramji. 1966. Developmental studies on *Cabomba caroliniana* Gray. *Proceedings of the Indian Academy of Science, Section B*, **64**:216–223.

Page, R. D. M. 1993. On islands of trees and the efficacy of different methods of branch swapping in finding most-parsimonious trees. *Systematic Biology* **42**:200–210 .

Page, V. M. 1979. Dicotyledonous wood from the Upper Cretaceous of central California. *Journal of the Arnold Arboretum* **60**:323–349.

Palser, B. F. 1976. The bases of angiosperm phylogeny: embryology. *Annals of the Missouri Botanical Garden* **62**:621–646.

Pandow, H. 1962. Histogenetische studien an den blüten einiger Phanerogamen. *Botanischen Studien* **13**:1–106.

Pant, D. D. and A. Choudhury. 1977. On the genus *Belemnopteris* Feistmantel. *Palaeontographica Beitrage zur Naturgeschichte der Vorzeit* **164B**:153–166.

Pant, D. D. and K. B. Singh. 1968. On the genus *Gangamopteris* McCoy. *Palaeontographica Beitrage zur Naturgeschichte der Vorzeit* **124B**:83–104.

Pant, D. D., and R. S. Singh. 1974. On the stem and attachment of *Glossopteris* and *Gangamopteris* leaves. Part II.--Structural Features. *Paleontographica Beitrage zur Naturgeschicte der Vorzeit.* **147B**:42–73.

Parrish, J. T. 1982. Atmospheric circulation, upwelling, and organic-rich rocks in the Mesozoic and Cenozoic Eras. *Paleogeography, Paleoclimatology, Paleoecology* **40**:31–66.

Parrish, J. T. 1987. Global paleogeography and paleoclimate of the late Cretaceous and early Tertiary. In *The Origins of Angiosperms and their Biological Consequences*, eds. E.M. Friis, W.G. Chaloner and P.R. Crane, pp.51–73. Cambridge University Press, Cambridge.

Payer, J.-B. 1857. *Traité d'organogénie comparée de la fleur.* Victor Masson, Paris.

Pearson, H. H. W. 1906. Some observations of *Welwitschia mirabilis* Hooker. *Philisophical Transactions of the Royal Society of London B* **198**:265–304.

Pearson, H. H. W. 1929. *Gnetales*, Cambridge University Press, Cambridge.

Pedersen, K. R., P. R. Crane, A.N. Drinnan, and E.M. Friis. 1991. Fruits from the mid-Cretaceous of North America with pollen grains of the *Clavatipollenites* type. *Grana* **30**:577–590.

Pedersen, K.R., P.R. Crane, and E.M. Friis. 1989a. Pollen organs and seeds with *Eucommiidites* pollen. *Grana* **28**:279–294.

Pedersen, K. R., P. R. Crane, and E. M. Friis. 1989b. The morphology and phylogenetic significance of *Vardekloeftia* Harris (Bennettitales). *Review of Palaeobotany and Palynology* **60**:7–24.

Penny, J. H. J. 1989. New early Cretaceous forms of the angiosperm pollen genus *Afropollis* from England and Egypt. *Review of Paleobotany and Palynology* **58**:289–299.

Philipson, W. R. 1993a. Amborellaceae. In *The Families and Genera of Vascular Plants.* Vol. 2. *Flowering Plants. Dicotyledons: Magnoliid, Hamamelid and Caryophyllid Families*, eds. K. Kubitzki, J. G. Rohwer, and V. Bittrich, pp. 92–93. Springer-Verlag, Berlin.

Philipson, W. R. 1993b. Monimiaceae. In *The Families and Genera of Vascular Plants. Vol. II. Flowering Plants. Dicotyledons: Magnoliid, Hamamelid and Caryophyllid Families*, eds. K. Kubitzki, J. G. Rohwer, and V. Bittrich, pp. 426–437. Springer-Verlag, Berlin.

Philipson, W. R. 1993c. Trimeniaceae. In *The Families and Genera of Vascular Plants. Vol. II. Flowering Plants. Dicotyledons: Magnoliid, Hamamelid and Caryophyllid Families*, eds. K. Kubitzki, J. G. Rohwer, and V. Bittrich, pp. 596–599. Springer-Verlag, Berlin.

Pigg, K. B. 1988. Anatomically preserved *Glossopteris* and *Dicroidium* from the central Transantarctic Mountains. Ph.D. dissertation, The Ohio State University, Columbus, Ohio.

Pigg, K. B. 1990. Anatomically preserved *Glossopteris* foliage from the central Transantarctic Mountains. *Review of Palaeobotany and Palynology* **66**:105–127.

Pigg, K. B. and S. McLoughlin. 1992. Comparison of anatomically preserved glossopterid leaves from the Bowen Basin, Queensland, and Sydney Basin, New South Wales,

Australia. *Resumes des Communications, Abstracts, Organisation internationale de Paleobotanique IVème Conference, Paris,* p. 123.

Pigg, K. B. and T. N. Taylor. 1990. Permineralized *Glossopteris* and *Dicroidium* from Antarctica. In *Antarctic Paleobiology: Its Role in the Reconstruction of Gondwana,* eds. T. N. Taylor and E. L. Taylor, pp. 16–172. Springer-Verlag, New York.

Pigg, K. B. and T. N. Taylor. 1993. Anatomically preserved *Glossopteris* stems with attached leaves from the central Transantarctic Mountains, Antarctica. *American Journal of Botany* **80**:500–516.

Pigg, K. B. and M. L. Trivett. 1994. Evolution of the glossopterid gymnosperms from Permian Gondwana. *Journal of Plant Research* **107**:461–478.

Plymale, E. L. and R. B. Wylie. 1944. The major veins of mesomorphic leaves. *American Journal of Botany* **31**:99–106.

Posluszny, U., and W. A. Charlton. 1993. Evolution of the helobial flower. *Aquatic Botany* **44**:303–324.

Prance, G. T. 1977. Floristic inventory of the tropics: where do we stand. *Annals of the Missouri Botanical Garden* **64**:659–684.

Pray, T. R. 1955. Foliar venation of angiosperms. II. Histogenesis of the venation of *Liriodendron. American Journal of Botany* **42**:18–27.

Pray, T. R. 1960. Ontogeny of the open dichotomous venation in the pinna of the fern *Nephrolepis. American Journal of Botany* **47**:319–328.

Pray, T. R. 1962. Ontogeny of the closed dichotomous venation of *Regnellidium. American Journal of Botany* **49**:464–472.

Pray, T. R. 1963. Origin of vein endings in angiosperm leaves. *Phytomorphology* **13**:60–81.

Pulle, A. 1938. The classification of the spermatophytes. *Chronica Botanica* **4**:92.

Qin, Hai-Ning. 1989. An investigation on carpels of Lardizabalaceae in relation to taxonomy and phylogeny. *Cathaya* **1**:61–82.

Qiu, Y.-L., M. W. Chase, D. H. Les, and C. R. Parks. 1993. Molecular phylogenetics of the Magnoliidae: cladistic analyses of nucleotide sequences of the plastid gene *rbcL. Annals of the Missouri Botanical Garden* **80**:507–606.

Queiroz, A., de. 1993. For consensus (sometimes). *Systematic Biology* **42**:368–372.

Quibell, C. H. 1941. Floral anatomy and morphology of *Anemopsis californica. Botanical Gazette* **102**:749–758.

Quinn, C. J., L. E. Equiarte, E. Golenburg, G. H. Learn, Jr., S. W. Graham, S. Dayanandan, and V. A. Albert. 1993. Phylogenetics of seed plants: an analysis of nucleotide sequences from the plastid gene *rbcl. Annals of the Missouri Botanical Garden* **80**:528–580.

Raciborski, M. 1894. Die Morphologie der Cabombaceen und Nymphaeaceen. *Flora* **76**:244–279.

Raju, M. V. S. 1961. Morphology and anatomy of the Saururaceae. I. Floral anatomy and embryology. *Annals of the Missouri Botanical Garden* **48**:107–124.

Raven, P. H. 1976. The bases of angiosperm phylogeny: cytology. *Annals of the Missouri Botanical Garden* **62**:724–764.

Read, C. B. and S. H. Mamay. 1964. Upper Paleozoic floral zones and floral provinces of the United States. *U.S. Geological Survey Professional Paper* **454K**:1–35.

Rees, M. 1993. Trade-offs among dispersal strategies in British plants. *Nature* **366**:150–152.

Regal, P. J. 1977. Ecology and evolution of flowering plant dominance. *Science* **196**:622–629.

Reihman, M. A. and J. T. Schabilion. 1978. Petrified neuropterid foliage from a Middle Pennsylvanian coal ball. *American Journal of Botany* **65**:834–844.

Remy, W. and R. Remy. 1959. *Pflanzenfossilen*, Akademie-Verlag, Berlin.

Retallack, G. and D. L. Dilcher. 1981a. A coastal hypothesis for the dispersal and rise to dominance of flowering plants. In *Paleobotany, Paleoecology and Evolution*, ed. K. J. Niklas, pp. 27–77. Praeger Publishers, New York.

Retallack, G. J. and D. L. Dilcher. 1981b. Arguments for a glossopterid origin of angiosperms. *Paleobiology* **7**:54–67.

Rice, K. A., M. J. Donoghue, and R. G. Olmstead. 1995. A reanalysis of the large *rbc*L dataset. *American Journal of Botany (Supplement)* **82**:157–158.

Rigby, J. F. 1984. The origin of the *Glossopteris* flora-some thoughts based on macrophyte remains. In *Evolutionary Botany and Biostratigraphy (A. K. Ghosh Commemoration Volume)*, eds. A. K. Sharma, G. C. Mitra, and M. Banerjee pp. 19–28. Today and Tomorrow's Printers and Publishers, New Delhi.

Robertson, R. E., and S. C. Tucker. 1979. Floral ontogeny of *Illicium floridanum*, with emphasis on stamen and carpel development. *American Journal of Botany* **66**:605–617.

Rodenburg, W. F. 1971. A revision of the genus *Trimenia* (Trimeniaceae). *Blumea* **19**:3–15.

Rodin, R. J. 1953. Seedling morphology of *Welwitschia*. *American Journal of Botany* **40**:371–378.

Rodin, R. J. 1958a. Leaf anatomy of *Welwitschia* I. Early development of the leaf. *American Journal of Botany* **45**:90–95.

Rodin, R. J. 1958b. Leaf anatomy of *Welwitschia* II. A study of mature leaves. *American Journal of Botany* **45**:96–103.

Rodin, R. J. 1966. Leaf structure and evolution in American species of *Gnetum*. *Phytomorphology* **16**:56–68.

Rodin, R. J. 1967. Ontogeny of foliage leaves in *Gnetum*. *Phytomorphology* **17**:118–128.

Rohwer, J. G. 1993. Lauraceae. In *The Families and Genera of Vascular Plants*. Vol. 2. *Flowering Plants. Dicotyledons: Magnoliid, Hamamelid and Caryophyllid Families*, eds. K. Kubitzki, J. G. Rohwer, and V. Bittrich, pp. 366–391. Springer-Verlag, Berlin.

Roth, I. 1977. *Fruits of Angiosperms*. Gebrüder Borntraeger, Berlin.

Rothwell, G. W. 1986. Classifying the earliest gymnosperms. In *Systematic and Taxonomic Approaches in Palaeobotany*, eds. R. A. Spicer and B. A. Thomas, pp. 137–162. Clarendon Press, Oxford.

Rothwell, G. W. 1988. Cordaitales. In *Origin and Evolution of Gymnosperms*, ed. C. B. Beck, pp. 273–297. Columbia University Press, New York.

Rothwell, G. W. and R. Serbet. 1993. The evolution of spermatophytes, a numerical cladistic analysis. *American Journal of Botany (Supplement)* **80**:174.

Rothwell, G. W. and R. Serbet. 1994. Lignophyte phylogeny and the evolution of spermatophytes: a numerical cladistic analysis. *Systematic Botany* **19**:443–482.

Rutishauser, R. 1993. Reproductive development in seed plants: research activities at the intersection of molecular genetics and systematic botany. *Progress in Botany* **54**:79–101.

Saitou, N. and M. Nei. 1987. The neighbor-joining method: a new method for reconstructing phylogenetic trees. *Molecular Biology and Evolution* **4**:406–425.

Sanderson, M. J. and M. J. Donoghue. 1994. Shifts in diversification rate with the origin of angiosperms. *Science* **264**:1590–1593.

Satina, S. 1944. Perclinal chimeras in *Datura* in relation to development and structure (A) of the style and stigma (B) of calyx and corolla. *American Journal of Botany* **31**:493–502.

Satina, S. 1945. Periclinal chimeras in *Datura* in relation to the development and structure of the ovule. *American Journal of Botany* **32**:72–81.

Satina, S. and A. F. Blakeslee. 1941. Periclinal chimeras in *Datura stramonium* in relation to development of leaf and flower. *American Journal of Botany* **28**:862–871.

Satina, S. and A. F. Blakeslee. 1943. Perclinal chimeras in *Datura* in relation to the development of the carpel. *American Journal of Botany* **30**:395–450.

Sattler, R. 1973. *Organogenesis of Flowers.* University of Toronto Press, Toronto.

Sattler, R. 1974. A new approach to gynoecial morphology. *Phytomorphology* **24**:22–34.

Sattler, R. and C. Lacroix. 1988. Development and evolution of basal cauline placentation: *Basella rubra*. *American Journal of Botany* **75**:918–927.

Sattler, R. and L. Perlin. 1982. Floral development of *Bougainvillea spectabilis* Willd., *Boerhaavia diffusa* L. and *Mirabilis jalapa* L. (Nyctaginaceae). *Botanical Journal of the Linnean Society* **84**:161–182.

Sauter, J. J., W. I. Iten, and M. H. Zimmermann. 1973. Studies on the release of sugar into the vessels of sugar maple (*Acer saccharum*). *Canadian Journal of Botany* **51**:1–8.

Scheckler, S. E. and H. P. Banks. 1971. Anatomy and relationships of some Devonian progymnosperms from New York. *American Journal of Botany* **58**:737–751.

Schmitz, F. 1872. Die Blüthen-Entwicklung der Piperaceen. *Botanische Abhandlungen aus dem Gebiet der Morphologie und Physiologie* **2**:1–74.

Schneider, E. L., and J. M. Jeter. 1982. Morphological studies of the Nymphaeaceae. XII. The floral biology of *Cabomba caroliniana*. *American Journal of Botany* **69**:1410–1419.

Schneider, E. L., and P. S. Williamson. 1993. Nymphaeaceae. In *The Families and Genera of Vascular Plants*. Vol. 2. *Flowering Plants. Dicotyledons: Magnoliid, Hamamelid and Caryophyllid Families*, eds. K. Kubitzki, J. G. Rohwer, and V. Bittrich, pp. 486–493. Springer-Verlag, Berlin.

Schulte, P. J., A. C. Gibson, and P. S. Nobel. 1989. Water flow in vessels with simple or compound perforation-plates. *Annals of Botany* **64**:171–178.

Schwarzwalder, R. N. and D. L. Dilcher. 1991. Systematic placement of the Platanaceae in Hamamelidae. *Annals of the Missouri Botanical Garden* **78**:962–969.

Scott, A. C. and T. N. Taylor. 1983. Plant/animal interactions during the Upper Carboniferous. *Botanical Review* **49**:259–307.

Scott, D. H. 1899. On the primary wood of certain araucarioxylons. *Annals of Botany* **13**:615–619.

Scott, D. H. 1923. *Studies in Fossil Botany*, 3rd ed., Vol. 2. *Spermatophyta*, A. and C. Black, London.

Seigler, D. S. 1977. Plant systematics and alkaloids. *Alkaloids* **16**:1–82.

Shannon, S. and D. R. Meeks-Wagner. 1993. Genetic interactions that regulate inflorescence development in *Arabidopsis*. *The Plant Cell* **5**:639–655.

Shenhav, H. 1970. Petrography, porosity and permeability of sandstone members in the Helez Formation (Lower Cretaceous). *Geological Survey of Israel, Report* OD/3/70, pp. 1–57.

Shenhav, H. and R. Shoresh. 1972. A possible source rock for oil, the Gevar'am Formation, Lower Cretaceous. *Geological Survey of Israel, Report* OD/1/72:1–34.

Simon, J. P. 1970. Comparative serology of the order Nymphaeales. I. Preliminary survey on the relationships of *Nelumbo*. *Aliso* **7**:243–261.

Simon, J. P. 1971. Comparative serology of the order Nymphaeales. II. Relationships of Nymphaeaceae and Nelumbonaceae. *Aliso* **7**:325–350.

Singh, H. 1978. *Embryology of Gymnosperms*. Gebrüder Borntraeger, Berlin.

Skipworth, J. P. 1970. Development of floral vasculature in the Magnoliaceae. *Phytomorphology* **20**:228–236.

Skipworth, J. P., and W. R. Philipson. 1967. The critical vascular system and the interpretation of the *Magnolia* flower. *Phytomorphology* **16**:463–469.

Slade, B. F. 1957. Leaf development in relation to venation as shown in *Cercis siliquastrum* L., *Prunus serrulata* Lindl., and *Acer pseudoplatanus* L. *New Phytology* **56**:281–300.

Slatyer, R. O. 1967. *Plant–Water Relationships*. Academic Press, New York.

Smith, A. C. 1947. The families Illiciaceae and Schisandraceae. *Sargentia* **7**:1–224.

Smith, J. F. and K. J. Sytsma. 1994. Molecules and morphology: congruence of data in *Columnea* (Gesneriaceae). *Plant Systematics and Evolution* **193**:37–52.

Snow, N. and P. Goldblatt. 1992. Chromosome number in *Ticodendron* (Fagales, Ticodendraceae). *Annals of the Missouri Botanical Garden* **79**: 906–907.

Sperry, J. S. 1985. Embolism formation in the palm *Rhapis excelsa*. *International Association of Wood Anantomists Bulletin, New Series* **6**:283–292.

Sporne, K. R. 1948. Correlation and classification in dicotyledons. *Proceedings of the Linnean Society of London* **160**:40–47.

Sporne, K. R. 1956. The phylogenetic classification of the angiosperms. *Biological Reviews of the Cambridge Philosophical Society* **31**:1–29.

Sporne, K. R. 1974. *The Morphology of Angiosperms*, Hutchinson, London.

Srivastava, L. M. 1970. The secondary phloem of *Austrobaileya scandens*. *Canadian Journal of Botany* **48**:341–359.

Stagner, H. R. 1941. Geology of the fossil leaf beds of the Petrified Forest National Monument. In Daugherty, L. H., *The Upper Triassic Flora of Arizona with a Discussion of its Geologic Occurrence*. Contributions to Paleontology, **526**:9–17. Carnegie Institution of Washington.

Stebbins, G. L. 1974. *Flowering Plants: Evolution above the Species Level*, The Belknap Press of Harvard University Press, Cambridge, Massachusetts.

Steele, K., K. E. Holsinger, R. K. Jansen, and D. W. Taylor. 1991. Assessing the reliability of 5S rRNA sequence data for phylogenetic analysis in green plants. *Molecular Biology and Evolution* **8**:240–248.

Steenis, C. G. G. J., van. 1948. *Gymnotheca*, a good genus of Saururaceae. *Blumea* **6**:244–245.

Steeves, T. A., M. W. Steeves, and A. R. Olson. 1991. Flower development in *Amelanchier alnifolia* (Maloideae). *Canadian Journal of Botany* **69**:844–857.

Sternberg, G. K. 1820–1838. *Versuch einer geognostichen botanischen Darstellung der Flora der Vorwelt*, Vol. 1; part 3, pp. 1–39 (1823); part 4, pp. 1–48 (1825), Kommission im Deutschen Museum, Leipzig.

Stevenson, D. 1993. Homology of the seed and associated structures in spermatophytes. *American Journal of Botany (Supplement)* **80**:125.

Stevenson, D. W. and J. I. Davis. 1994. Cladistics of monocot families: morphologists too use exemplars. *American Journal of Botany (Supplement)* **81**:189.

Stevenson, D. W. and H. Loconte. in press. Cladistic analysis of monocot families. In *Monocotyledons*, eds. P. Rudall and C. Humphries. Royal Botanic Gardens, Kew, England.

Stewart, C.-B. 1993. The powers and pitfalls of parsimony. *Nature* **361**:603–607.

Stewart, C.-B., J. W. Schilling, and A. C. Wilson. 1987. Adaptive evolution in the stomach enzymes of foregut fermenters. *Nature* **330**:401–404.

Stewart, R. N. and L. G. Burk. 1970. Independence of tissues derived from apical layers in ontogeny of the tobacco leaf and ovary. *American Journal of Botany* **57**:1010–1016.

Stewart, R. N., P. Semeniuk, and H. Dermen. 1974. Competition and accommodation between apical layers and their derivatives in the ontogeny of chimeral shoots of *Pelargonium X hortorum*. *American Journal of Botany* **61**:54–67.

Stewart, W. N. and G. W. Rothwell. 1993. *Paleobotany and the Evolution of Plants*, 2nd ed., Cambridge University Press, Cambridge.

Stidd, B. A., L. L. Oestry, and T. L. Phillips. 1975. On the frond *Sutcliffia insignis* var. *tuberculata*. *Review of Palaeobotany and Palynology* **20**:55–66.

Stone, D. E. 1989. Biology and evolution of temperate and tropical Juglandaceae. In *Evolution, Systematics, and Fossil History of the Hamamelidae*, eds. P. R. Crane and S. Blackmore, Vol. 2, pp. 117–145. Clarendon Press, Oxford.

Stout, S. A. 1992. Aliphatic and aromatic triterpenoid hydrocarbons in a Tertiary angiospermous lignite. *Organic Geochemistry* **18**:51–66.

Stubblefield, S. P. and H. P. Banks. 1978. The cuticle of *Drepanophycus spinaeformis*, a long-ranging Devonian lycopod from New York and eastern Canada. *American Journal of Botany* **65**:110–118.

Suzuki, M., L. Joshi, and S. Noshiro. 1991. *Tetracentron* wood from the Miocene of Noto Peninsula, central Japan, with a short revision of homoxylic fossil woods. *Botanical Magazine (Tokyo)* **104**:37–48.

Swamy, B. G. L. 1949. Further contributions to the morphology of the Degeneriaceae. *Journal of the Arnold Arboretum* **30**:10–38.

Swamy, B. G. L. 1952. Some aspects in the embryology of *Zygogynum bailloni* v. Tiegh. *Proceedings of the National Institiute of Science India* **18**:399–406.

Swamy, B. G. L. 1953a. *Sarcandra irvingbaileyi*, a new species of vesselless angiosperm from south India. *Proceedings of the National Institiute of Science India* **19**:301–306.

Swamy, B. G. L. 1953b. Some observations on the embryology of *Decaisnea insignis* Hook. et Thoms. *Proceedings of the National Institiute of Science India* **19**:307–310.

Swamy, B. G. L. 1953c. A taxonomic revision of the genus *Ascarina* Forst. *Proceedings of the National Institiute of Science India* **19**:371–388.

Swamy, B. G. L. 1953d. The morphology and relationships of the Chloranthaceae. *Journal of the Arnold Arboretum* **34**:375–408.

Swamy, B. G. L. and I. W. Bailey. 1950. *Sarcandra*: a vesselless genus of the Chloranthaceae. *Journal of the Arnold Arboretum* **31**:117–129.

Swofford, D. L. 1991a. *PAUP: Phylogenetic Analysis Using Prsimony. Version 3.0.* Illinois Natural History Survey, Champaign, Illinois.

Swofford, D. L. 1991b. When are phylogeny estimates from molecular and morphological data incongruent? In *Phylogenetic Analysis of DNA Sequences*, eds. M. M. Miyamoto and J. Cracraft, pp. 295–333. Oxford University Press, New York.

Swofford, D. L. 1993. *PAUP: Phylogenetic Analysis Using Parsimony. Version 3.1.1* Illinois Natural History Survey, Champaign, Illinois.

Sytsma, K. J. 1990. DNA and morphology: inference of plant phylogeny. *Trends in Evolution and Ecology* **5**:104–110.

Sytsma, K. J. and W. J. Hahn. 1994. Molecular systematics: 1991–1993. *Progress in Botany* **55**:307–333.

Sytsma, K. J. and J. F. Smith. 1992. Molecular systematics of Onagraceae: examples from *Clarkia* and *Fuchsia*. *Molecular Systematics of Plants*, eds. P. S. Soltis, D. E. Soltis, and J. J. Doyle, pp. 295–323. Chapman & Hall, New York.

Szymkowiak, E. J. and I. M. Sussex. 1992. The internal meristem layer (L3) determines floral meristem size and carpel number in tomato periclinal chimeras. *The Plant Cell* **4**:1089–1100.

Takahashi, A. 1988. Morphology and ontogeny of stem xylem elements in *Sarcandra glabra* (Thunb.) Nakai (Chloranthaceae) additional evidence for the occurrence of vessels. *Botanical Magazine (Tokyo)* **101**:387–395.

Takaso, T. 1985. A developmental study of the integument in gymnosperms. 3. *Ephedra distachya* L. and *E. equisetina* Bge. *Acta Botanica Neerlandica* **34**:33–48.

Takaso, T. and F. Bouman. 1986. Ovule and seed ontogeny in *Gnetum gnemon* L. *Botanical Magazine (Tokyo)* **99**:241–266.

Takhtajan, A. L. 1953. Phylogenetic principles of the system of higher plants. *The Botanical Review* **19**:1–45.

Takhtajan, A. L. 1969. *Flowering Plants. Origin and Dispersal*, Oliver and Boyd, Edinburgh.

Takhtajan, A. L. 1976. Neoteny and the origin of flowering plants. In *Origin and Early Evolution of Angiosperms*, C. B. Beck, ed., pp. 207–219. Columbia University Press. New York.

Takhtajan, A. L. 1980. Outline of the classification of flowering plants (Magnoliophyta). *The Botanical Review* **46**:225–359.

Takhtajan, A. L. 1987. *Systema Magnoliophytorum*. Nauka, Lenningrad.

Takhtajan, A. L. 1991. *Evolutionary Trends in Flowering Plants*, Columbia University Press, New York.

Tamura, M. 1972. Morphology and phyletic relationship of the Glaucidiaceae. *Botanical Magazine (Tokyo)* **85**:29–41.

Tamura, M. 1993. Ranunculaceae. In *The Families and Genera of Vascular Plants. Vol. 2. Flowering Plants. Dicotyledons: Magnoliid, Hamamelid and Caryophyllid Families*, eds. K. Kubitzki, J. G. Rohwer, and V. Bittrich, pp. 563–583. Springer-Verlag, Berlin.

Taylor, D. W. 1990. Paleobiogeographic relationships of angiosperms from the Cretaceous and early Tertiary of the North American area. *The Botanical Review* **56**:279–417.

Taylor, D. W. 1991a. Angiosperm ovules and carpels: their characters and polarities, distribution in basal clades and structural evolution. *Postilla* **208**:1–40.

Taylor, D. W. 1991b. Structure of the female reproductive axes in *Cycadeoidea* (Bennettitales): implications for polarizing ancestral angiosperm states. *American Journal of Botany (Supplement)* **78**:126.

Taylor, D. W. and L. J. Hickey. 1990a. An Aptian plant with attached leaves and flowers: implications for angiosperm origin. *Science* **247**:702–704.

Taylor, D. W. and L. J. Hickey. 1990b. A new hypothesis of the morphology of the ancestral angiosperm. *American Journal of Botany (Supplement)* **77**:159.

Taylor, D. W. and L. J. Hickey. 1992. Phylogenetic evidence for the herbaceous origin of angiosperms. *Plant Systematics and Evolution* **180**:137–156.

Taylor, D. W., J. M. Moldowan, and L. J. Hickey. 1992. Investigation of the terrestrial occurrence and biological source of the petroleum geochemical biomarker oleanane. In *Fifth North American Paleontological Convention, Abstracts and Programs*, eds. S. Lidgard and P. R. Crane, pp. 286. The Paleontological Society by The University of Tennessee, Knoxville.

Taylor, S. A. 1967. The comparative morphology and phylogeny of the Lardizabalaceae. Ph.D dissertation. Indiana University, Bloomington, Indiana.

Taylor, T. N. and E. L. Taylor. 1993. *The Biology and Evolution of Fossil Plants*. Prentice-Hall, Englewood Cliffs, New Jersey.

Taylor, T. N. and M. A. Millay. 1979. Pollination biology and reproduction in early seed plants. *Review of Palaeobotany and Palynology* **27**:239–355.

Tebbs, M. C. 1993. Piperaceae. In *The Families and Genera of Vascular Plants.* Vol. 2. *Flowering Plants. Dicotyledons: Magnoliid, Hamamelid and Caryophyllid Families,* eds. K. Kubitzki, J. G. Rohwer, and V. Bittrich, pp. 516–520. Springer-Verlag, Berlin.

Teichman, I. von and A. E. van Wyk. 1991. Trends in the evolution of dicotyledonous seeds based on character associations, with special reference to pachychalazy and recalcitrance. *Botanical Journal of the Linnean Society of London* **105**:211–237

Tepfer, S. S. 1953. Floral anatomy and ontogeny in *Aquilegia formosa* var. *truncata* and *Ranunculus repens. University of California Publications in Botany* **25**:513–648.

Thomas, H. H. 1925. The Caytoniales a new group of angiospermous plants from the Jurassic rocks of Yorkshire. *Philosophical Transactions of the Royal Society of London,* **213B**:299–363.

Thomas, H. H. 1957. Plant morphology and the evolution of flowering plants. *Botanical Journal of the Linnean Society of London,* **168**:125–133.

Thomas, H. H. and N. Bancroft. 1913. On the cuticles of some Recent and fossil cycadean fronds. *The Transactions of the Linnean Society of London. Second Series, Botany* **8**:155–204.

Thompson, W. P. 1918. Independent evolution of vessels in Gnetales and angiosperms. *Botanical Gazette* **69**:83–90.

Thorne, R. F. 1963. Some problems and guiding principles of angiosperm phylogeny. *American Naturalist* **97**:287–305.

Thorne, R. F. 1968. Synopsis of a putatively phylogenetic system of flowering plants. *Aliso* **6**:57–66.

Thorne, R. F. 1974. A phylogenetic classification of the Annoniflorae. *Aliso* **8**:147–209.

Thorne, R. F. 1976. A phylogenetic classification of the Angiospermae. *Evolutionary Biology* **9**:35–106.

Thorne, R. F. 1981. Phytochemistry and angiosperm phylogeny: a summary statement. In *Phytochemistry and Angiosperm Phylogeny,* eds. D. A. Young and D. S. Seigler, pp. 233–295. Praeger Scientific, New York.

Thorne, R. F. 1983. Proposed new realignments in the angiosperms. *Nordic Journal of Botany* **3**:85–117.

Thorne, R. F. 1992a. Classification and geography of the flowering plants. *The Botanical Review* **58**:225–348.

Thorne, R. F. 1992b. An updated classification of the flowering plants. *Aliso* **13**:365–389.

Tidwell, W. D., A. D. Simper, and G. F. Thayn. 1977. Additional information concerning the controversial Triassic plant: *Sanmiguelia. Palaeontographica Beitrage zur Natur-geschichte der Vorzeit* **163B**:143–151.

Tiffney, B. H. 1984. Seed size, dispersal syndromes, and the rise of the angiosperms: evidence and hypothesis. *Annals of the Missouri Botanical Garden* **71**:551–576.

Tiffney, B. H. and K. J. Niklas. 1985. Clonal growth in land plants: a paleobotanical perspective. In *Population Biology and Evolution of Clonal Organisms*, eds. J. B. C. Jackson, L. W. Buss, and R. E. Cook, pp. 35–66. Yale University Press, New Haven.

Tilman, D. 1988. *Plant Strategies and the Dynamics and Structure of Plant Communities*, Princeton University Press, Princeton, N.J.

Tilney-Bassett, R. A. E. 1986. *Plant Chimeras*, Edward Arnold, London.

Tilton, V. R. 1980. On micropyles. *Canadian Journal of Botany* **58**:1872–1884.

Tobe, H. 1991. Reproductive morphology, anatomy, and relationships of *Ticodendron*. *Annals of the Missouri Botanical Garden* **78**:135–142.

Tobe, H. and R. C. Keating. 1985. The morphology and anatomy of *Hydrastis* (Ranunculales): systematic reevaluation of the genus. *Botanical Magazine (Tokyo)* **98**:291–316.

Todzia, C. A. 1988. Chloranthaceae: *Hedyosmum*. *Flora Neotropica, Monograph* 48: 139. The New York Botanical Garden, New York.

Todzia, C. A. 1993. Chloranthaceae. In *The Families and Genera of Vascular Plants. Vol. 2. Flowering Plants. Dicotyledons: Magnoliid, Hamamelid and Caryophyllid Families*, eds. K. Kubitzki, J. G. Rohwer, and V. Bittrich, pp. 281–287. Springer-Verlag, Berlin.

Todzia, C. A. and R. C. Keating. 1991. Leaf architecture of the Chloranthaceae. *Annals of the Missouri Botanical Garden* **78**:476–496.

Traverse, A. 1988. *Paleopalynology*. Unwin Hyman, Boston.

Trivett, M. L. and G. W. Rothwell. 1991. Diversity among Paleozoic Cordaitales. *Neues Jahrbuch für Geologie Paläontologie* **183**:289–305.

Troitsky, A. V., Y. F. Melekhovets, G. M. Rakhimova, V. K. Bobrova, K. M. Valiego-Roman, and A. S. Antonov. 1991. Angiosperm origin and early statges of seed plant evolution deduced from rRNA sequence comparisons. *Journal of Molecular Evolution* **32**:253–261.

Troll, W. 1939a. *Vergleichende Morphologie der höheren Pflanzen*. Vol.1, Part 2, Section 1. Gebrüder Borntraeger, Berlin.

Troll, W. 1939b. Die morphologische Natur der Karpelle. *Chronica Botanica* **5**:38–41.

Tryon, R. M. and A. F. Tryon. 1982. *Ferns and Allied Plants with Special Reference to Tropical America*, Springer-Verlag, New York.

Tucker, S. C. 1975. Floral development in *Saururus cernuus* (Saururaceae). 1. Floral initiation and stamen development. *American Journal of Botany* **62**:993–1007.

Tucker, S. C. 1976. Floral development in *Saururus cernuus* (Saururaceae). II. Carpel initiation and floral vasculature. *American Journal of Botany* **63**:289–301.

Tucker, S. C. 1979. Ontogeny of the inflorescence of *Saururus cernuus* (Saururaceae). *American Journal of Botany* **66**:227–236.

Tucker, S. C. 1980. Inflorescence and flower development in the Piperaceae. I. *Peperomia*. *American Journal of Botany* **67**:686–702.

Tucker, S. C. 1981. Inflorescence and floral development in *Houttuynia cordata* (Saururaceae). *American Journal of Botany* **68**:1017–1032.

Tucker, S. C. 1982a. Inflorescence and flower development in the Piperaceae. II. Inflorescence development of *Piper. American Journal of Botany* **69**:743–752.

Tucker, S. C. 1982b. Inflorescence and flower development in the Piperaceae. III. Floral ontogeny of *Piper. American Journal of Botany* **69**:1389–1401.

Tucker, S. C. 1985. Initiation and development of inflorescence and flower in *Anemopsis californica. American Journal of Botany* **72**:20–31.

Tucker, S. C., A. W. Douglas, and L. Han-Xing. 1993. Utility of ontogenetic and conventional characters in determining phylogenetic relationships of Saururaceae and Piperaceae (Piperales). *Systematic Botany* **18**:614–641.

Uhl, N. W. 1969. Anatomy and ontogeny of the cincinni and flowers in *Nannorhops ritchiana* (Palmae). *Journal of the Arnold Arboretum* **50**:411–431.

Uhl, N. W. 1988. Floral organogenesis in palms. In *Aspects of Floral Development*, eds. P. Leins, S. C. Tucker, and P. K. Endress, pp. 25–44. J. Cramer, Berlin.

Uhl, N. W. and J. Dransfield. 1987. *Genera Palmarum. A Classification of Palms Based on the Work of Harold E. Moore, Jr.* Allen Press, Lawrence, Kansas.

Upchurch, G. R., Jr. and J. A. Wolfe. 1987. Mid-Cretaceous to Early Tertiary vegetation and climate: evidence from fossil leaves and woods. In *The Origins of Angiosperms and their Biological Consequences*, eds. E. M. Friis, W. G. Chaloner, and P. R. Crane, pp. 75–105. Cambridge University Press, Cambridge.

Van Konijnenburg-van Cittert, J. H. A. 1992. An enigmatic Liassic microsporophyll, yielding *Ephedripites* pollen. *Review of Palaeobotany and Palynology* **71**:239–254.

Van Fink, W. 1982. Histochemische Untersuchungen uber Starkverteilung und Phosphatase Aktivität im Holz einiger tropischer Baumarten. *Holzforschung* **36**:295–302.

Verdcourt, B. 1985. Notes on Malasian Chloranthaceae. *Kew Bulletin* **40**:213–224.

Vijayaraghavan, R. 1964. Morphology and embryology of a vesselless dicotyledon--*Sarcandra irvingbaileyi* Swamy, and systematic position of the Chloranthaceae. *Phytomorphology* **14**:429–441.

Vink, W. 1993. Winteraceae. In *The Families and Genera of Vascular Plants*. Vol. 2. *Flowering Plants. Dicotyledons: Magnoliid, Hamamelid and Caryophyllid Families*, eds. K. Kubitzki, J. G. Rohwer, and V. Bittrich, pp. 630–638. Springer-Verlag, Berlin.

Vishnu-Mittre. 1957. Studies on the fossil flora of Nipania (Rajmahal Series), India--Pentoxyleae. *Palaeobotanist* **6**:31–46.

Vokes, H.E. 1946. Contribution to the paleontology of the Lebanon Mountains, Republic of Lebanon. Part 3: The pelecypod fauna of the "Olive Locality" (Aptian) at Abeih. *Bulletin of the American Museum of Natural History* **87**:1–25.

Wagner, R. H. 1964. Stephanian B flora from the Cinera-Matallana Coalfield (León) and neighboring outliers. II. *Mixoneura, Reticulopteris, Linopteris,* and *Odontopteris. Notas y Comunicaciones del Instituto Geológico y Minero de España* **75**:5–56.

Wagner, W. H., Jr. 1979. Reticulate veins in the systematics of modern ferns. *Taxon* **28**:87–95.

Wagner, W. H., Jr., J. M. Beitel, and F. S. Wagner. 1982. Complex venation patterns in the leaves of *Selaginella*: megaphyll-like leaves in lycophytes. *Science* **218**:793–794.

Weissbrod, T., G. Gvirtzman and Z. Lewy. 1990. Recent advances in correlation of marine and nonmarine sequences in southern Israel. In *Abstracts, International Cretaceous Field Conference, IGCP*, p. 11. Project 245 and 265, Jerusalem.

Walker, J. W. 1971. Pollen morphology, phytogeography and phylogeny of the Annonaceae. *Contributions of the Gray Herbium* **202**:1–131.

Walker, J. W. 1972. Contributions to the pollen morphology and phylogeny of the Annonaceae II. *Botanical Journal of the Linnean Society of London* **65**:173–178.

Walker, J. W. 1974. Aperture evolution in the pollen of primitive angiosperms. *American Journal of Botany* **61**:1112–1136.

Walker, J. W. 1976a. Comparative pollen morphology and phylogeny of the Ranalean complex. In *Origin and Early Evolution of Angiosperms*, ed. C. B. Beck, pp. 241–299. Columbia University Press, New York,

Walker, J. W. 1976b. Evolutionary significance of the exine in pollen of primitive angiosperms. In *The Evolutionary Significance of the Exine*, eds. I. K. Ferguson and J. Muller, pp. 251–308. Linnean Society Symposium Series No. 1, London.

Walker, J. W., G. J. Brenner, G. J. and A. G. Walker. 1984. Winteraceous pollen in the Lower Cretaceous of Israel: early evidence of a magnolialean angiosperm family. *Science* **220**:1273–1275.

Walker, J. W. and J. A. Doyle. 1975. The bases of angiosperm phylogeny: palynology. *Annals of the Missouri Botanical Garden* **62**:664–723.

Walker, J. W. and J. J. Skvarla. 1975. Primitively columellaless pollen: a new concept in the evolutionary morphology of angiosperms. *Science* **187**:445–447.

Walker, J. W. and A. G. Walker. 1984. Ultrastructure of Lower Cretaceous angiosperm pollen and origin and early evolution of flowering plants. *Annals of the Missouri Botanical Garden* **71**:464–521.

Ward, J. V., J. A. Doyle, and C. L. Hotton. 1988. Probable granular magnoliid angiosperm pollen from the Early Cretaceous. *American Journal of Botany* **75**(6):118–119.

Wardrop, A. B. 1951. Cell wall organization and the properties of xylem. I. Cell wall organization and the variation of breaking load of xylem in conifer stems. *Australian Journal of Research and Science Series B* **4**:391–414.

Watson, J. and C. A. Sincock. 1992. *Bennettitales of the English Wealden*, The Palaeontographical Society, London.

Weberling, F. 1988. Inflorescence structure in primitive angiosperms. *Taxon* **37**:657–690.

Weberling, F. 1989. *Morphology of Flowers and Inflorescences*. Cambridge University Press, New York.

Wellwood, R. W. 1962. Tensile testing of small wood samples. *Pulp Paper Magazine Canada* **63**(2):T61–T67.

West, W. C. 1969. Ontogeny of oil cells in the woody Ranales. *Bulletin of the Torrey Botanical Club* **96**:329–344.

Westoby, M. and B. Rice. 1982. Evolution of the seed plants and inclusive fitness of plant tissues. *Evolution* **36**:713–724.

Wettstein, R. R., von. 1901. *Handbuch der systematischen Botanik.* Franz Deuticke, Vienna.

Wettstein, R. R., von. 1907a. *Handbuch der systematischen Botanik,* 2nd ed., Franz Deuticke, Leipzig.

Wettstein, R. R., von. 1907b. Über des vorkommen zweigeschlectiger Infloreszenzen bei *Ephedra. Festschrifter Naturwissenschaften Vereins Universität Wien* **25**:21–28.

Wettstein, R. R., von. 1911. *Handbuch der Systematischen Botanik,* 3rd ed., Franz Deuticke, Leipzig.

Wettstein, R. R., von. 1935. *Handbuch der systematischen Botanik,* 4th ed., Franz Deuticke, Leipzig.

Wheeler, E. A. 1983. Intervascular pit membranes in *Ulmus* and *Celtis* native to the United States. *International Association of Wood Anatomists Bulletin, New Series* **4**:79–88.

Wheeler, E. A. and P. Baas. 1991. A survey of the fossil record for dicotyledonous wood and its significance for evolutionary and ecological wood anatomy. *International Association of Wood Anantomists Bulletin, New Series* **12**:275–332.

Wheeler, E. A. and P. Baas. 1993. The potentials and limitations of dicotyledonous wood anatomy for climatic reconstruction. *Paleobiology* **19**:487–498.

Wheeler, E. A. and P. S. Herendeen. 1993. Wood structure in extant archaic dicotyledons and the Cretaceous fossil record. *American Journal of Botany (Supplement)* **80**:127.

Wheeler, E. A., M. Lee, and L. C. Matten. 1987. Dicotyledonous woods from the Upper Cretaceous of southern Illinois. *Botanical Journal of the Linnean Society of London* **95**:77–100.

Whitehouse, H. L. K. 1950. Multiple-allelomorph incompatibility of pollen and style in the evolution of the angiosperms. *Annals of Botany* **14**:198–216.

Wiegrefe, S. J., E. Conti, K. J. Systma, and R. P. Guries. 1993. The Ulmaceae, one family or two? The molecular prospective. *American Journal of Botany (Supplement)* **80**:182.

Wieland, G. R. 1906. *American Fossil Cycads,* Vol. 1. Carnegie Institution of Washington, Washington, DC.

Wieland, G. R. 1916. *American Fossil Cycads,* Vol. 2. *Taxonomy,* Carnegie Institution of Washington, Washington, DC.

Wilkinson, H. P. 1979. The plant surface (mainly leaf). In *Anatomy of the Dicotyledons,* eds. C. R. Metcalfe and L. Chalk, Vol. I, pp. 97–165. Clarendon Press, Oxford.

Willense, M. T. M. and A. W. Franssen-Verheijen. 1992. Pollen tube growth and ovule penetration in *Gasteria verrucosa* (Mill.) H. Duval. In *Angiosperm Pollen and Ovules,* eds. E. Ottaviano, D. L. Mulcahy, M. S. Goria, and G. B. Mulcahy, pp. 168–173. Springer-Verlag, New York.

Williamson, P. S. and E. L. Schneider. 1993a. Cabombaceae. In *The Families and Genera of Vascular Plants.* Vol. 2. *Flowering Plants. Dicotyledons: Magnoliid, Hamamelid and Caryophyllid Families,* eds. K. Kubitzki, J. G. Rohwer, and V. Bittrich, pp.157–160. Springer-Verlag, Berlin.

Williamson, P. S. and E. L. Schneider. 1993b. Nelumbonaceae. In *The Families and Genera of Vascular Plants*. Vol. 2. *Flowering Plants. Dicotyledons: Magnoliid, Hamamelid and Caryophyllid Families*, eds. K. Kubitzki, J. G. Rohwer, and V. Bittrich, pp. 470–472. Springer-Verlag, Berlin.

Wilson, T. K. 1960. The comparative morphology of the Canellaceae. I. Synopsis of genera in wood anatomy. *Tropical Woods* 112:1–27.

Wilson, T. K. 1964. The ethereal oil cells of *Canella*. *American Journal of Botany* 51:676.

Wilson, T. K. 1965. The comparative morphology of the Canellaceae. II. Anatomy of the young stem and node. *American Journal of Botany* 52:369–378.

Wilson, T. K. and L. M. Maculans. 1967. The morphology of the Myristicaceae. I. Flowers of *Myristica fragrans* and *M. malabarica*. *American Journal of Botany* 54:214–220.

Wilson, M. A., B. S. Gaut, and M. Clegg. 1990. Chloroplast DNA evolves slowly in the palm family (Arecaceae). *Molecular Biology and Evolution* 7:303–314.

Wing, S. L., L. J. Hickey, and C. C. Swisher. 1993. Implications of an exceptional fossil flora from Late Cretaceous vegetation. *Nature* 363:342–344.

Wing, S. L. and B. H. Tiffney. 1987a. Interactions of angiosperms and herbivorous tetrapods through time. In *The Origins of Angiosperms and Their Biological Consequences*, eds. E. M. Friis, W. G. Chaloner, and P. R. Crane, pp. 203–224. Cambridge University Press, Cambridge.

Wing, S. L. and B. H. Tiffney. 1987b. The reciprocal interaction of angisoperm evolution and tetrapod herbivory. *Review of Palaeobotany and Palynology* 50:179–210.

Withner, C. I. 1941. Stem anatomy and phylogeny of the Rhoipteleaceae. *American Journal of Botany* 28:872–878.

Wolfe, J. A. 1989. Leaf-architectural analysis of the Hamamelidae. In *Evolution, Systematics, and Fossil History of the Hamamelidae*, eds. P. R. Crane and S. Blackmore, Vol. 1, pp. 75–104. Clarendon Press, Oxford.

Wolfe, J. A., J. A. Doyle, and V. M. Page. 1976. The bases of angiosperm phylogeny: paleobotany. *Annals of the Missouri Botanical Garden* 62:801–824.

Wolfe, K. H., M. Gouy, Y.-W. Yang, P. M. Sharp, and W.-H. Li. 1989. Date of the monocot-dicot divergence estimated from chloroplast DNA sequence data. *Proceedings of the National Academy of Sciences of the United States of America* 86:6201–6205.

Wood, C. E., Jr. 1971. The Saururaceae in the southeastern United States. *Journal of the Arnold Arboretum* 52:479–485.

Wu, C.-Y. and K. Kubitzki. 1993a. Circaeasteraceae. In *The Families and Genera of Vascular Plants*. Vol. 2. *Flowering Plants. Dicotyledons: Magnoliid, Hamamelid and Caryophyllid Families*, eds. K. Kubitzki, J. G. Rohwer, and V. Bittrich, pp. 288–289. Springer-Verlag, Berlin.

Wu, C.-Y. and K. Kubitzki. 1993b. Lardizabalaceae. In *The Families and Genera of Vascular Plants*. Vol. 2. *Flowering Plants. Dicotyledons: Magnoliid, Hamamelid and

Caryophyllid Families, eds. K. Kubitzki, J. G. Rohwer, and V. Bittrich, pp. 361–365. Springer-Verlag, Berlin.

Wu, C.-Y. and K. Kubitzki. 1993c. Saururaceae. In *The Families and Genera of Vascular Plants*. Vol. 2. *Flowering Plants. Dicotyledons: Magnoliid, Hamamelid and Caryophyllid Families*, eds. K. Kubitzki, J. G. Rohwer, and V. Bittrich, pp. 586–588. Springer-Verlag, Berlin.

Wylie, R. B. 1939. Relations between tissue organization and vein distribution in dicotyledon leaves. *American Journal of Botany* **26**:219–225.

Young, D. A. 1981. Are the angiosperms primitively vesselless? *Systematic Botany* **6**:313–330.

Young, D. A. and D. S. Seigler (eds.). 1981. *Phytochemistry and Angiosperm Phylogeny*. Praeger Scientific, New York.

Zavada, M. S. and D. L. Dilcher. 1986. Comparative pollen morphology and its relationship to phylogeny of pollen in the Hamamelidae. *Annals of the Missouri Botanical Garden* **73**:348–381.

Zavada, M. S. and T. N. Taylor. 1986. The role of self-incompatibility and sexual selection in the gymnosperm-angiosperm transition: a hypothesis. *The American Naturalist* **128**:538–550.

Zhang, S.-Y. 1992. Systematic wood anatomy of the Rosaceae. *Blumea* **37**:81–152.

Zhang, S.-Y., L.-G. Lei, H.-Q. Li, and Q.-U. Su. 1990. A preliminary study on tracheary elements in the endemic species *Sarcandra hainanensis* from China--evidence for the occurrence of vessels in *Sarcandra*. *Acta Botanica Borealis Sinica* **10**:95–98.

Zimmer, E. A., R. K. Hamby, M. L. Arnold, D. A. Leblanc, and E. C. Theriot. 1989. Ribosomal RNA phylogenies amd flowering plant evolution. In *The Hierarchy of Life*, B. Fernholm, K. Bremer, and H. Jornvall, pp. 205–214. Elsevier Science Publishers, New York.

Zimmermann, M. H. 1983. *Xylem Structure and the Ascent of Sap*. Springer-Verlag, Berlin.

APPENDIX

A stratigraphic time scale of the important time periods for angiosperm evolutions. Dates from W. B. Harland et al., *A Geologic Time scale* (Cambridge University Press, New York, 1990).

				Millions of years before present
Cenozoic	Tertiary			
Mesozoic	Cretaceous	Late	Maestrichtian	65
			Campanian	74
			Santonian	83
			Coniacian	87
			Turonian	89
			Cenomanian	90
		Early	Albian	97
			Aptian	112
			Barremian	125
			Hauterivian	132
			Valanginian	135
			Berriasian	141
	Jurassic	Late (Malm)	Tithonian	146
			Kimmeridgian	152
			Oxfordian	155
		Middle (Dogger)		157
		Early (Lias)		178
	Triassic	Late	Rhaetian	208
			Norian	209
			Carnian	223
		Middle		235
				235

Index